Fabio Altomare and Albert M. Chang

One-Dimensional Superconductivity in Nanowires

Related Titles

Bezryadin, A.

Superconductivity in Nanowires
Fabrication and Quantum Transport

2012
ISBN: 978-3-527-40832-0

Bhattachariya, R., Parantharam, M.P.

High Temperature Superconductors

2010
ISBN 978-3-527-40827-6

Luryi, S., Xu, J., Zaslavsky, A.

Future Trends in Microelectronics
Up the Nano Creek

2007
ISBN: 978-0-470-08146-4

Wolf, E. L.

Nanophysics and Nanotechnology
An Introduction to Modern Concepts in Nanoscience

2006
ISBN: 978-3-527-40651-7

Wong, B., Mittal, A., Cao, Y., Starr, G. W.

Nano-CMOS Circuit and Physical Design

2004
ISBN: 978-0-471-46610-9

Buckel, W., Kleiner, R.

Superconductivity
Fundamentals and Applications

2004
ISBN: 978-3-527-40349-3

Reich, S., Thomsen, C., Maultzsch, J.

Carbon Nanotubes
Basic Concepts and Physical Properties

2004
ISBN: 978-3-527-40386-8

Fabio Altomare and Albert M. Chang

One-Dimensional Superconductivity in Nanowires

WILEY-VCH

WILEY-VCH Verlag GmbH & Co. KGaA

The Authors

Dr. Fabio Altomare
D-Wave Systems Inc.
100 - 4401 Still Creek Drive
Burnaby, BC, V5C 6G9
Canada

Prof. Albert M. Chang
Duke University
Department of Physics
Physics Building
Science Drive
Durham, NC 27708-0305
USA

Cover Figure
Artistic rendering of an 8 nm wide, 20 μm long nanowire fabricated on an InP semiconducting stencil. Electrical contacts to the nanowire (left and right pads) are realized during the nanowire deposition. At the center, a QPS junction is current biased by a current source: The number of windings of the phase of the order parameter decreases because of a phase slippage event. The diagram does not indicate the actual nanowire connection in a circuit. Rendering of the windings courtesy of A. Del Maestro.

Library of Congress Card No.:
applied for

British Library Cataloguing-in-Publication Data:
A catalogue record for this book is available from the British Library.

Bibliographic information published by the Deutsche Nationalbibliothek
The Deutsche Nationalbibliothek lists this publication in the Deutsche Nationalbibliografie; detailed bibliographic data are available on the Internet at http://dnb.d-nb.de.

Composition le-tex publishing services GmbH, Leipzig
Printing and Binding betz-druck GmbH, Darmstadt
Cover Design Adam-Design, Weinheim

Print ISBN 978-3-527-40995-2
ePDF ISBN 978-3-527-64907-5
ePub ISBN 978-3-527-64906-8
mobi ISBN 978-3-527-64905-1
oBook ISBN 978-3-527-64904-4

Printed in the Federal Republic of Germany
Printed on acid-free paper

To Maria Elena, Mattia, Gabriele, Giulia and to my parents:
Thank you for your love and support.

Fabio

To my family: Ying, Emily and Austin, my sister Margaret, my brother Winston, my father Bertrand Tsu-Shen, and my late mother, Virginia Pang-Ying.

Albert

Contents

Preface

This book is devoted to the topic of superconductivity in very narrow metallic wires. Interest in such wires is driven by the continuing drive for miniaturization in the electronics industry, where the reduction of heat dissipation by the use of interconnects which are superconducting may be necessary. This has led to the invention of new methods of producing very narrow wires with good quality and uniformity in dimensions, and opened up the possibility of novel device paradigms.

The superconducting state is a state of coherent pairs of electrons, held together by the mechanism of the Cooper instability. When the lateral dimensions are small, a new pathway opens up for the superconductor to produce dissipation, associated with the enhanced rate at which fluctuations can occur in the complex order parameter – a quantity which describes the magnitude and phase of the Cooper pairs. In many regimes, the fluctuations of the phase of the Cooper pairs is often the dominant process, in analogy with what occurs in dissipative Josephson junctions. This new pathway is related with the motion of vortices in type-II superconductors, which gives rise to dissipation in 2D (two-dimensional) and bulk superconductors, in which a rapid change of phase occurs as a vortex passes by. Here, because of the physical dimension of the system (comparable to 10 nm), the vortex quickly passes across the entire narrow wire, producing dissipation in the process.

Although nanowires are small in their transverse dimensions, they are still much larger than the Fermi wavelength (λ_F) of the electronic system, which is of the order of a few angstrom in all metals. In conventional superconductors such as Al, Pb, Sn, Nb, MoGe, and so on, in thin film or nanowire form, the coherence length is $\xi \approx 5$–100 nm, typically 10–1000 times the Fermi wavelength. The system we will consider is therefore in a sort of mixed-dimensional regime. From the condensate perspective, the system is 1D (one-dimensional), in the sense that the transverse dimensions are smaller than the Cooper pair size: for this reason, the wave function, or alternatively, the order parameter, describing the Cooper pairs, is uniform, and thus position-independent, in the transverse directions. From the fermionic quasiparticle excitation perspective, the system is effectively 3D (three-dimensional; λ_F is much smaller than the lateral dimension) and there is a large number of transverse channels (from $\approx$ 100 in a multiwalled carbon nanotube, to $\approx$ 3000 in an $\approx$ 8 nm diameter aluminum nanowire), analogous to transverse modes in a wave guide.

In this limit, the dominant collective excitations are no longer pair-breaking excitations across the superconducting energy gap, but rather "phase-slips," which are topological defects in the ground state configuration. Phase slips are related to the motion of vortices in type-II superconductors which give rise to dissipation in 2D and bulk superconductors: in a 1D system, they produce a sudden change of phase by 2π across a core region of reduced superconducting correlation which gives rise to a voltage pulse.

Another intriguing aspect arises from the fact that even in wires which are not ballistic along the wire length, the typical level spacing in the transverse direction can significantly exceed the superconducting gap energy scale. Thus, the possibility of singularity in the density of the electronic state in the normal system associated with each transverse channel can cause oscillatory behaviors in the superconducting properties.

From the above discussion, it is clear that the regime of interest is delineated by the condition that the size of the Cooper pairs, or the superconducting coherence length, be larger than the transverse directions perpendicular to the length of the wire. In this limit the order behavior of the Cooper pair is largely uniform across the wire length, and only variations along the wire length need to be considered. At temperatures well below the superconducting transition temperature, the scale for the observation of dimensionality effects is in the 10–100 nm range (5–50 nm in radius): in this regime, new physics have been predicted, including universal scaling laws in the conductance of the wire.

To access the regime where quantum processes become dominant, however, a more stringent requirement is necessary, that of a sufficiently large probability of fluctuations to occur: this more stringent criterion places the scale requirement in the 10 nm (5 nm in radius) range. Nanowires in this regime may either be a single monolithic wire, such as nanowires made of MoGe, Al, Nb, In, PbIn, Sn, and so on, or coupled wires such as in carbon nanotube bundles.

It should be noted that, in the strict sense, only systems in 2D or 3D have a true, sharp, thermodynamic phase transition into the superconducting state at a finite temperature T_c. In the 2D case, it is a Berenzinskii–Thouless–Kosterlitz type of second-order phase transition, while in 3D, it is a second-order continuous phase transition in the Ginzburg–Landau sense, at least for type-II superconductors supporting the existence of vortices within the bulk. In contrast, in a 1D nanowire, the fluctuations cause the transition to become smeared, so that the resistance remains finite, albeit small, below the transition. Early theoretical analyses were motivated by experimental observations of such behavior, and attempted to quantify the amount of residual resistance. Thus, naturally, questions arise as to whether the resistance vanishes at zero T, and whether novel excitations are able to limit the supercurrent below the depairing limit.

From a broader point of view, 1D superconducting nanowires are interesting from a variety of perspectives, including many body physics, quantum phase transition (QPT), macroscopic quantum tunneling (MQT) processes, and device applications. The field, by nature, involves rather technically sophisticated methods. This is true from either the experimental or theoretical side.

On the experimental side, there are many technical challenges, and thus it is not for the faint of heart. Such challenges include nanowire fabrication, the delicate nature of nanowires with respect to damage–they act as excellent fuses, and their sensitivity to environmental interference from external noise sources, and so on. At the level of 10 nm in transverse dimensions, corresponding roughly to 40 atoms across, even width fluctuations of a few atoms can have significant influence on the energetics and properties. Thus, to obtain intrinsic behaviors of relevance to a uniform wire, rather than behavior limited to weak-links or a very thin region in a nonuniform wire, fabrication is exceedingly demanding. Arguably, only in recent years has the emergence of fabrication techniques come into existence with sufficient precision for producing unusually uniform nanowires. Thus, substantial progress is occurring on the experimental front.

On the theoretical side, analyses invariably involve sophisticated quantum field theory (QFT), quantum phase transition (QPT), Bogoliubov–de Gennes BdG, Ginzburg–Landau (GL), Gorkov–Eilenberger–Usadel self-consistent (including nonlinearities) techniques, all of which require a rather advanced level of understanding of the theoretical machinery. This is often compounded by the fact that the concept of the quantum-mechanical tunneling in the phase of the superconducting order-parameter is difficult to motivate from a classical perspective, since, unlike coordinates or momenta, the phase is a concept born out of wave mechanics. Instead, the tunneling in the phase degree of freedom is usually introduced via the Feynman path integral type of formulation, as one finds a more natural description, for this phenomenon, in terms of instantons in field theory language.

The goal of this book is to produce a relatively self-contained introduction to the experimental and theoretical aspects of the 1D superconducting nanowire system. The aim is to convey what the important issues are, from experimental, phenomenological, and theoretical aspects. Emphasis is placed on the basic concepts relevant and unique to 1D, on identifying novel behaviors and concepts in this unique system, as well as on the prospects for potentially new device applications based on such new concepts and behaviors. The latest experimental techniques and results in the field are summarized. On the theoretical side, much of the field theoretic methods for analyzing the various quantum phase transitions, such as superconductor–metal transitions, superconductor–insulation transitions, and so on, brought about by disorder, are highly technical and the details are beyond the scope of this book. Nevertheless, an attempt is made to summarize the relevant issues and predictions, to pave the way for understanding the formalisms and issues addressed in available journal literature. It is the hope of the authors that this book will serve as a starting point for those interested in joining this exciting field, as well as serving as a useful reference for active researchers.

To this end, our philosophy is to present the field as an active, exciting, and ongoing discourse, rather than one that is fully established. Thus, many of the concepts and experiments are still fraught with a certain degree of healthy controversy. Thus, an attempt is made to convey a sense of openness to the discourse in the field.

The book is organized as follows: Chapter 1 contains a brief history of the field, and a succinct summary of the various theoretical methodologies for understanding conventional superconductivity. These methods are widely used in analyzing 1D superconducting nanowire systems. Chapter 2 is devoted to the basic concepts of 1D superconductivity, including size quantization and its influence on superconducting properties, leading to the phenomenon of shape resonances, the phase-slip phenomenon, which originated from an attempt to explain the broadened temperature transition and the finite voltage along the wire below but near the transition, as well as the conditions and relevant energy scales in molecular systems such as carbon nanotubes. In Chapter 3, the quantum theory based on path integral formulation is summarized. The various types of quantum phase transitions and competing physical scenarios are described. Chapter 4 explores new concepts and potentially new devices based on the idea of a duality between Cooper pairs and the phase slip. Novel QPS junctions are described. These are believed to offer new venues for a current version of the Shapiro steps, as well as a platform for qubits. Nonlinear and nonequilibrium effects based on the Usadel equations are described in Chapter 5.

On the experimental side, the all-important description of the state-of-the-art fabrication methodologies is presented in Chapter 6. Experimental techniques, such as filtering to remove external environmental noise are summarized in Chapter 7. Finally, in Chapter 8, we discuss the current state of experimental progress, and the many open questions, as well as future prospects. To conclude, in Chapter 9, we describe recent experimental results in superconducting nanowire single-photon detector that are now approaching the 1D superconducting limit and devices that are related to 1D superconductivity via the proximity effect: in this class, we find nanotubes and semiconducting nanowires, which have recently indicated of the presence of Majorana fermions.

The authors are indebted to Sergei Khlebnikov (Purdue University), in particular for sharing his insight and for providing helpful comments on the entire theory section. The authors acknowledge the help of (in alphabetical order) G. Berdiyorov (University of Antwerp), E. Demler (Harvard University), A. Del Maestro (University of Vermont), D. Golubev (Karlsruhe Institute of Technology), F. Peeters (University of Antwerp), G. Refael (Caltech), S. Sachdev (Harvard University), A. Zaikin (Lebedev Inst. of Physics and Karlsruhe Institute of Technology) for critically reading the manuscript. The authors would also like to thank M.R. Melloch (Purdue University), C.W. Tu (University of California at San Diego), P. Li, P.M. Wu, G. Finkelstein, Y. Bomze, I. Borzenets (Duke University), and Li Lu (Institute of Physics, CAS, Beijing) for their help and fruitful discussions.

September, 2012 *F. Altomare and A.M. Chang*

Abbreviations and Symbols

Acronyms

APE	Anti-proximity Effect
BdG	Bogoliubov–de–Gennes
BCS	Bardeen–Cooper–Schreiffer
CPR	Current–Phase Relation
cQPS	Coherent Quantum Phase-Slip
DCR	Dark Current Rate
DQM	Dissipative Quantum Mechanics
e-beam	Electron beam
GIO	Giordano expression for the resistance due to quantum phase slips, or to QPS and TAPS
GL	Ginzburg–Landau
HQS	Silsesquioxane, a type of negative electron beam resist
IRFP	Infinite-Randomness Fixed Point
JJ	Josephson Junction
KQPS	Khlebnikov Quantum Phase-Slip
KTB	Kosterlitz–Thouless–Berezinskii
LA	Langer–Ambegaokar
LAMH	Langer–Ambegaokar–McCumber–Halperin
MBE	Molecular-Beam-Epitaxy
MH	McCumber–Halperin
PMMA	Polymethylmethacrylate: probably the most common electron beam resist, mainly used as positive tone resist
PSC	Phase-Slip Center
QPS	Quantum Phase-Slip
RTFIM	Random-Transverse-Field Ising Model
SC	Superconducting
SG	sine-Gordon
SEM	Scanning Electron Microscope
SIT	Superconductor-Insulator Transition

SmNW	Semiconducting Nanowire
SMT	Superconductor-Metal Transition
SNAP	Superlattice Nanowire Pattern Transfer
SNAP	Superconducting Nanowire Avalanche Photo Detector
SNSPD	Superconducting Nanowire Single-Photon Detector
SQUID	Superconducting Interference Device
SSPD	Superconducting Single-Photon Detector
SWNT	Single-Walled Nanotube
SWCNT	Single-Walled Carbon Nanotube
TAP or TAPS	Thermally-Activated Phase Slip

Symbols

w	Width of nanowire or wire
h	Height of nanowire or thin film
d	Diameter of nanowire
L	Length of nanowire
L_x	Length in x-direction of a thin film
L_y	Length in y-direction of a thin film
A	Cross sectional area of a nanowire
V	Volume
T	Temperature
$k_B T$	Thermal energy scale; k_B is the Boltzmann's constant
β	$\frac{1}{k_B T}$
e	fundamental unit of charge: $+1.602 \times 10^{-19}$ C $=$ $+ 4.80 \times 10^{-10}$ esu
μ	(i) Chemical potential (ii) parameter that controls the KTB (Kosterlitz–Thouless–Berezinskii) phase transition
E_F	Fermi energy
v_F	Fermi velocity (speed)
$N(0) \equiv D(E_F)$	Density of states at the Fermi level of both spins, in the normal state
$-e$	Charge of the electron
$(-2e)$	Charge of the Cooper-pair of electrons
m	electron mass
M	Cooper-pair mass $M = 2m$
$D = \frac{1}{3} v_F l_{mfp}$	Diffusion coefficient
l_{mfp}	Mean-free-path
T_c	Critical temperature, or normal to superconducting transition temperature
Δ	Superconducting gap
Δ_0	Zero temperature superconducting gap

ω_D	Debye frequency
n_s	Superconducting carrier density
g	Gorkov coupling constant, take to be positive $g = VV$
u	Amplitude of electron-like component of a Bogoliubov quasiparticle
v	Amplitude of hole-like component of a Bogoliubov quasiparticle
ξ	Superconducting coherence length, usually in the dirty limit of $\xi_{cln} \sim \xi_{bulk} \sim \xi_{BCS} \gg l_{mfp}$
ξ_0	Zero temperature superconducting coherence length
ξ_{bulk}	Superconducting coherence length is bulk 3D material
ξ_{cln}	Superconducting coherence length in the clean limit
ξ_{BCS}	The BCS superconducting coherence length
λ_L	London penetration depth
λ_n or λ_i	Eigenvalues of index n or index i
E_C	Coulomb charging energy $E_C = (2e)^2/C$, where C is the capacitance
E_J	Josephson energy
C	(i) Capacitance per unit length of a nanowire (ii) capacitance of a Josephson junction
L_{kin}	Kinetic inductance per unit length of a nanowire
c_{pl}	Mooij–Schön plasmon mode propagation speed
ψ or $\psi^\dagger$	Electron annihilation or creation operator
ψ	(i) Ginzburg–Landau superconducting order parameter (dimensionless) (ii) Gross–Pitaevskii superconducting order parameter
Ψ	Ginzburg–Landau superconducting order parameter
τ_{GL}	Ginzburg–Landau relaxation time: $\tau_{GL} = \frac{\pi}{8}\frac{\hbar}{T_c - T}$
ϕ_0	Superconducting flux quantum: $\phi_0 = hc/(2e)$ (cgs); $\phi_0 = h/(2e)$ (SI)
R_Q	Superconducting quantum resistance $R_Q \equiv h/(2e)^2 \approx 6.453\,\mathrm{k\Omega}$
A	Vector potential
V	(i) electrostatic potential (ii) strength of electron-phonon coupling in the Gorkov coupling constant $g = VV$
$\mathbf{E}$	Electric field
$\mathbf{B}$	Magnetic field
J	Electrical current density
J_s	Electrical current density of the superfluid
J_c	Critical electrical current density
j	Reduced electrical current density
j_c	Reduced critical electrical current density
I	Electrical current
I_c	Critical electrical current

I_s	Switching current, at which a nanowire switches from a superconducting state to a normal state, typically somewhat smaller than the I_c in the depairing limit
I_{bias}	Externally applied bias current
a	(i) Lattice constant (ii) Proportionality constant in the Giordano expression for the quantum phase-slip rate or quantum phase-slip resistance
φ	Phase of the complex superconducting order-parameter
H_c	Thermodynamic critical (magnetic) field
R_{QPS}	The quantum phase-slip contribution to the resistance
R_{TAP}	The thermally-activated phase-slip contribution to the resistance
Γ_{QPS}	The quantum phase-slip (tunneling) rate
Γ_{TAP}	The phase-slip rate due to thermal-activation
$\Gamma^{\pm}$	The phase-slip rate: + corresponds to increasing the phase-winding by one unit, − to decreasing by one unit
Γ_{inst}	Quantum tunneling rate of an instanton
Ω	Attempt frequency for the phase-slip process
$\Omega^{\pm}$	Attempt frequency for the phase-slip process: + corresponds to increasing the phase-winding by one unit, − to decreasing by one unit
Z	(i) Partition function (ii) Mooij–Schön plasmon propagation impedance
S	Action, in almost all cases, the Euclidean action in the imaginary-time formulation
S_D	Drude contribution to the action
S_J	Josephson contribution to the action
S_L	London contribution to the action
S_{em}	Electromagnetic field contribution to the action
S_{diss}	Dissipation contribution to the action
S_{bias}	Contribution to the action from a biasing current
S_{bdry}	Boundary contribution to the action
S_{1D}	Action for a 1D superconducting nanowire
S_{QPS}	Action due to a single or multiple quantum phase-slip
F	Free energy
ΔF	Free energy barrier for the creation of a phase-slip
$\Delta F^{\pm}$	Free energy barrier for the creation of a phase-slip: + corresponds to increasing the phase-winding by one unit, − to decreasing by one unit
$\hat{G}$	The Gorkov Green's functions in matrix form in Nambu space
$\check{G}$	The Gorkov Green's functions in the Keldysh formulation, with both the advanced and retarded, as well as the Keldysh component. The matrix is in the direct product space of forward and backward branches of the Keldysh contour, with the Nambu space.

G	The normal components of the Gorkov Green's functions. They represent the diagonal components of the $\hat{G}$ matrix
F	The superconducting components of the Gorkov Green's functions, representing the off-diagonal components of the $\hat{G}$ matrix (sometimes with an extra sign change, and or hermitian conjugation)
$G^{R,A,K}$	In the Keldysh formulation: Retarded (R) G, Advanced (A) G, and Keldysh (K) G
$F^{R,A,K}$	In the Keldysh formulation: Retarded (R) F, Advanced (A) F, and Keldysh (K) F
$\hat{g}$	The Eilenberger–Larkin–Ovchinnikov Green's functions, which are the Gorkov ones averaged over an energy variable
$\check{g}$	The Eilenberger–Larkin–Ovchinnikov Green's functions, which are the Gorkov ones averaged over an energy variable, in the Keldysh formulation, with both the advanced and retarded, as well as the Keldysh component. The matrix is in the direct product space of forward and backward branches of the Keldysh contour, with the Nambu space.
g	The normal components of the Eilenberger–Larkin–Ovchinnikov Green's functions, which are the Gorkov ones averaged over an energy variable. They represent the diagonal components of the $\hat{g}$ matrix
f	The superconducting components of the Eilenberger–Larkin–Ovchinnikov Green's functions, which are the Gorkov ones averaged over an energy variable, representing the off-diagonal components of the $\hat{g}$ matrix
σ	Electron spin index
$\sigma_{x,y,z}$	x: x-component of Pauli matrix; y: y-component, z: z-component
$g^{R,A,K}$	In the Keldysh formulation: Retarded (R) g, Advanced (A) g, and Keldysh (K) g
$f^{R,A,K}$	In the Keldysh formulation: Retarded (R) f, Advanced (A) f, and Keldysh (K) f
f_{QPS}	QPS fugacity
x_0	Core size of a quantum phase-slip, typically taken to be of order ξ_0
τ_0	Time-scale of a quantum phase-slip, typically $\gtrsim \hbar/\Delta$
D_N	Inverse of the compressibility of normal electron fluid
D_S	Inverse of the compressibility of superconducting electron fluid
V_N	Voltage of the normal fluid
V_S	Voltage of the super fluid
r	Normal electron to Cooper-pair conversion resistance
γ	Parameter proportional to the rate of normal electron to Cooper-pair conversion
λ_Q	Length scale for the normal electron to Cooper–pair conversion process

$f_{\mathrm{dpl,p}}$	Fugacity of QPS–anti-QPS dipoles, separated by p lattice sites
K or K_{s}	The superconducting stiffness
$K_{\mathrm{s,w}}$ or K_{w}	The superconducting stiffness for a 2D strip of width w
α_{dis}	Dimensionless disorder parameter in the Khlebnikov–Pryadko theory of quantum phase-slips
A_{dis}	A parameter, which characterizes the correlation integral of the disorder potential
C_{w}	Capacitance per unit area of a narrow 2D superconducting strip
$L_{\mathrm{kin,w}}$	Kinetic inductance for a unit area of a narrow 2D superconducting strip
q	Vortex charge-vector
$q^{0;1,1}$	The 0-th, 1st and 2nd components of the (2+1)D q-vector
J_{vor}	Vortex current density-vector
$f_{\mu\nu}$	The field tensor components of an effective photon field
n_{cp}	Cooper-pair number
n_{flux}	The flux number penetration a superconducting loop
ω_{pl}	Plasma frequency, e.g. in a Josephson junction, or that characterizing the vibrations within the local minimum of the free-energy
V_{c}	The critical voltage in a QPS-junction. It is the dual to the critical current I_{c} in a conventional Josephson Junction

Color Plates

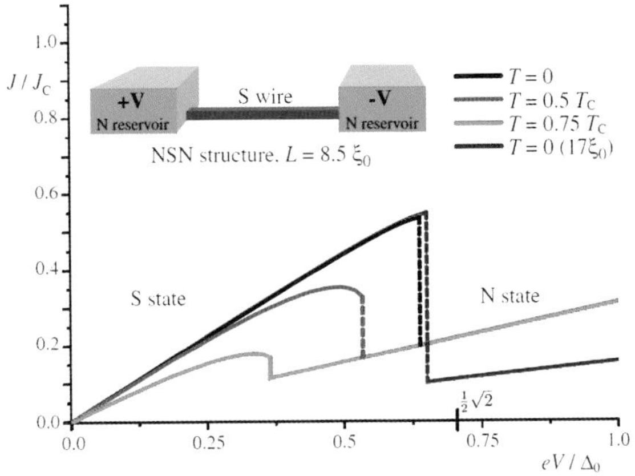

Figure 5.4 The calculated current (J)–voltage (V) relation of a superconducting wire of length $L = 8.5\xi_0$ between normal metallic reservoirs (see inset) at several temperatures, and for a wire of length $17\xi_0$. J_c is the critical current density, and Δ_0 the bulk gap energy. From [8].

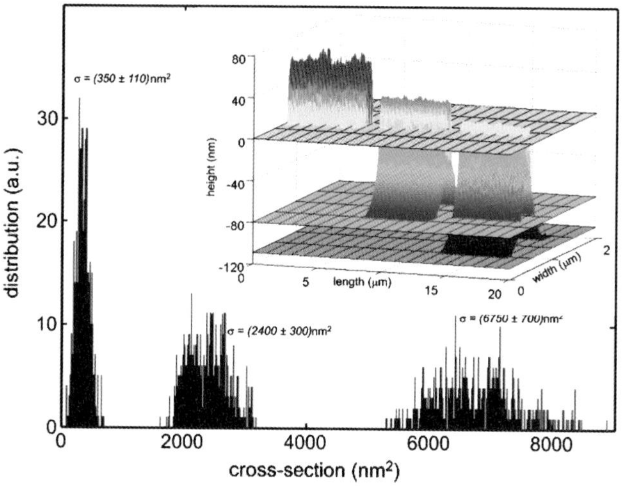

Figure 6.7 Histograms showing the distribution of the wire cross section before and after sputtering. To collect statistics, about 500 SPM (scanning probe microscope) scans were taken across the wire with the step along its axis ~ 12 nm, which is comparable to the radius of curvature of the SPM tip. Narrowing of the histograms is due to the "polishing" effect of ion sputtering. The inset shows the evolution of the sample shape while sputtering measured by SPM. The bright color above the gray plane corresponds to Al and the blue below the plane to Si. Planes (gray, green, orange) indicate Si substrate base levels after successive sessions of sputtering. As Si is sputtered faster than Al, the wire is finally situated at the top of the Si pedestal. Gray plane (height = 0) separates Si from Al. From [39].

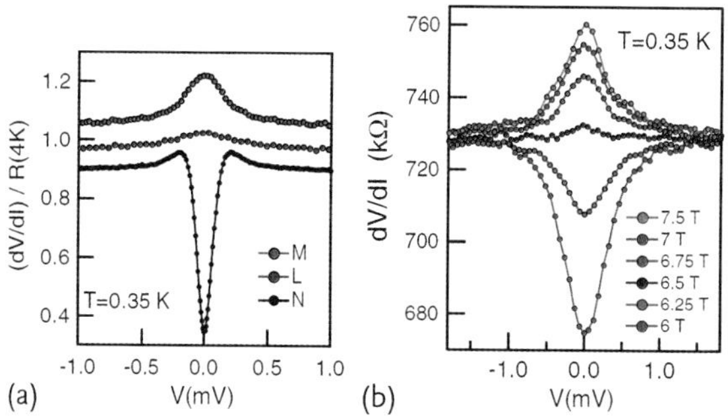

Figure 7.6 (a) Normalized differential resistance at $T = 0.35$ K for indicated MoGe wires (M, L insulating, N superconducting). Data for wires L and N are downshifted by 0.05 and 0.1, respectively. (b) Differential resistance as a function of bias voltage at $T = 0.35$ K in the transitional regime of the SIT for superconducting nanowires F. Adapted from [41].

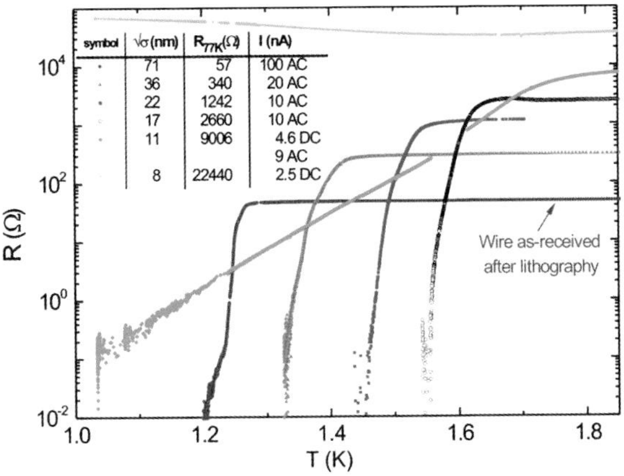

Figure 7.7 Resistance vs. temperature for the same aluminum wire of length $L = 10\,\mu\text{m}$ after several sputtering sessions. The sample and the measurement parameters are listed in the table. For low-Ohmic samples, lock-in AC measurements with the front-end preamplifier with input impedance $100\,\text{k}\Omega$ were used; for resistance above $\sim 500\,\Omega$, a DC nanovolt preamplifier with input impedance $\sim 1\,\text{G}\Omega$ was used. The absence of data for the 11 nm sample at $T \sim 1.6$ K is due to switching from a DC to AC setup. Note the qualitative difference of $R(T)$ dependencies for the two thinnest wires from the thicker ones. From [34].

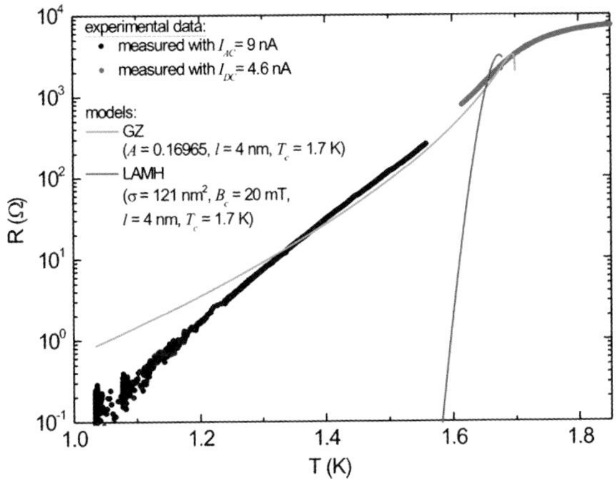

Figure 7.8 $R(T)$ dependence for the $\sqrt{\sigma} \sim 11$ nm aluminum sample. The green line shows the result of fitting to a renormalization theory (Reference 18 in [34]) with A, l, and T_c being the fitting parameters. The same set of parameters together with the critical magnetic field B_c measured experimentally is used to show the corresponding effect of thermally activated phase slips on the wire's $R(T)$ transition (red line) (References 1,2 in [34]). The parameter σ is obtained from the normal state resistance value and the known sample geometry (Reference 12 in [34]). The estimation for σ is in reasonable agreement with SPM analysis as well as with evolution of σ over all sputtering sessions (see Figure 7.7). From [34].

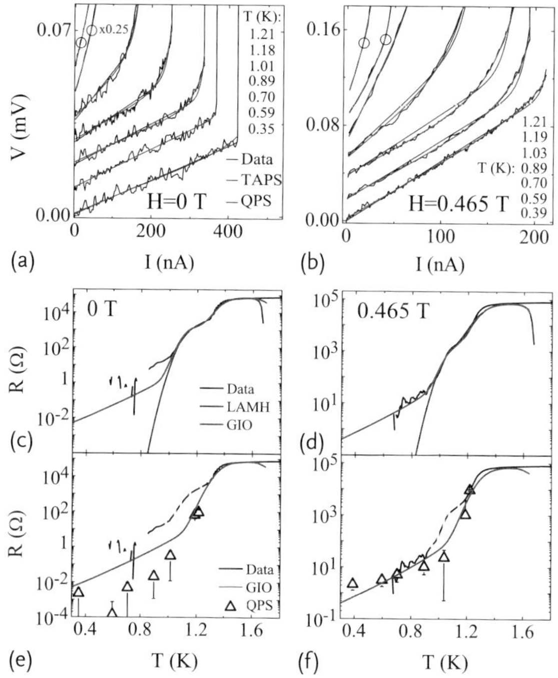

Figure 7.9 Nonlinear I–V curves and linear resistance for Al nanowire sample at different magnetic fields (H): Black curves – data; red curves – fits to the GIO ($\equiv$ GIO + LAMH; GIO denotes Giordano) expressions for QPS ($\equiv$ TAPS + QTPS; TAPS denotes thermal-activated phase-slips and QTPS quantum-tunneling of phase-slips), and blue curves – fits to the LAMH expressions for TAPS alone. (a) and (b) show I–V curves offset for clarity. The fits to QPS are of higher quality compared to TAPS; each fit includes a series resistance term V_S. These I–V curves can be fitted equally well by a power law form $V = V_S + (I/I_k)^{\nu}$ where $12 < \nu < 3.2$ for $0 \leq H \leq 1.05$ T [21, 37]. (c) and (d) show linear resistance after background subtraction (see Figure 1a in [24] and [37]). The LAMH fits are poor at low T. (e) and (f) show the resistance contribution due to phase slips (R_{QPS}) extracted in (a) and (b) from fits to the I–V curves using the GIO expressions (discrete points, Δ). RQPS and the linear resistance from (c) and (d) are refitted using the GIO expressions (red) with the same $a_{GIO}(= 1.2)$ while disregarding the irrelevant shoulder feature (dashed line). Adapted from [24].

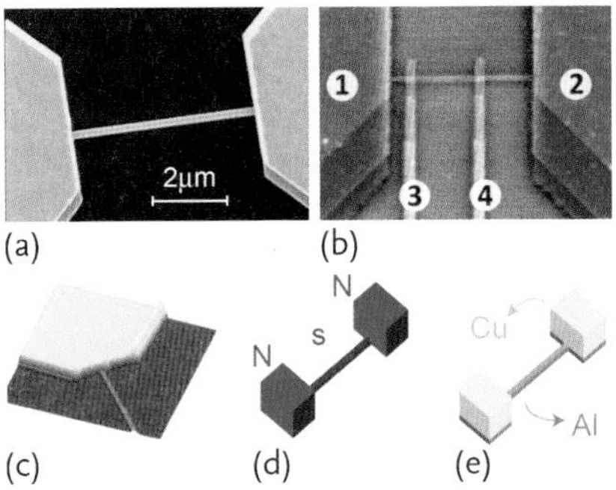

Figure 7.11 A superconducting Al nanowire connected to two massive normal reservoirs, consisting of the same Al, covered by a normal metal Cu layer: (a) SEM-picture, (c) AFM-picture, and (d) and (e) show a schematic representation. The thin Al of the pads is driven normal by the inverse proximity effect of the thick normal Cu. Normal tunneling probes are attached for local measurements (b). From [53].

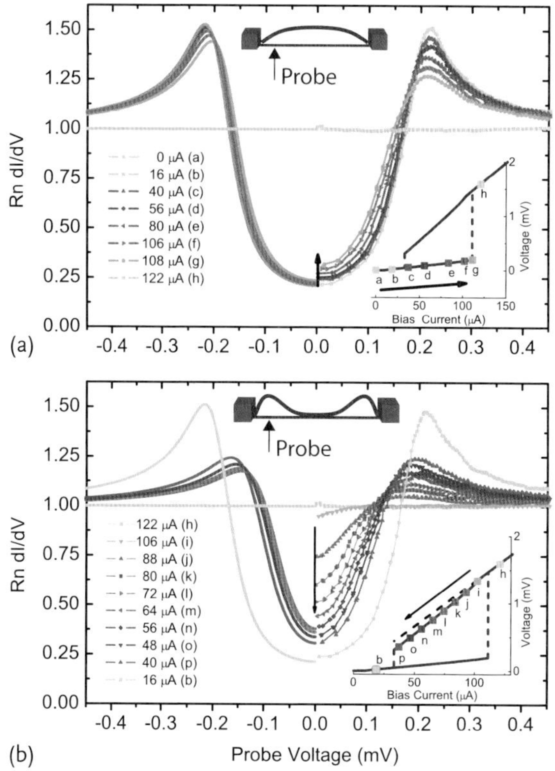

Figure 7.12 The local density of states (DOS $\propto dI/dV$) (a) the global superconducting state and (b) the bimodal state for different bias currents I_{12} of a 100 nm wider Al nanowire, measured at 200 mK. For the global superconducting state, the gap is only weakly dependent on the bias current, while for the bimodal state, for which only regions adjacent to the normal contacts are superconducting while the middle of the nanowire is in the normal state, one observes a DOS gradually changing from a normal into a superconducting state. From [53].

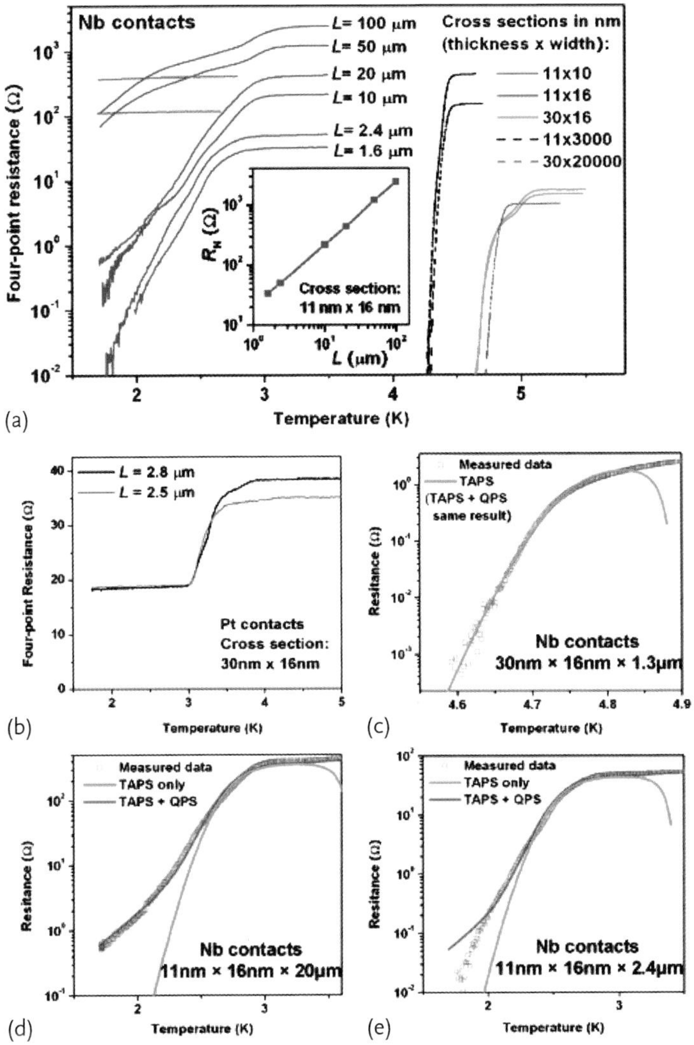

Figure 7.13 Temperature dependence for the four-point resistance of Nb NW arrays and films. (a) Superconducting Nb contacted NW arrays and films. Red lines: arrays of 12 NWs of cross section 11 nm × 10 nm and length L (from top to bottom) = 3 and 0.9 μm. Blue lines: arrays of 100 NWs of cross section 11 nm × 16 nm and L = 100, 50, 20, 10, 2.4, and 1.6 μm. Green lines: arrays of 250 NWs of cross section 30 nm × 16 nm and L = 1.5 and 1.3 μm. Black dashed lines: 11 nm thick films with width of 3 μm, L = 60 and 20 μm. Purple dashed line: a 30 nm thick film with a width of 20 μm, L = 2.5 μm. Inset: Length dependence of the normal-state resistance for arrays of 100 NWs of cross section 11 nm × 16 nm. (b) Normal-state Pt contacted NW arrays of cross section 30 nm × 16 nm. (c) 30 nm × 16 nm × 1.3 μm data in (a) fitted to TAPS theory (TAPS + QPS gives an indistinguishable result). (d) 11 nm × 16 nm × 20 μm data fitted to the theories. (e) 11 nm × 16 nm × 2.4 μm data fitted to the theories. From [40].

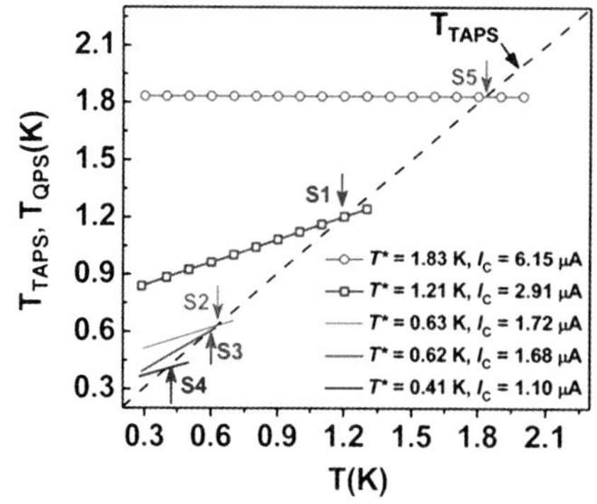

Figure 7.17 The best-fit effective temperature for fluctuations at different bath temperatures for five different $Mo_{1-x}Ge_x$ nanowire samples (S1–S5). For all TAPS rate calculations, the effective temperature is chosen as the bath temperature (shown by the black dashed line). For the QPS rates, the effective temperature TQPS, used in the corresponding QPS fits, similar to the blue-line fits of b [4], is shown by the solid lines. For each sample, below the crossover temperature T^* (indicated by arrows), QPS dominates the TAPS. They find that the T^* decreases with decreasing critical depairing current of the nanowires, which is the strongest proof of QPS. The trend indicates that the observed behavior of TQPS below T^* is not due to extraneous noise in the setup or granularity of wires, but, indeed, is due to QPS. Adapted from [4].

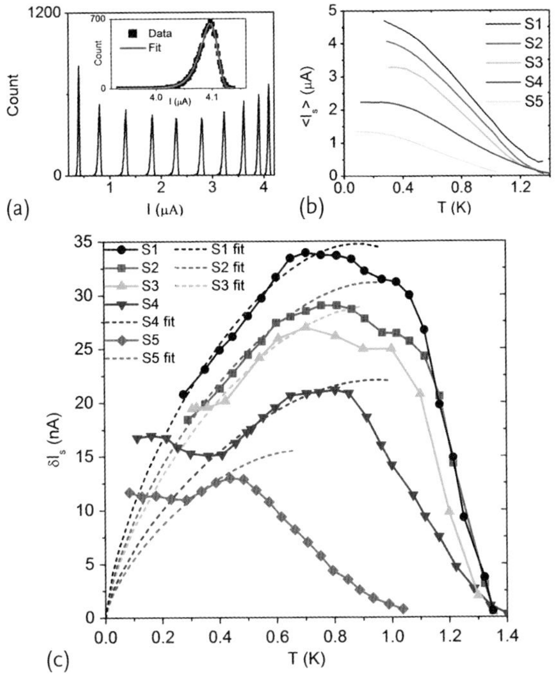

Figure 7.18 (a) I_s distribution for Al nanowire S2 at different temperatures: right to left: 0.3–1.2 K in 0.1 K increments. The inset shows the 0.3 K distribution, fitted by the Gumbel distribution (Reference 20 in [28]). (b) $\langle I_s \rangle$ vs. temperature – top to bottom S1–S5. (c) Symbols: δI_s vs. temperature. Dashed lines: fittings in the single TAPS regime using (Eq. 2 in [28]). An additional scale factor of 1.25, 1.11, 1.14, 0.98, and 1.0, for S1–S5, respectively (average 1.1 ± 0.1), is multiplied to match the data. Alternatively, a $\approx$ 6% adjustment in the exponent fits the data without the scale factor. Adapted from [28].

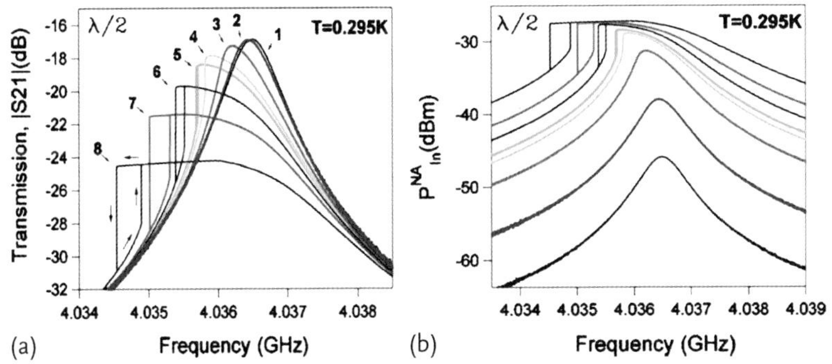

Figure 7.19 (a) ($Mo_{1-x}Ge_x$ sample S2) Transmission amplitude S_{21} (dB) in forward and backward frequency sweep for various driving powers. The graph shows Duffing bifurcation occurring at higher driving powers. The curves correspond to different driving powers: 1: $P_{out}^{NA} = -29$ dBm (black); 2: -21 dBm (blue); 3: -14 dBm (red); 4: -11 dBm (orange); 5: -10 dBm (green); 6: -8 dBm (black); 7: -6 dBm (violet); and 8: -3 dBm (black). (b) (sample S2) Replotting of the data from (a) as the transmitted power P_{in}^{NA} measured at the network analyzer input vs. frequency. From [60].

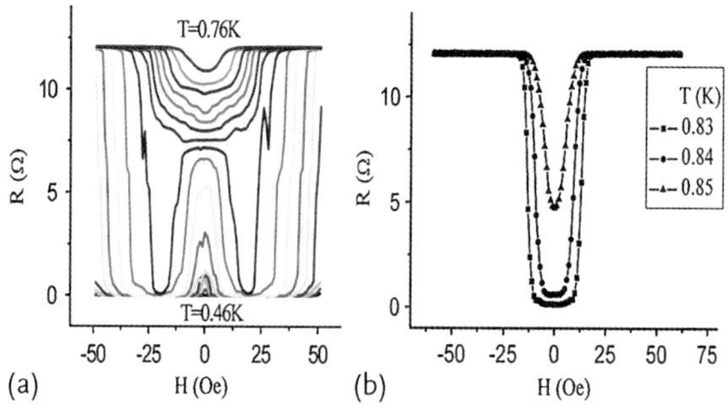

Figure 7.25 (a) Magnetic-field dependence of Zn wire resistance, at a high current $I \approx 4.4\ \mu A$, with temperatures ranging from 0.46 to 0.76 K, every 0.02 K. (b) Magnetic-field dependence of wire resistance, at a low current $I = 0.4\ \mu A$, with temperatures ranging from 0.83 to 0.85 K, every 0.01 K. From [31].

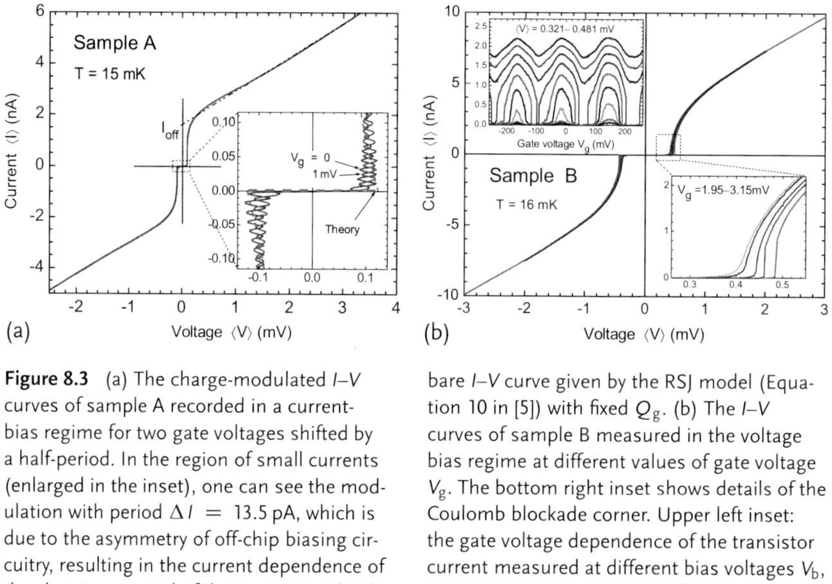

Figure 8.3 (a) The charge-modulated I–V curves of sample A recorded in a current-bias regime for two gate voltages shifted by a half-period. In the region of small currents (enlarged in the inset), one can see the modulation with period $\Delta I = 13.5$ pA, which is due to the asymmetry of off-chip biasing circuitry, resulting in the current dependence of the electric potential of the transistor island, $\delta V = (R_{\text{bias1}} - R_{\text{bias2}})I$ and, therefore, of the effective gate charge, $\delta Q_g = (C_g + C_0)\delta V$. The green dashed line shows the shape of the bare I–V curve given by the RSJ model (Equation 10 in [5]) with fixed Q_g. (b) The I–V curves of sample B measured in the voltage bias regime at different values of gate voltage V_g. The bottom right inset shows details of the Coulomb blockade corner. Upper left inset: the gate voltage dependence of the transistor current measured at different bias voltages V_b, providing a steady increase of $\langle V \rangle$ from 0.321 up to 0.481 mV in 20 μV steps (from bottom to top). Adapted from [5].

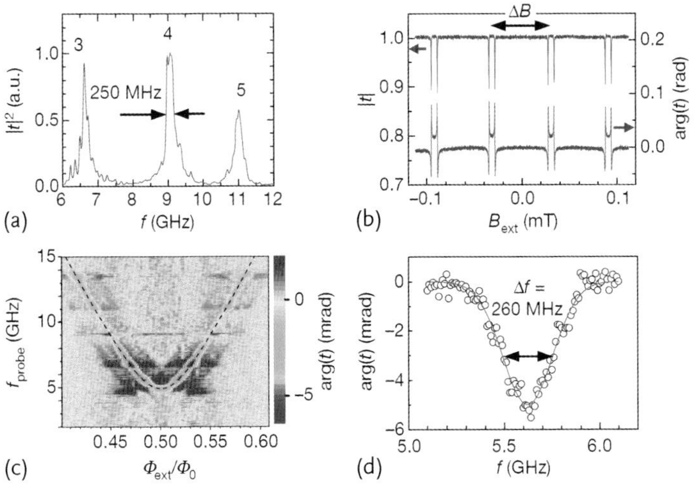

Figure 8.8 Experimental data. (a) Power transmission through the resonator measured within the bandwidth of our experimental setup. Peaks in transmission power coefficient, $|t|^2$, correspond to resonator modes, with mode number m indicated for each peak (a.u., namely, arbitrary units). (b) Transmission through the resonator as function a of external magnetic field B_{ext} at $m = 4$ ($f_4 = 9.08\,\text{GHz}$). The periodic structure in amplitude ($|t|$) and phase ($\arg(t)$) corresponds to the points where the lowest-level energy gap $\Delta E/h$ matches f_4. The period $\Delta B = 0.061\,\text{mT}$ ($= \Phi_0/S$) indicates that the response comes from the loop (shown in Figure 1b) with the effective loop area $S = 32\,\mu\text{m}^2$. (c) The two-level spectroscopy line obtained in two-tone measurements. The phase of transmission, $\arg(t)$, through the resonator at f_4 is monitored, while another tone with frequency f_{probe} from an additional microwave generator, and B_{ext}, are independently swept. The plot is filtered to eliminate the contribution of other resonances ($2 < m < 6$), visible as horizontal red features. The dashed line is the fit to the energy splitting, with $\Delta E/h = 4.9\,\text{GHz}$, $I_{\text{p}} = 24\,\text{nA}$. (d) The resonant dip is measured at $\Phi/\Phi_0 = 0.52$. The red curve is the Gaussian fit. From [14].

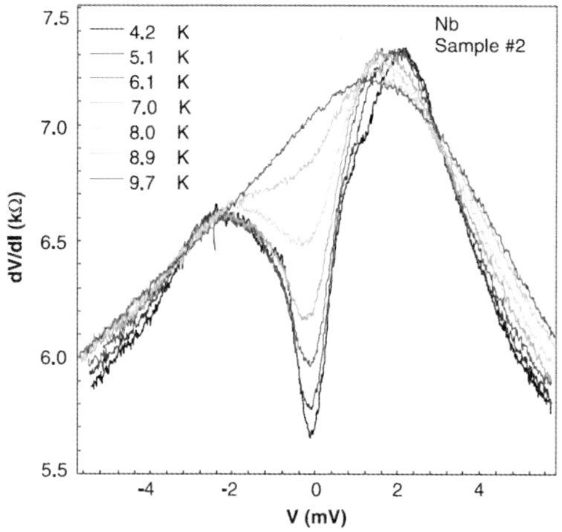

Figure 9.3 Differential resistance dV/dI vs. bias voltage V for a SWNT sample #2 of diameter $D < 1.8$ nm with Nb electrodes over a range of temperatures around T_c of Nb (9.2 K). The magnitude of the Andreev dip decreases with increasing T and disappears above T_c. From [5].

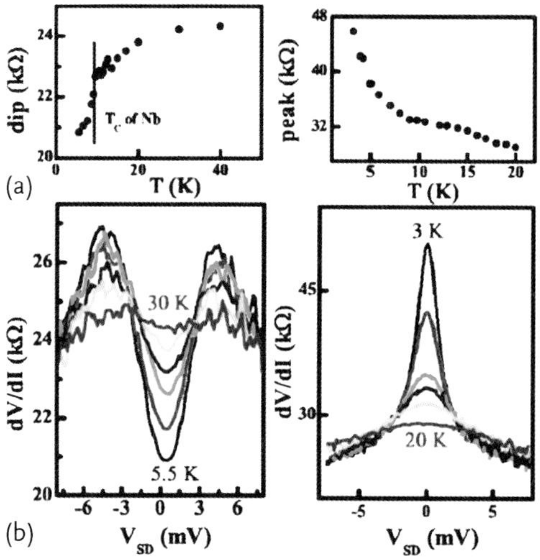

Figure 9.5 Differential resistance as a function of bias and temperature for CNFET (carbon nanotube field-effect-transistor) no. 1. $V_g = 47$ V for the dip on the left. $V_g = 48$ V for the peak on the right. The temperatures for the curves shown in (c) are 5.5, 8.5, 9.5, 15, 20, and 30 K. The temperatures for the curves shown in (d) 3, 4, 7, 9, 15, and 20 K. From [10].

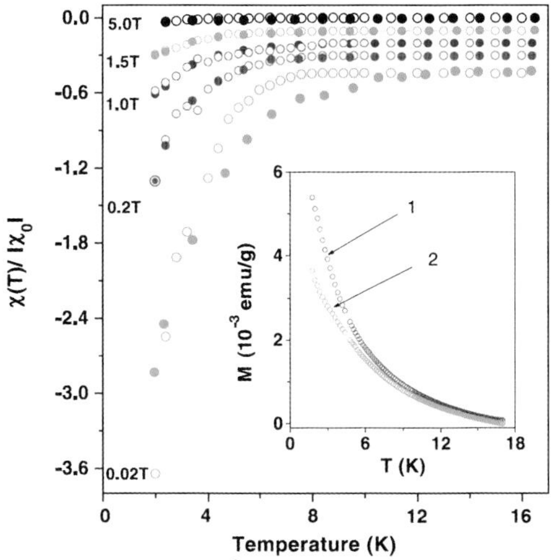

Figure 9.7 Normalized magnetic susceptibility of the SWNTs plotted as a function of temperature for five values of the magnetic field. The curves are displaced vertically for clarity. Values shown are for theory (open symbols) and experiment (filled symbols). χ_0 denotes the value of the susceptibility at $T = 1.6$ K and magnetic field $= 0.2$ T. The experimental value of χ_0, when normalized to the volume of the SWNTs, is -0.015 (in units where $B = 0$ denotes $\chi_0 = -1$). The scatter in the theory points reflects statistical fluctuations inherent in the Monte Carlo calculations. (Inset) Temperature dependence of magnetization density for zeolite AFI crystallites (curve 1) and for AFI crystallites with SWNTs in their channels (curve 2). Both curves are measured at 2000 Oe. From [13].

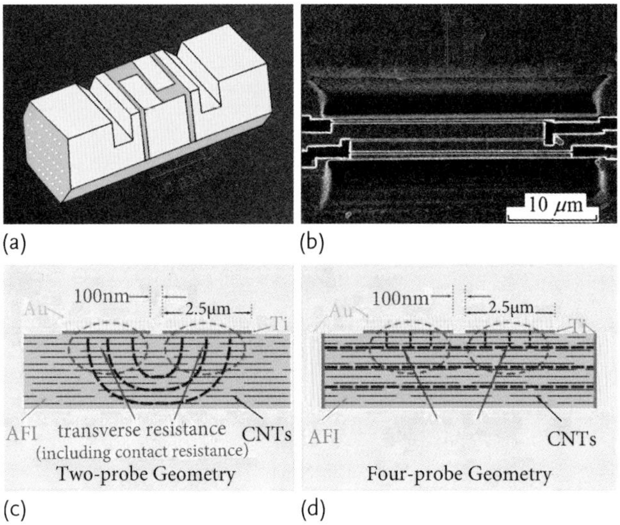

Figure 9.8 (a) Cartoon of the sample contain an array of SWNT in an AFI zeolite crystal. Yellow denotes gold and green denotes the AFI crystal surface exposed by FIB etching. Nanotubes are delineated schematically by open circles. (b) SEM image of one sample. The *c*-axis is along the N-S direction. The thin, light, horizontal line in the middle is the 100 nm separation between the two surface voltage electrodes that are on its two sides. The dark regions are the grooves cut by the FIB and sputtered with Au/Ti to serve as the end-contact current electrodes. (c) and (d) show schematic drawings of the two-probe and four-probe geometries, respectively. Blue-dashed lines represent the current paths. In (d), the two end-contact current pads are 4 mm in depth and 30 mm in width. The difference between the two-probe and the four-probe measurements is the transverse resistance, delineated by the red circles in (c). From [17].

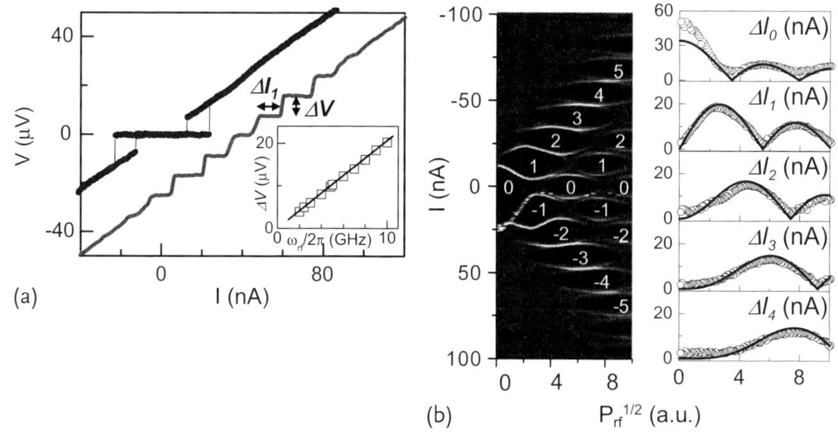

Figure 9.10 (a) $V(I)$ characteristics for device no. 3 at 40 mK, with (red) and without (black) an externally applied 4-GHz radiation (this device has $R_N = 860\,\Omega$ and $I_C = 26$ nA at $T = 40$ mK). The red trace is horizontally offset by 40 nA. The applied microwave radiation results in voltage plateaus (Shapiro steps) at integer multiples of $\Delta V = 8.3\,\mu$V. (Inset) Measured voltage spacing ΔV (symbol) as a function of microwave angular frequency ω_{RF}. The solid line (theory) shows the agreement with the AC Josephson relation $\Delta V = \hbar\omega_{RF}/(2e)$. From [30].

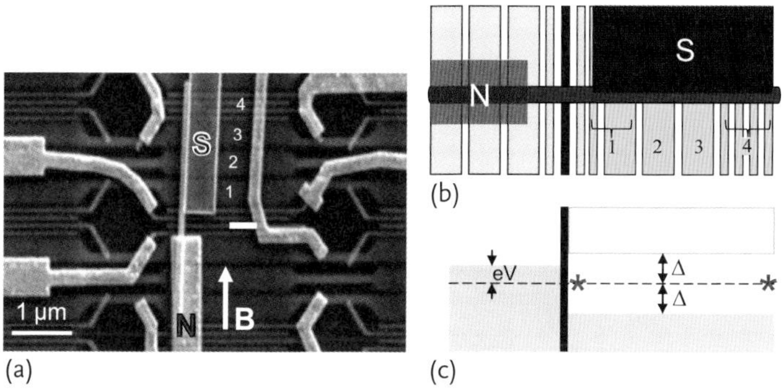

Figure 9.11 (a) Implemented version of theoretical proposals, using InSb SmNW and Nb superconductor. Scanning electron microscope image of the device with normal (N) and superconducting (S) contacts. The S contact only covers the right part of the InSb nanowire. The underlying gates, numbered 1 to 4, are covered with a dielectric. (Note that gate 1 connects two gates, and gate 4 connects four narrow gates; see (b).) (b) schematic of our device. (c) an illustration of energy states. The black rectangle indicates the tunnel barrier separating the normal part of the nanowire on the left from the wire section with induced superconducting gap, Δ. (In (a), the barrier gate is also shown in white.) An external voltage, V, applied between N and S drops across the tunnel barrier. (Red) stars, again, indicate the idealized locations of the Majorana pair. Only the left Majorana is probed in this experiment. Adapted from [24].

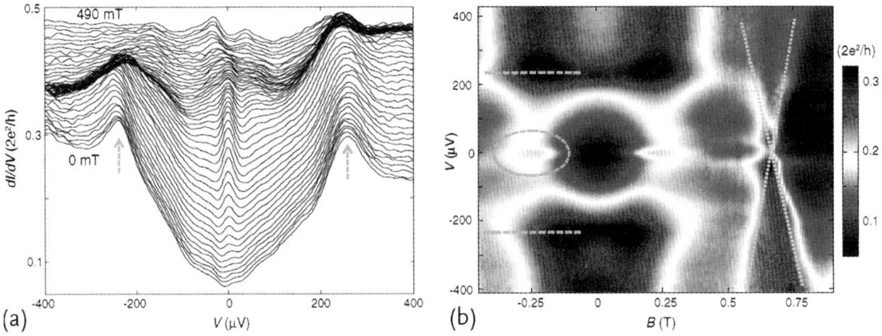

Figure 9.12 Magnetic field-dependent spectroscopy. (a) dI/dV vs. V at 70 mK taken at different B fields (from 0 to 490 mT in 10 mT steps; traces are offset for clarity, except for the lowest trace at $B = 0$). Data are from device 1. Arrows indicate the induced gap peaks. (b) Colorscale plot of dI/dV vs. V and B. The ZBP is highlighted by a dashed oval; green dashed lines indicate the gap edges. At ≈ 0.6 T, a non-Majorana state is crossing zero bias with a slope equal to $\approx$ 3 meV/T (indicated by sloped yellow dotted lines). Traces in (a) are extracted from (b). From [24].

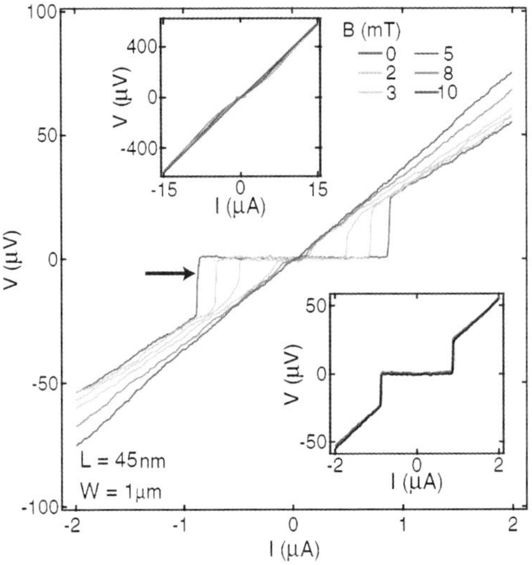

Figure 9.13 (main) V vs. I for Al-Bi_2Se_3-Al S-TI-S devices of dimensions $(L, W) =$ (45 nm, 1 μm) for $B = 0, 2, 3, 5, 8, 10$ mT and at a temperature of 12 mK. At $B = 0$, I_C is 850 nA, which is reduced upon increasing B. For this device, the product $I_C R_N = 30.6$ μV, which is much lower than theoretically expected for conventional JJs. (Upper-left inset) I–V curves overlap for all values of B at $V \geq 2\Delta/e \sim 300$ μV. (Lower-right inset) Sweeps up and down in I show little hysteresis, indicating that the junction is in the overdamped regime. Adapted from [36].

Part One Theoretical Aspects of Superconductivity in 1D Nanowires

1
Superconductivity: Basics and Formulation

1.1
Introduction

The Bardeen–Cooper–Schrieffer (BCS) theory captures the essential physics that gives rise to the condensation of the Cooper pairs into a coherent superconducting state. It specifically deals with a homogeneous and clean system in three dimensions and is essentially a self-consistent, mean field treatment. Following its success, subsequent theories are based on a mean field treatment on the BCS or related models. In order to account for inhomogeneous systems, such as those occurring in multilayer systems of superconductor- insulators, or normal metals, de Gennes, and independently Bogoliubov, derived a system of coupled equations between electrons and holes that yield the solutions for the fermionic quasiparticles above the superconducting condensate, which are separated from the condensate by the superconducting energy gap, Δ. Such a formalism is useful for computing the properties of ultrathin SC films and clean SC nanowires, as will be discussed in Chapter 2.

Prior to the BCS microscopic theory, Ginzburg and Landau sought to generalize the Landau theory of second order phase transitions to the superconductivity problem. The result is the celebrated Ginzburg–Landau (GL) theory of superconductivity. The free energy is written in terms of the superconducting order parameter, which in this case is a complex quantity. This theory is expected to be valid near T_c, and was derived from a microscopic BCS model by Gorkov.

Gorkov invented a powerful methodology by deducing the equations of motion for the Green's functions. An anomalous Green's function, F, accounting for pair-correlations was introduced in addition to the normal electron Green's function. The two Green's functions form a closed set of equations, the solutions of which yield all the results of the BCS theory, and moreover, can be readily extended to incorporate dirty systems with impurities, as well as deal with nonlinearities, dynamics, and so on. Thus, type-II superconductors can readily be described. From the perspective of this text, the central importance of the Gorkov equations is the ultimate deduction of the Usadel equation for related Green's functions (to the Gorkov GF's), in the limit of dirty systems. The derivation of the Usadel diffusion equation was based on the works of Eilenberger and of Larkin and Ovchinnikov,

One-Dimensional Superconductivity in Nanowires, First Edition. F. Altomare and A.M. Chang

who independently applied the quasiclassical approximation to the Gorkov equations, and identified an energy-integrated version of the Gorkov Green's functions. These approaches led to a simplification of the Gorkov equations into Boltzmann transport-like equations for these modified Green's functions.

The Usadel diffusion equation is much more tractable and amenable to numerical implementations, enabling realistic experimental geometries and situations to be analyzed. In particular, issues of quasiparticle injection at the normal-superconductor interface, nonequilibrium quasiparticle distribution, and so on, are readily computed. These methodologies based on the Usadel equation are naturally suited to analyzing systems with 1D SC nanowires. In fact, Dr. Pauli Virtanen provides downloadable C++ programs for such computations on his website: http://ltl.tkk.fi/~theory/usadel1/!

Beyond these standard methodologies, a powerful technique for analyzing complex superconducting phenomena has been developed over the past 20 years, which is particularly well-suited for understanding quantum phase transitions, such as those in dissipative Josephson junction arrays of 2D and 1D varieties, dissipative single Josephson junctions, vortex states and associated transitions, as well as transitions in the 1D superconducting nanowire system. This technique is based on the path integral formulation. Thus, introducing such a formulation is important in bridging the gap to enable the study of quantum-tunneling processes, such as the macroscopic quantum tunneling of the phase slip, which is central to the understanding of the unique behaviors in 1D superconducting nanowires.

1.2 BCS Theory

Conventional superconductivity has, as its key components, the binding of the Cooper pairs, and the Bose condensation of such Cooper pairs. The binding arises from an attractive interaction between electrons mediated by phonons or other excitations, for example, plasmons. The essential behaviors and outstanding characteristics are captured by a self-consistent mean field approximation. The celebrated BCS theory, proposed by Bardeen, Cooper, and Schrieffer in 1957 [1], provides both a microscopic model, which provides the underpinning of our current understanding of conventional superconductivity and a basic formalism, which is most conveniently cast in terms of the second-quantized formalism, that is, in terms of the creation and destruction operators of the single particle electronic states.

The Cooper instability [2] indicates that the Fermi sea is unstable to the formation of bound-pairs (Cooper pairs) of electrons, in the presence of an effective, attractive interaction. In the original formulation, and in most conventional superconductors to date, the attractive interaction is a time-retarded interaction mediated by lattice vibration phonons. A passing electron, whose velocity, given by the Fermi velocity, is much larger than the phonon propagation speed, polarizes the local ions: before the lattice can relax, a second electron arrives and feels the attraction from the still-polarized positive ions. This attraction produces a pairing of electrons which gives rise to an entity which is bosonic in character. The two electrons thus form a Cooper pair whose size-scale is denoted by the superconducting coherence length ξ.

The Bose condensation of such Cooper pairs into a macroscopic, coherent quantum ground state is the source of the unusual physical properties associated with superconductivity. In a weak-coupling superconductor, where the electron–phonon coupling is much smaller than the Fermi energy, the size of the Cooper pair, or ξ, is typically much larger than the Fermi wavelength λ_{F}. In a homogeneous system, the attraction is maximal when the relative total momentum of the two electrons is zero, that is, when the center of mass is stationary. This gives the requirement of pairing between $+\boldsymbol{k}$ and $-\boldsymbol{k}$ states. At the same time, the s-wave channel has the largest attractive potential, leading to an s-wave superconductor, for which the energy gap created by the condensation is isotropic in k-space. As the spatial wavefunction is symmetric, in order to satisfy Pauli's exclusion principle, the spin must be in an antisymmetric, singlet state.

To simplify the calculations, the BCS model assumes that the attractive phonon-mediated potential is a delta-function in real space (a contact interaction). This is sensible since the interaction occurs on the length scale of the Fermi wavelength λ_{F}, and the length scale of the phonons associated with the Debye frequency, of order the lattice constant, whereas the Cooper pair has a much larger scale given by the superconducting coherence length, ξ. The coherence length has its minimum value at $T = 0$, with a value typically exceeding 10 nm.

The second quantization formulation is then equivalent to a formulation based on the occupation of single-particle states – in this case of a homogeneous, disorder-free system, the $\boldsymbol{k}$-states. The Pauli exclusion in the occupation dictating that a given state to be either singly occupied or unoccupied, with multi-occupation forbidden, is encoded in the anticommutation relation of the creation and annihilation operators. In the $\boldsymbol{k}$-representation, the model Hamiltonian is given by

$$H = \sum_{k\sigma} \epsilon_k c^{\dagger}_{k\sigma} c_{k\sigma} - \sum_{k,k':|\epsilon_k|,|\epsilon_{k'}|\leq \hbar\omega_{\mathrm{D}}} V c^{\dagger}_{k\uparrow} c^{\dagger}_{-k\downarrow} c_{-k'\downarrow} c_{k'\uparrow} . \tag{1.1}$$

Here, $c^{\dagger}_{k\sigma}$ is the creation operator for an electron in momentum-state $\boldsymbol{k}$ and spin σ. $\epsilon_k = \hbar^2 \boldsymbol{k}^2/(2m) - E_{\mathrm{F}}$ is the energy of the single-particle $\boldsymbol{k}$-state measured from the Fermi energy of the system, $V > 0$ for an attractive interaction, and the scattering is between $(\boldsymbol{k}, -\boldsymbol{k})$ – and $(\boldsymbol{k}', -\boldsymbol{k}')$-states. The pairing in real space is an s-wave, which is symmetric under exchange, while the spin is an antisymmetric singlet, in order to satisfy Pauli exclusion. Because phonons mediate the interaction, the scattering term is restricted to states within $\hbar\omega_{\mathrm{D}}$ of the Fermi surface, where ω_{D} is the Debye frequency. This form of the phonon-mediated attractive interaction between electrons is equivalent to a contact potential interaction in real space, that is,

$$V(\boldsymbol{r}_i, \boldsymbol{r}_j) = -V\mathsf{V}\delta(\boldsymbol{r}_i - \boldsymbol{r}_j) = -g\delta(\boldsymbol{r}_i - \boldsymbol{r}_j) , \tag{1.2}$$

between electrons at positions $\boldsymbol{r}_i$, $\boldsymbol{r}_j$, where V is the interacting strength, V is the system volume, and g the volume-independent Gorkov coupling.

To find the ground state energy of the system, a variational wavefunction in the form is assumed

$$|\Psi_{\text{BCS}}\rangle = \prod_k \left(u_k + v_k c_{k\uparrow}^\dagger c_{-k\downarrow}^\dagger\right)|0\rangle \,, \tag{1.3}$$

where $|0\rangle$ is the vacuum, and u_k and v_k are the amplitude of electron-like Cooper pairs and hole-like Cooper pairs, respectively, chosen to be real, and subjected to the normalization condition $u_k^2 + v_k^2 = 1$. By varying the expectation value of the ground state E_{g} energy and minimizing $E_{\text{g}} = \langle\Psi_{\text{BCS}}|H|\Psi_{\text{BCS}}\rangle$ with respect to v_k, one finds the condition for

$$2\epsilon_k u_k v_k = \sum_{k':|\epsilon_{(k,k')}|\le\hbar\omega_{\text{D}}} V\left(u_k^2 - v_k^2\right) u_{k'} v_{k'} \tag{1.4}$$

for the ground state energy

$$\langle\Psi_{\text{BCS}}|H|\Psi_{\text{BCS}}\rangle = 2\sum_{k'} v_k^2 \epsilon_k - \sum_{k,k':|\epsilon_{(k,k')}|\le\hbar\omega_{\text{D}}} V u_k v_k u_{k'} v_{k'} \,. \tag{1.5}$$

Changing to the customary variables of E_k and Δ_k, where

$$u_k = \frac{1}{\sqrt{2}}\sqrt{1 + \frac{\epsilon_k}{E_k}} \tag{1.6}$$

and

$$v_k = \frac{1}{\sqrt{2}}\sqrt{1 - \frac{\epsilon_k}{E_k}} \tag{1.7}$$

with

$$E_k = \sqrt{\epsilon_k^2 + \Delta_k^2} \,, \tag{1.8}$$

the minimization condition yields the self-consistency expression

$$\Delta_k = \frac{1}{2}\sum_{k':|\epsilon_{k'}|\le\hbar\omega_{\text{D}}} \frac{V\Delta_{k'}}{\sqrt{\epsilon_{k'}^2 + \Delta_{k'}^2}} \,. \tag{1.9}$$

This is the BCS gap equation for the superconducting gap of a quasi-particle of momentum $\boldsymbol{k}$. The restriction in $\boldsymbol{k}'$ leads to a $\boldsymbol{k}$-independent gap

$$\Delta_k = \Delta \,, \quad |\epsilon_k| < \hbar\omega_{\text{D}} \,, \tag{1.10}$$

and

$$\Delta_k = 0 \,, \quad \text{otherwise} \,. \tag{1.11}$$

Converting to an integral using the normal electron density of states, and approximating the density of states with its value at the Fermi energy $N(0)$ gives

$$\frac{1}{N(0)g} = \int_0^{\hbar\omega_D} \frac{d\epsilon}{\sqrt{\epsilon^2 + \Delta^2}} . \tag{1.12}$$

The solution yields the celebrated BCS expression for the energy gap Δ

$$\Delta = 1.14\hbar\omega_D e^{-1/N(0)g} = 1.14\hbar\omega_D e^{-1/[N(0)VV]} , \tag{1.13}$$

with $g \equiv VV$ the Gorkov coupling constant. The nonanalytic dependence on g indicates that this result cannot be obtained by a perturbation calculation in the small parameter V/E_F!

At finite temperatures $T > 0$, the gap equation is modified by the Fermi–Dirac occupation of the fermionic excitations across the gap ($\beta = 1/(k_B T)$)

$$\frac{1}{N(0)g} = \int_0^{\hbar\omega_D} \frac{d\epsilon}{\sqrt{\epsilon^2 + \Delta^2}} \left[1 - 2\frac{1}{1 + \exp(\beta\sqrt{\epsilon^2 + \Delta^2})} \right] . \tag{1.14}$$

At T_c, the gap vanishes, giving the condition

$$\frac{1}{N(0)g} = \int_0^{\hbar\omega_D} \frac{d\epsilon}{\epsilon} \tanh \frac{\epsilon}{2k_B T} \tag{1.15}$$

and the relation

$$\Delta = 1.76 k_B T_c . \tag{1.16}$$

The BCS gap, $\Delta(T)$, occupies a central role in the phenomenon of conventional superconductivity. From the existence of the gap, many important properties can be derived, such as the quasiparticle excitation spectrum

$$E_k = \sqrt{\epsilon_k^2 + \Delta_k^2} \tag{1.17}$$

and the density of states

$$D_{qp}(E_k) = D(E_F)\frac{E_k}{\sqrt{E_k^2 - \Delta_k^2}} = N(0)\frac{E_k}{\sqrt{E_k^2 - \Delta_k^2}} ; \quad D(E_F) \equiv N(0) , \tag{1.18}$$

the London equations [3] accounting for the Meissner effect of flux expulsion, flux quantization, specific heat, critical current, and so on. Our purpose here is to introduce the essential aspect of BCS superconductivity, and to wet the appetite of those readers unfamiliar with this subject matter. We refer to the excellent standard textbooks, for example, de Gennes [4], Tinkham [5], Ashcroft and Mermin [6], Fetter and Walecka [7], and so on, for a comprehensive treatment.

1.3
Bogoliubov–de Gennes Equations – Quasiparticle Excitations

The BCS theory in the previous section is for a 3D, homogeneous system without disorder. An essential feature of the theory is the pairing between time-reversed states. The role of time-reversed states was highlighted by Anderson's well-known theoretical analysis, demonstrating that nonmagnetic impurities do not suppress the superconducting gap [8]. To go beyond a translational invariant system, and account for boundaries, and for the possibility of tunnel junctions or SNS bridges, the theories of Bogoliubov [9, 10] and de Gennes [11] are required.

In the method of Bogoliubov and de Gennes, a solution is sought for the fermionic electron-like and hole-like quasiparticle excitations above the gap, which separates the condensed Cooper pairs from these unbound, quasiparticles. The theory was developed in part to address situations where there are boundaries or coupling to normal metals [11]. Because the presence of boundaries is properly accounted for, this formalism lends itself readily to deduce the behaviors of clean, 1D superconducting nanowires in which the boundaries are in the two lateral directions, perpendicular to the length of the nanowire, as was done in the work of Shanenko, Croitoru, Peeters, and collaborators [12–14] (please see Chapter 2).

The fermionic electron- and hole-like quasiparticles are each linear combinations of the electron and hole wavefunctions in the normal state. However, because of the spatial dependences introduced by the boundaries, the linear combinations are in general not as simple as in the homogeneous case. The point contact attractive interaction between electrons may be spatially dependent: $-V(\boldsymbol{r})\mathrm{V}\delta(\boldsymbol{r}) \equiv -g(\boldsymbol{r})\delta(\boldsymbol{r})$ (see (1.2)).

The starting point is the equation of motion for the field operator $\psi_\sigma^\dagger(\boldsymbol{r}, t)$, which creates an electron of spin σ at position $\boldsymbol{r}$. In the special case of a homogeneous, clean system, it is related to the creation operator $c_{\boldsymbol{k}\sigma}^\dagger$ for the $\boldsymbol{k}\sigma$ state by the (inverse-)Fourier transform

$$\psi^\dagger(\boldsymbol{r}, t) = \frac{1}{\sqrt{2\pi}} \sum_{\boldsymbol{k}} e^{i\boldsymbol{k}\cdot\boldsymbol{r}} c_{\boldsymbol{k}}^\dagger(t) . \tag{1.19}$$

The usual commutation with the BCS Hamiltonian in real space gives the equation of motion:

$$\begin{aligned} -i\hbar\frac{\partial\psi_\sigma^\dagger(\boldsymbol{r}, t)}{\partial t} &= \left[\frac{\boldsymbol{p}^2}{2m} + U_0(\boldsymbol{r})\right]\psi_\sigma^\dagger(\boldsymbol{r}, t) \\ &\quad - g(\boldsymbol{r})\sum_{\sigma'}\psi_\sigma^\dagger(\boldsymbol{r}, t)\psi_{\sigma'}^\dagger(\boldsymbol{r}, t)\psi_{\sigma'}(\boldsymbol{r}, t) ; \quad g(\boldsymbol{r}) = V(\boldsymbol{r})\mathrm{V} . \end{aligned} \tag{1.20}$$

Here, the one-electron potential $U_0(\boldsymbol{r})$ includes the boundary and static impurity potentials. Using a Hartree–Fock mean field approximation, the term cubic in the field operators is replaced in the anomalous channel by

$$\left\langle \psi_\sigma^\dagger(\boldsymbol{r}, t)\psi_{\sigma'}^\dagger(\boldsymbol{r}, t)\right\rangle \psi_{\sigma'}(\boldsymbol{r}, t) , \tag{1.21}$$

and, in the normal channel by

$$\left\langle \psi^\dagger_{\sigma'}(\boldsymbol{r},t)\psi_\sigma(\boldsymbol{r},t)\right\rangle \psi^\dagger_\sigma(\boldsymbol{r},t)\,. \tag{1.22}$$

Here, the averaging denotes a thermal average. For the anomalous channel, in anticipation of a coupling similar to s-wave coupling in the homogeneous case, only the dominant terms, which come from the down- and up-spin correlators (and up- and down-), are kept, that is,

$$g(\boldsymbol{r})\left\langle \psi^\dagger_\downarrow \psi^\dagger_\uparrow\right\rangle = V(\boldsymbol{r})V\left\langle \psi^\dagger_\downarrow \psi^\dagger_\uparrow\right\rangle = \Delta^\dagger(\boldsymbol{r})\,, \tag{1.23}$$

which defines the partial potential $\Delta^\dagger(\boldsymbol{r})$. For the normal channel, the self-consistent potential contributes a term

$$U_{\text{sc}}(\boldsymbol{r}) = -g(\boldsymbol{r})\left\langle \psi^\dagger_{\sigma'}\psi_{\sigma'}\right\rangle \tag{1.24}$$

to the single-particle potential $U(\boldsymbol{r}) = U_{\text{o}}(\boldsymbol{r}) + U_{\text{sc}}(\boldsymbol{r})$, where U_{o} represents the contributions from static background ions, impurities, and boundaries. New fermionic operators $\gamma^\dagger_{n\tau}$, which carry the usual anticommutation relationships, are introduced, namely,

$$\begin{aligned}
\psi^\dagger_\uparrow(\boldsymbol{r},t) &= \sum_n \left[u^\dagger_n(\boldsymbol{r})e^{iE_nt/\hbar}\gamma^\dagger_{n\uparrow} + v_n(\boldsymbol{r})e^{-iE_nt/\hbar}\gamma_{n\downarrow}\right]\,,\\
\psi^\dagger_\downarrow(\boldsymbol{r},t) &= \sum_n \left[u^\dagger_n(\boldsymbol{r})e^{iE_nt/\hbar}\gamma^\dagger_{n\downarrow} - v_n(\boldsymbol{r})e^{-iE_nt/\hbar}\gamma_{n\uparrow}\right]\,,
\end{aligned} \tag{1.25}$$

where the energies E_n are positive. The eigenfunctions $u_n(\boldsymbol{r})$ and $v_n(\boldsymbol{r})$ satisfy the following coupled equations, namely,

$$\begin{aligned}
E_n u_n(\boldsymbol{r}) &= \left[\frac{\boldsymbol{p}^2}{2m} + U(\boldsymbol{r})\right] u_n(\boldsymbol{r}) + \Delta(\boldsymbol{r})v_n(\boldsymbol{r})\\
E_n v_n(\boldsymbol{r}) &= -\left[\frac{\boldsymbol{p}^2}{2m} + U(\boldsymbol{r})\right] v_n(\boldsymbol{r}) + \Delta(\boldsymbol{r})^* u_n(\boldsymbol{r})\,.
\end{aligned} \tag{1.26}$$

The pair potential satisfies the self-consistency condition

$$\Delta(\boldsymbol{r}) = g(\boldsymbol{r})\sum_n v^\dagger_n(\boldsymbol{r})u_n(\boldsymbol{r})\left\{1 - 2\frac{1}{1+\exp[E_n/(k_\text{B}T)]}\right\}\,. \tag{1.27}$$

In addition,

$$\begin{aligned}
U_{\text{sc}}(\boldsymbol{r}) = -g(\boldsymbol{r})\sum_n &\left\{|u_n(\boldsymbol{r})|^2\frac{1}{1+\exp[E_n/(k_\text{B}T)]}\right.\\
&\left.+|v^\dagger_n(\boldsymbol{r})|^2\left[1 - \frac{1}{1+\exp[E_n/(k_\text{B}T)]}\right]\right\}.
\end{aligned} \tag{1.28}$$

In the presence of an external magnetic field, these equations are supplemented, of course, by the London equations. As expected, in the homogeneous case in three dimensions, they reproduce the BCS results.

1.4
Ginzburg–Landau Theory

The Ginzburg–Landau (GL) [15] approach to the description of conventional superconductivity is based on the notion of an order parameter, and follows Landau's phenomenological theory of second order phase transitions. The free energy of the superconducting state is envisioned to differ from the normal state by a contribution from the condensate when the order parameter becomes nonzero below a transition temperature T_c. The thermodynamic state is given by the solution which minimizes the free energy with respect to variations in the order parameter. In the case of a charged superconductor, which couples to the electromagnetic scalar and vector field, minimization with respect to the potentials leads to the London equations. The GL equations therefore represent a mean field treatment and are valid below, but near T_c. Very close to T_c, additional fluctuations may arise. Gorkov showed that the GL equations (see below) can be derived from a microscopic theory starting from a BCS model Hamiltonian [16].

For a superconductor, the order parameter Ψ is a complex quantity, having a magnitude, $|\Psi|$, and a phase, φ. The physical interpretation of this order parameter is that its modulus squared $2|\Psi|^2$ yields the density fraction of superfluid component. Thus, Ψ may be thought of as the "wavefunction" of the Cooper pairs; these pairs are coherent below T_c and are described by a single "wavefunction." The GL theory was used by Little [17], and Langer and Ambegaokar to analyze the phase-slip process, to account for the generation of a finite voltage below T_c in thin superconducting whiskers, at current levels below the expected critical current [18]. Thus, in some sense, the GL approach has been essential in providing a physical picture of a phase-slip defect. A time-dependent version was used by McCumber and Halperin to more accurately estimate the prefactor in the rate for the thermal generation of phase slips for an activated process just below T_c [19]. In this approach, relaxation to equilibrium from an external perturbation is characterized by a relaxation time, and modeled by the addition of a relaxation term which is proportional to the variation of the free energy density with respect to the order parameter.

The phenomenological GL free energy, in cgs units, has the following form:

$$F = F_n + \int d^3x \left[\alpha|\Psi|^2 + \frac{\beta}{2}|\Psi|^4 + \frac{1}{2M}\left|\left(\frac{\hbar}{i}\nabla + \frac{2e}{c}\boldsymbol{A}\right)\Psi\right|^2 + \frac{|\boldsymbol{B}|^2}{8\pi}\right] . \tag{1.29}$$

Here, M refers to the mass of the Cooper pair, and its charge is $-2e$. The inclusion of the gradient term with the covariant derivative $(\hbar/i\nabla + 2e/c\boldsymbol{A})$ accounts for the kinetic energy term when the superfluid is in motion, such as when a current is flowing.

The coefficients are dependent on temperature. Near T_c,

$$\alpha = a(T - T_c) + O((T - T_c)^2) , \tag{1.30}$$

$$\beta = b + O(T - T_c) , \tag{1.31}$$

where a and b are positive constants, and

$$M = M_o + O(T - T_c) . \tag{1.32}$$

Note that α changes sign and is negative below T_c. This leads to a minimum energy state which has a nonzero value for the order parameter.

Minimization with respect to the order parameter Ψ^* yields the GL equations

$$\frac{\delta F}{\delta \Psi^*} = \alpha(T)\Psi + \beta(T)|\Psi|^2\Psi - \frac{\hbar^2}{2M}\left(\nabla + \frac{2ie}{\hbar c}A\right)^2 \Psi = 0 . \tag{1.33}$$

Near T_c, the solution for a uniform state is given by

$$|\Psi|^2 = \frac{a(T_c - T)}{b} . \tag{1.34}$$

Thus, the superfluid density, n_s, goes as

$$n_s = 2|\Psi|^2 \propto \left(1 - \frac{T}{T_c}\right) . \tag{1.35}$$

Variation with respect to the vector potential gives

$$\nabla \times \boldsymbol{B} = \frac{4\pi}{c}\boldsymbol{J} , \tag{1.36}$$

with

$$\begin{aligned} \boldsymbol{J} &= \frac{-2e}{2M}\left[\Psi^*\left(\frac{\hbar}{i}\nabla + \frac{2e}{c}\boldsymbol{A}\right)\Psi - \Psi\left(\frac{\hbar}{i}\nabla - \frac{2e}{c}\boldsymbol{A}\right)\Psi^*\right] \\ &= -\frac{2e\hbar}{M}|\Psi|^2\left(\nabla\phi + \frac{2e}{\hbar c}\boldsymbol{A}\right) . \end{aligned} \tag{1.37}$$

These are the phenomenological London equations, which were put forth by London [3] to account for the complete expulsion of magnetic flux from the interior of a bulk SC sample, beyond the London penetration depth λ.

In the Ginzburg–Landau theory, the coherence length ξ characterizing the size of the Cooper pair is given by

$$\xi(T) = \sqrt{\frac{\hbar^2}{2M|\alpha(T)|}} , \quad \xi(0) = \sqrt{\frac{\hbar^2}{2M|\alpha(T=0)|}} = \sqrt{\frac{\hbar^2}{2MaT_c}} , \tag{1.38}$$

while the penetration depth

$$\begin{aligned} \lambda(T) &= \sqrt{\frac{Mc^2}{16\pi e^2|\Psi|^2}} = \sqrt{\frac{Mc^2\beta}{16\pi e^2|\alpha(T)|}} , \quad \lambda(0) = \sqrt{\frac{Mc^2 b}{16\pi e^2 a T_c}} ; \\ \lambda(T) &= \sqrt{\frac{M}{4\mu_o e^2|\Psi|^2}} \quad \text{[SI units]} . \end{aligned} \tag{1.39}$$

1.4.1
Time-Dependent Ginzburg–Landau Theory

In equilibrium, the minimization of the free energy yields the lowest energy state, for which the free energy does not change, to first order, with any variation of the order parameter or the fields. If, however, the system is not in an equilibrium state, the system should relax toward the equilibrium state. To account for the time-dependence of this relaxation process, the time-dependent Ginzburg–Landau equation is often used to describe the temporal behavior. The relaxation rate is assumed proportional to the variation of the free energy density with the order parameter ψ^*. The TDGL equation is written as

$$\left(i\hbar\frac{\partial}{\partial t} - 2\mu\right)\Psi = -\frac{i}{\tau_{\mathrm{GL}}}\frac{\hbar}{|\alpha(T)|}\frac{\partial f}{\partial \Psi^*}$$
$$= -\frac{i}{\tau_{\mathrm{GL}}}\frac{\hbar}{|\alpha(T)|}\left[\alpha(T) + \beta(T)|\Psi|^2 - \frac{\hbar^2}{2M}\left(\nabla + \frac{2ie}{\hbar c}A\right)^2\right]\Psi . \quad (1.40)$$

Here, τ_{GL} is the Ginzburg–Landau relaxation time $\tau_{\mathrm{GL}} = (\pi/8)\hbar/(T_c - T)$, μ the chemical potential, and f is the free energy density. The last expression, of course, equals zero in equilibrium, but is nonzero when displaced from equilibrium.

This form of the TDGL is somewhat controversial and is believed to be accurate when close to T_c [20, 21]. It was used by McCumber and Halperin to estimate the attempt frequency to surmount the free energy barrier due to a phase slip in the vicinity of T_c. Review papers on applying the TDGL approach to the phase-slip phenomena in narrow superconducting devices are readily available in the literature [22, 23]. In addition, the generalized TDGL approach of Kramer and coworkers [24, 25] has been widely used to describe the property of both 1D [26, 27] and 2D [28, 29] superconducting systems, for example.

1.5
Gorkov Green's Functions, Eilenberger–Larkin–Ovchinnikov Equations, and the Usadel Equation

Gorkov developed a powerful method for understanding superconductivity by introducing a set of coupled equations for the dynamics (time evolution) of the Green's functions [30]. The equations couple the normal Green's functions, G, and the anomalous Green's, F, functions relevant to Cooper pairing. The equations of motion follow from the time evolution of the fermion field operators (see (1.20)), in which the interaction terms are again treated in a mean field manner. From such equations, essentially all interesting physical quantities can be obtained.

However, these coupled equations are difficult to solve. Starting from these equations, Eilenberger [31], and separately Larkin and Ovchinnikov [32], developed transport-like equations for a set of Green's functions closely related to Gorkov's Green's functions; these Green's functions are related to Gorkov's via an integration over the energy variable. By exploiting the fact that most quantities of interest are derivable from such Green's functions integrated over energy, and exploiting the quasiclassical approximation (which amounts to the neglect of the second order spatial derivatives relative to terms involving k_{F} times the first order derivatives), after averaging over the position of impurities, the number of Green's functions was reduced from four to two, and a transport-like equation for these Green's functions integrated over energy emerged.

Going one step further, Usadel noted that in the dirty limit, the Eilenberger–Larkin–Ovchinnikov Green's functions are nearly isotropic in space [33]. By exploiting this condition, the Eilenberger equations are further simplified to the transport-like Usadel equations describing diffusive motion of the Cooper pairs and the normal electrons. These equations now form the corner stone of many analyses of the dynamics of superconducting nanowires and SNS (superconductor–normal metal–superconductor) bridges. The purpose of this section is to provide a brief summary and some intuitive understanding of the transport-like Usadel equations, starting from the Gorkov formulation. These equations will be essential to understanding 1D nanowires, when either parts of the nanowires are driven into the normal state in a nonequilibrium situation, or else are connected to normal leads at one or multiple points along the nanowire.

The BCS Hamiltonian, which contains a contact interaction for the attraction between electrons, is written in the form

$$H = \int \left(\sum_{\sigma} \hat{\psi}_{\sigma}^{\dagger} \left[\frac{\left(\frac{\hbar}{i}\nabla + \frac{eA}{c}\right)^2}{2m} + u(\boldsymbol{r}) \right] \hat{\psi}_{\sigma} - \frac{g}{2} \sum_{\sigma \neq \sigma'} \hat{\psi}_{\sigma}^{\dagger} \hat{\psi}_{\sigma'}^{\dagger} \hat{\psi}_{\sigma'} \hat{\psi}_{\sigma} \right) d^3 \boldsymbol{r} \,, \tag{1.41}$$

where $-e$ is the charge of the electron, σ denotes the spin (up or down), $u(\boldsymbol{r})$ is a one-particle potential, which includes impurity and boundary effects, $g = VV$ is the Gorkov electron phonon coupling, and as before, the interaction between electrons is nonzero only within a region of the Debye energy $\hbar\omega_{\mathrm{D}}$ of the Fermi surface. The Heisenberg field operators

$$\psi_{\sigma}(\boldsymbol{r}, t) = e^{iHt/\hbar} \hat{\psi}_{\sigma}(\boldsymbol{r}) e^{-iHt/\hbar} \,, \tag{1.42}$$

and

$$\psi_{\sigma}^{\dagger}(\boldsymbol{r}, t) = e^{iHt/\hbar} \hat{\psi}_{\sigma}^{\dagger}(\boldsymbol{r}) e^{-iHt/\hbar} \,, \tag{1.43}$$

satisfy the following equations of motion:

$$
\begin{aligned}
-i\hbar\frac{\partial \psi_\sigma(\mathbf{r},t)}{\partial t} &= -\left[\frac{\left(\frac{\hbar}{i}\nabla + \frac{eA}{c}\right)^2}{2m} + u(\mathbf{r}) - \mu\right]\psi_\sigma(\mathbf{r},t) \\
&\quad + g\psi^\dagger_{\sigma'}(\mathbf{r},t)\psi_{\sigma'}\mathbf{r},t)\psi_\sigma(\mathbf{r},t) \\
-i\hbar\frac{\partial \psi^\dagger_\sigma(\mathbf{r},t)}{\partial t} &= +\left[\frac{\left(\frac{\hbar}{i}\nabla - \frac{eA}{c}\right)^2}{2m} + u(\mathbf{r}) - \mu\right]\psi^\dagger_\sigma(\mathbf{r},t) \\
&\quad - g\psi^\dagger_\sigma(\mathbf{r},t)\psi^\dagger_{\sigma'}(\mathbf{r},t)\psi_{\sigma'}(\mathbf{r},t)\ ,
\end{aligned}
\tag{1.44}
$$

where $\sigma' \neq \sigma$ and the terms $-\mu$ have been added to measure the energy relative to the chemical potential μ. Defining the normal Green's function, G, and anomalous Green's function, F, which measures pair correlations

$$
\begin{aligned}
G_1 &\equiv G_{\uparrow\uparrow}(\mathbf{r},t;\mathbf{r}',t') = -i\left\langle T(\psi_\uparrow(\mathbf{r},t)\psi^\dagger_\uparrow(\mathbf{r}',t'))\right\rangle\ , \\
G_2 &\equiv G^\dagger_{\downarrow\downarrow}(\mathbf{r},t;\mathbf{r}',t') = -i\left\langle T(\psi^\dagger_\downarrow(\mathbf{r},t)\psi_\downarrow(\mathbf{r}',t'))\right\rangle\ , \\
F_1 &\equiv F_{\uparrow\downarrow}(\mathbf{r},t;\mathbf{r}',t') = -i\left\langle T(\psi_\uparrow(\mathbf{r},t)\psi_\downarrow(\mathbf{r}',t'))\right\rangle\ , \\
F_2 &\equiv F^\dagger_{\downarrow\downarrow}(\mathbf{r},t;\mathbf{r}',t') = -i\left\langle T(\psi^\dagger_\downarrow(\mathbf{r},t)\psi^\dagger_\uparrow(\mathbf{r}',t'))\right\rangle\ .
\end{aligned}
\tag{1.45}
$$

Here, the chemical potential μ is given by the relationship $2\mu = 2(\partial E/\partial N) = E_{N+2} - E_N$. Equation (1.45) represents the thermodynamic Green's functions, where the brackets around an operator O denote $\langle O\rangle \equiv \mathrm{Tr}(e^{-\beta H}O)$.

For these Green's functions, their dynamical equations of motion form a closed set of coupled equations. The coupled equations are derived from the equations of motion for ψ and $\psi^\dagger$, making use of the same mean field approximation employed in the BCS solution as well as the BdG formulations to account for pair correlations. Defining a Green's function matrix $\hat{G}$,

$$
\hat{G} \equiv \begin{pmatrix} G_1 & F_1 \\ -F_2 & G_2 \end{pmatrix}\ ,
\tag{1.46}
$$

and going over to imaginary time with $t \to -i\tau$, the dynamical equations of motion for the Green's functions in matrix notation are found to be [30–32]

$$
\begin{pmatrix}
\left[-\hbar\frac{\partial}{\partial\tau} - \frac{\left(\frac{\hbar}{i}\nabla + \frac{eA}{c}\right)^2}{2m} - u(\mathbf{r}) + \mu\right] & \Delta(\mathbf{r}) \\
-\Delta^*(\mathbf{r}) & \left[\hbar\frac{\partial}{\partial\tau} - \frac{\left(\frac{\hbar}{i}\nabla - \frac{eA}{c}\right)^2}{2m} - u(\mathbf{r}) + \mu\right]
\end{pmatrix}\hat{G}
= \delta(\mathbf{r}-\mathbf{r}')\delta(\tau-\tau')\hat{1}\ .
\tag{1.47}
$$

Here,

$$\Delta^*(\boldsymbol{r}, \boldsymbol{\tau}) = -gF^\dagger_{\downarrow\uparrow}(\boldsymbol{r}, \tau; \boldsymbol{r}, \tau) \tag{1.48}$$

is the gap function.

The potential $u(\boldsymbol{r})$ may include the random impurity potentials, as well as that from an applied electric potential ϕ. From these equations, many properties, including nonequilibrium properties, can be obtained. However, the equations are difficult to solve in realistic situations of relevance to experiment, except for special cases, such as thin films at weak perturbation of external drive.

Eilenberger, and independently Larkin and Ovchinnikov, applied a quasiclassical approximation to the Fourier transformed Green's functions, $\hat{G}(\boldsymbol{k}; \boldsymbol{r}, \tau, \tau')$, where the transform is taken in the relative coordinates $\boldsymbol{r}'' = \boldsymbol{r} - \boldsymbol{r}'$ [31, 32, 34]. These Green's functions, in matrix notation, are given by

$$\begin{aligned}\hat{G}(\boldsymbol{k}; \boldsymbol{R}, \tau, \tau') &= \int \hat{G}(\boldsymbol{r}, \tau; \boldsymbol{r}', \tau')e^{-i\boldsymbol{k}\cdot\boldsymbol{r}''} d^3r'' \\ &= \int \hat{G}\left(\boldsymbol{R} + \frac{\boldsymbol{r}''}{2}, \tau; \boldsymbol{R} - \frac{\boldsymbol{r}''}{2}, \tau'\right) e^{-i\boldsymbol{k}\cdot\boldsymbol{r}''} d^3r'' .\end{aligned} \tag{1.49}$$

Here, the choice of singling out the relative coordinate is predicated on the weak dependence of the Green's functions on the center-of-mass coordinate $(\boldsymbol{R} \equiv (\boldsymbol{r} + \boldsymbol{r}')/2)$, once averaged over impurity positions. (Note that Eilenberger introduced extra gauge potential phase factors in his definition of the Green's functions, and thus these matrix equations are in a slightly different but equivalent form [34].) Following an average over the impurity positions, one arrives at a set of equations which are now first order in the covariant derivatives, rather than second order. In the absence of paramagnetic impurities that flip spins, the Gorkov equations become (from here on, we denote $\boldsymbol{R}$ by $\boldsymbol{r}$)

$$\begin{aligned}&\begin{pmatrix} -\hbar\frac{\partial}{\partial\tau} - \xi(\boldsymbol{k}) - \boldsymbol{v}(\boldsymbol{k}_\mathrm{F})\cdot\frac{\hbar}{i}\nabla - \frac{e\hbar}{mc}\boldsymbol{k}_\mathrm{F}\cdot\boldsymbol{A} & \Delta(\boldsymbol{r}) \\ -\Delta^*(\boldsymbol{r}) & \hbar\frac{\partial}{\partial\tau} - \xi(\boldsymbol{k}) - \boldsymbol{v}(\boldsymbol{k}_\mathrm{F})\cdot\frac{\hbar}{i}\nabla + \frac{e\hbar}{mc}\boldsymbol{k}_\mathrm{F}\cdot\boldsymbol{A} \end{pmatrix} \hat{G}(\boldsymbol{k}; \boldsymbol{r}, \tau, \tau') \\ &\quad - \int d\tau_1 \hat{\Sigma}(\boldsymbol{k}_\mathrm{F}; \boldsymbol{r}, \tau, \tau_1)\hat{G}(\boldsymbol{k}; \boldsymbol{r}, \tau_1, \tau') = \delta(\tau - \tau')\hat{1} ,\end{aligned} \tag{1.50}$$

where

$$\xi(\boldsymbol{k}) = \frac{\hbar^2 \boldsymbol{k}^2}{2m} - \mu , \tag{1.51}$$

and

$$\begin{aligned}\hat{\Sigma}(\boldsymbol{k}_\mathrm{F}; \boldsymbol{r}, \tau, \tau_1) &= \frac{1}{2}\int d^3q\, W(\boldsymbol{k}_\mathrm{F}, \boldsymbol{q})\hat{G}(\boldsymbol{k}; \boldsymbol{r}, \tau_1, \tau') \\ &\approx \frac{1}{4i}\int_{S_\mathrm{F}} d^2q_\mathrm{F}\rho(\boldsymbol{q}_\mathrm{F})\, W(\boldsymbol{k}_\mathrm{F}, \boldsymbol{q}_\mathrm{F})\frac{i}{\pi}\int d\xi'\hat{G}(\boldsymbol{q}; \boldsymbol{r}, \tau_1, \tau') .\end{aligned} \tag{1.52}$$

The evaluation of the integral over ξ' requires care to remove an unphysical divergence [31]. Here, $W(\boldsymbol{k}, \boldsymbol{q})$ is the probability of scattering from the $\boldsymbol{k}$-state to the $\boldsymbol{q}$-state. The approximation follows with the assumption that W varies slowly in the energy regions of relevance: $\xi(\boldsymbol{k})$ and $\xi'(\boldsymbol{q})$ (with $\xi'(\boldsymbol{q}) = \hbar^2 q^2/2m - \mu$). In the Born approximation

$$W(\boldsymbol{k}_\mathrm{F}, \boldsymbol{q}_\mathrm{F}) \approx 2\pi N(0) n_\mathrm{i} |u(\boldsymbol{k}_\mathrm{F} - \boldsymbol{q}_\mathrm{F})|^2 , \tag{1.53}$$

where n_i is the impurity concentration, and $u(\boldsymbol{k})$ is the Fourier transform of the impurity potential $u(\boldsymbol{r})$. The quasiclassical approximation amounts to neglecting the terms of second order in the covariant derivatives in comparison to the first order derivatives, that is, $(\nabla \pm eA/i\hbar)^2 \ll \boldsymbol{k}_\mathrm{F} \cdot (\nabla + e\boldsymbol{A}/i\hbar)$, a consequence of the fact that the length scales of relevance for superconductivity: λ, the penetration depth, and ξ, the coherence length, on which Δ and $\boldsymbol{A}$ vary, far exceeds the Fermi wavelength $\lambda_\mathrm{F} = 2\pi/k_\mathrm{F}$.

A key observation significantly simplifies the situation. As it turns out, physical quantities of interest, such as the self-consistent gap equation

$$\frac{\Delta(\boldsymbol{r})}{g} = \langle \psi_\downarrow(\boldsymbol{r}) \psi_\uparrow(\boldsymbol{r}) \rangle , \tag{1.54}$$

and the supercurrent density, and so on, can be written in terms of the energy integrated Green's functions

$$\hat{g}(\boldsymbol{k}_\mathrm{F}; \boldsymbol{r}, \tau, \tau') = \frac{i}{2\pi} \int d\xi\, \hat{G}(\boldsymbol{k}; \boldsymbol{r}, \tau_1, \tau') . \tag{1.55}$$

Here, the wave vector $\boldsymbol{k}$ is to be taken to indicate the direction on the Fermi surface, or in essence $\boldsymbol{k}_\mathrm{F}$, which points in the same direction as the vector $\boldsymbol{k}$, and the energy integration must be taken as the principal value for large energies ξ. The energy-variable integrated Green's functions satisfy the normalization condition

$$\int_0^{\hbar\beta} \hat{g}(\boldsymbol{k}_\mathrm{F}; \boldsymbol{r}, \tau, \tau_1) \hat{g}(\boldsymbol{k}_\mathrm{F}; \boldsymbol{r}, \tau_1, \tau') d\tau_1 = \delta(\tau - \tau') \hat{1} , \tag{1.56}$$

with $\beta = 1/(k_\mathrm{B} T)$. In the simplest case of time-independent perturbations, written in terms of the individual components

$$\begin{aligned} g(\boldsymbol{k}_\mathrm{F}, \omega; \boldsymbol{r}) &= \hat{g}_{11} = \hat{g}_{22} , \\ f(\boldsymbol{k}_\mathrm{F}, \omega; \boldsymbol{r}) &= i\hat{g}_{12} , \\ f^\dagger(\boldsymbol{k}_\mathrm{F}, \omega; \boldsymbol{r}) &= i\hat{g}_{21} , \end{aligned} \tag{1.57}$$

where the normalization condition now reads

$$g(\boldsymbol{k}_\mathrm{F}, \omega; \boldsymbol{r}) = \left[1 - f(\boldsymbol{k}_\mathrm{F}, \omega; \boldsymbol{r}) f^\dagger(\boldsymbol{k}_\mathrm{F}, \omega; \boldsymbol{r})\right]^{1/2} , \tag{1.58}$$

the resultant equations of motion of these Green's functions that are integrated over the energy variable are given by

$$\begin{aligned}
&\left[-2i\hbar\frac{\partial}{\partial\tau} + \boldsymbol{v}(\boldsymbol{k}_\mathrm{F})\cdot\left(\nabla + i\frac{2e}{\hbar c}\boldsymbol{A}\right)\right] f(\boldsymbol{k}_\mathrm{F},\omega;\boldsymbol{r}) \\
&\quad = 2\Delta(\boldsymbol{r})g(\boldsymbol{k}_\mathrm{F},\omega;\boldsymbol{r}) + \int_{S_\mathrm{F}} d^2 q_\mathrm{F}\rho(\boldsymbol{q}_\mathrm{F})\,W(\boldsymbol{k}_\mathrm{F},\boldsymbol{q}_\mathrm{F}) \\
&\qquad \times\left[g(\boldsymbol{k}_\mathrm{F},\omega;\boldsymbol{r})\,f(\boldsymbol{q}_\mathrm{F},\omega;\boldsymbol{r}) - f(\boldsymbol{k}_\mathrm{F},\omega;\boldsymbol{r})g(\boldsymbol{q}_\mathrm{F},\omega;\boldsymbol{r})\right] \\
&\left[2i\hbar\frac{\partial}{\partial\tau} - \boldsymbol{v}(\boldsymbol{k}_\mathrm{F})\cdot\left(\nabla - i\frac{2e}{\hbar c}\boldsymbol{A}\right)\right] f^\dagger(\boldsymbol{k}_\mathrm{F},\omega;\boldsymbol{r}) \\
&\quad = 2\Delta^*(\boldsymbol{r})g(\boldsymbol{k}_\mathrm{F},\omega;\boldsymbol{r}) + \int_{S_\mathrm{F}} d^2 q_\mathrm{F}\rho(\boldsymbol{q}_\mathrm{F})\,W^*(\boldsymbol{k}_\mathrm{F},\boldsymbol{q}_\mathrm{F}) \\
&\qquad \times\left[g(\boldsymbol{k}_\mathrm{F},\omega;\boldsymbol{r})\,f^\dagger(\boldsymbol{q}_\mathrm{F},\omega;\boldsymbol{r}) - f^\dagger(\boldsymbol{k}_\mathrm{F},\omega;\boldsymbol{r})g(\boldsymbol{q}_\mathrm{F},\omega;\boldsymbol{r})\right] ,
\end{aligned} \tag{1.59}$$

where S_F denotes the Fermi surface.

Defining the Fourier transform

$$\hat{g}(\boldsymbol{k}_\mathrm{F},\omega;\boldsymbol{r}) = \frac{1}{2}\int_{-\hbar/(k_\mathrm{B}T)}^{\hbar/(k_\mathrm{B}T)} \hat{g}(\boldsymbol{k}_\mathrm{F};\boldsymbol{r},\tau,0)e^{i\omega\tau}d\tau , \tag{1.60}$$

and the inverse transform

$$\hat{g}(\boldsymbol{k}_\mathrm{F};\boldsymbol{r},\tau,0) = i\beta\sum_{n=-\infty}^{n=\infty} e^{-i\omega_n\tau}\hat{g}(\boldsymbol{k}_\mathrm{F},\omega_n;\boldsymbol{r}) , \quad \beta = \frac{1}{k_\mathrm{B}T} , \tag{1.61}$$

the Eilenberger–Larkin–Ovchinnikov equations read

$$\begin{aligned}
&\left[2\omega + \boldsymbol{v}(\boldsymbol{k}_\mathrm{F})\cdot\left(\nabla + i\frac{2e}{\hbar c}\boldsymbol{A}\right)\right] f(\boldsymbol{k}_\mathrm{F},\omega;\boldsymbol{r}) \\
&\quad = 2\frac{\Delta(\boldsymbol{r})}{\hbar}g(\boldsymbol{k}_\mathrm{F},\omega;\boldsymbol{r}) + \int_{S_\mathrm{F}} d^2 q_\mathrm{F}\rho(\boldsymbol{q}_\mathrm{F})\,W(\boldsymbol{k}_\mathrm{F},\boldsymbol{q}_\mathrm{F}) \\
&\qquad \times\left[g(\boldsymbol{k}_\mathrm{F},\omega;\boldsymbol{r})\,f(\boldsymbol{q}_\mathrm{F},\omega;\boldsymbol{r}) - f(\boldsymbol{k}_\mathrm{F},\omega;\boldsymbol{r})g(\boldsymbol{q}_\mathrm{F},\omega;\boldsymbol{r})\right] \\
&\left[2\omega - \boldsymbol{v}(\boldsymbol{k}_\mathrm{F})\cdot\left(\nabla - i\frac{2e}{\hbar c}\boldsymbol{A}\right)\right] f^\dagger(\boldsymbol{k}_\mathrm{F},\omega;\boldsymbol{r}) \\
&\quad = 2\frac{\Delta^*(\boldsymbol{r})}{\hbar}g(\boldsymbol{k}_\mathrm{F},\omega;\boldsymbol{r}) + \int_{S_\mathrm{F}} d^2 q_\mathrm{F}\rho(\boldsymbol{q}_\mathrm{F})\,W^*(\boldsymbol{k}_\mathrm{F},\boldsymbol{q}_\mathrm{F}) \\
&\qquad \times\left[g(\boldsymbol{k}_\mathrm{F},\omega;\boldsymbol{r})\,f^\dagger(\boldsymbol{q}_\mathrm{F},\omega;\boldsymbol{r}) - f^\dagger(\boldsymbol{k}_\mathrm{F},\omega;\boldsymbol{r})g(\boldsymbol{q}_\mathrm{F},\omega;\boldsymbol{r})\right] .
\end{aligned} \tag{1.62}$$

These equations are reminiscent of Boltzmann transport equations, albeit for quantities which are complex, and thus can account for quantum-mechanical interference. The second term on the right-hand side has the appearance of a collision integral. An additional equation for $g(\boldsymbol{k}_\mathrm{F},\omega;\boldsymbol{r})$ is redundant.

These equations are supplemented by the self-consistent equations for the gap $\Delta(\boldsymbol{r})$ and for the relation between the magnetic field and the supercurrent, which now become

$$\begin{aligned}
&\Delta(\boldsymbol{r})N(0)\ln\frac{T}{T_c} + 2\pi N(0)k_B T \\
&\quad\times\sum_{n=0}^{\infty}\left[\frac{\Delta(\boldsymbol{r})}{\omega_n} - \int_{S_F} d^2k_F\rho(\boldsymbol{k}_F)f(\boldsymbol{q}_F,\omega_n;\boldsymbol{r})\right] = 0 \\
&\boldsymbol{J}_s(\boldsymbol{r}) = \frac{1}{4\pi}\nabla\times i\left[\boldsymbol{B}(\boldsymbol{r}) - \boldsymbol{B}_e(\boldsymbol{r})\right] \\
&\quad = i\frac{2e}{\hbar c}2\pi N(0)k_B T\sum_{n=0}^{\infty}\int_{S_F} d^2k_F\rho(\boldsymbol{k}_F)\boldsymbol{v}(\boldsymbol{k}_F)g(\boldsymbol{q}_F,\omega_n;\boldsymbol{r})\,,
\end{aligned} \tag{1.63}$$

with $\boldsymbol{J}_s$ the supercurrent density and $\boldsymbol{B}_e$ the magnetic field generated by normal electrons. Using (1.62) and (1.63), all the important physical quantities can be derived. Written in terms of the f and $f^\dagger$ only using the relationship $g = (1-|f|^2)^{1/2}$, these equation are nonlinear, and can therefore go beyond the linear approximation used by de Gennes [11]. For instance, they readily reproduce the higher order corrections cubic in $|\Delta|$ calculated by Maki for dirty superconductors [35–37].

Although these equations are substantially more convenient than the Gorkov equations, and are more amenable to numerical implementation for computing the physical quantities under realistic experimental conditions, one further development rendered this entire theoretical machinery far more tractable. Usadel, following the earlier work by Lüders [38–41], recognized that a crucial further simplification can be obtained in the case of a very dirty superconductor. Starting with the Eilenberger–Larkin–Ovchinnikov equations in the form of (1.62), Usadel noted that the large amount of scattering in a dirty system renders the Green's functions nearly isotropic in space, and they can thus be written as the sum of a dominant isotropic part, plus a smaller part dependent on the direction on the Fermi surface. Based on this idea, and keeping the leading terms, Usadel transformed the equations into a diffusion equation valid in this dirty limit.

Separating to the dominant isotropic term and the Fermi velocity-dependent term

$$\begin{aligned}
f(\boldsymbol{k}_F,\omega;\boldsymbol{r}) &= F(\omega;\boldsymbol{r}) + \boldsymbol{k}_F\cdot\boldsymbol{F}(\omega;\boldsymbol{r}) \\
g(\boldsymbol{k}_F,\omega;\boldsymbol{r}) &= G(\omega;\boldsymbol{r}) + \boldsymbol{k}_F\cdot\boldsymbol{G}(\omega;\boldsymbol{r})\,,
\end{aligned} \tag{1.64}$$

the quantities for g are expressible in terms of those for f via the normalization condition, which now, to leading order, reads

$$G(\omega;\boldsymbol{r}) = \left[1-|F(\omega;\boldsymbol{r})|^2\right]^{1/2}\,, \tag{1.65}$$

yielding, in addition,

$$\boldsymbol{G}(\omega;\boldsymbol{r}) = \frac{1}{2}\frac{F(\omega;\boldsymbol{r})\boldsymbol{F}^*(\omega;\boldsymbol{r}) - F^*(\omega;\boldsymbol{r})\boldsymbol{F}(\omega;\boldsymbol{r})}{G(\omega;\boldsymbol{r})}\,. \tag{1.66}$$

Under the conditions

$$G \gg 2\tau_{tr}\omega\,, \qquad F \gg 2\tau_{tr}\Delta\,, \tag{1.67}$$

and of course $|F| \gg |\hat{k}_F \cdot F|$, the resultant diffusion equation is given by

$$\begin{aligned} &2\omega F(\omega; r) - D\left(\nabla + \frac{2ie}{\hbar c}A\right) \\ &\quad\cdot\left[G(\omega; r)\left(\nabla + \frac{2ie}{\hbar c}A\right)F(\omega; r) - F(\omega; r)\nabla G(\omega; r)\right] \\ &= 2\frac{\Delta(r)}{\hbar}G(\omega; r) \end{aligned} \tag{1.68}$$

with the diffusion constant $D = (1/3)\tau_{tr}v_F^2$, where v_F is the Fermi velocity. The transport time is given by the average of the scattering rate over the Fermi surface S_F

$$\frac{1}{\tau_{tr}} = \int_{S_F} q_F \rho(q_F)\, W(k_F, q_F)\,, \qquad W(k_F, q_F) \approx 2\pi N(0) n_i |u(k_F - q_F)|^2\,, \tag{1.69}$$

where the approximation denotes the first Born approximation. Note that the frequencies ω are to be taken as the Matsubara frequency and are positive: $\omega = \omega_n = (2n+1)\pi k_B T/\hbar > 0$.

1.6 Path Integral Formulation

The techniques described thus far are powerful and have been extremely successful in describing most of the properties of conventional superconductors and the phenomena associated with them as well. However, in order to formulate a fully quantum theory that captures the physics of 1D superconducting nanowires in the dirty limit, in which the mean free path $l_{mfp} \ll \xi_0$, where ξ_0 is the clean limit coherence length, a path integral formulation has proven to be a convenient starting point. This formulation enables to go beyond their description close to T_c, and is also naturally suited to describe issues pertaining to quantum phase transitions, such as those occurring in Josephson junctions coupled to a dissipative environment [42–44], such as those occurring in arrays of Josephson junctions in 2D and 1D, or in vortex matter [45].

In the path integral formulation, the action plays a central role. Theoretical developments in the 1970s [42] enabled one to make a connection from the computation of quantum evolution in time, to the partition function, with the introduction of the imaginary time. The nonequilibrium case, such as under current or voltage bias, is also readily incorporated through the generalization of the Keldysh formulation, with the ordered imaginary-time integrals in the expansion of the path integral [42, 46–49]. More to the point, the computation of the low temperature behaviors of a 1D superconducting nanowire has required the use of path integral techniques based on instantons [50–52]. Instantons are saddle point solutions of the action and describe the quantum tunneling processes. Thus, such calculations yield the quantum tunneling rates for the phase slip.

Here, we summarize a formulation of conventional superconductors in a path integral formulation put forth by Otterlo, Golubev, Zaikin, and Blatter [49], which has developed out of earlier works [42, 46–48]. Such an approach not only reproduces the static results of the BCS theory, but enables the treatment of dynamical responses, including relaxation, collective modes, particularly the Carlson-Goldman [53] and Mooij–Schön modes [54], the latter being of direct relevance to 1D superconducting nanowires. Topological defects, such as the motion and the tunneling of vortices, as well as the quantum tunneling of phase slips in one dimension, are also included within this formalism. The Mooij–Schön mode will emerge to play a central role in the dynamics of phase slips and will be discussed in Chapter 3. The standard starting point to compute the partition function in the BCS model is written in terms of the path integral in the imaginary-time formulation [42, 49]

$$
\begin{aligned}
Z &= \int \mathcal{D}\psi \mathcal{D}\psi^* \mathcal{D}V \mathcal{D}^3 \boldsymbol{A} \exp\left(-\frac{S}{\hbar}\right) \\
\frac{S}{\hbar} &= \frac{1}{\hbar}\int d\tau d^3x \left\{ \psi_\uparrow^\dagger \left[\hbar\partial_\tau + ieV + \xi\left(\nabla + \frac{ie}{\hbar c}\boldsymbol{A}\right)\right]\psi_\uparrow \right. \\
&\quad + \psi_\downarrow^\dagger \left[\hbar\partial_\tau + ieV + \xi\left(\nabla + \frac{ie}{\hbar c}\boldsymbol{A}\right)\right]\psi_\downarrow - g\psi_\uparrow^\dagger\psi_\downarrow^\dagger\psi_\downarrow\psi_\uparrow \\
&\quad \left. - ien_\mathrm{i}V + \frac{\boldsymbol{E}^2 + \boldsymbol{B}^2}{8\pi} \right\} ,
\end{aligned}
\tag{1.70}
$$

where $\xi(\nabla) \equiv -\nabla^2/2m - \mu$. The action includes the action of the electromagnetic field, and the coupling between the electrons and electromagnetic field is contained in the covariant derivative: $\partial/\partial_r = \nabla + (ie/\hbar c)\boldsymbol{A}$. The background ion charge density is n_i.

In order to perform the integral over the fermion Grassman fields ψ_σ and $\psi_\sigma^\dagger$, it is useful to decouple quartic attractive interaction term into terms bilinear in the ψs by the introduction of the superconducting gap $\Delta = |\Delta|e^{i\varphi}$ via the Hubbard–Stratonovich transformation, that is,

$$
\begin{aligned}
\exp\left(\frac{g}{\hbar}\int d\tau d^3x \psi_\uparrow^\dagger\psi_\downarrow^\dagger\psi_\downarrow\psi_\uparrow\right) &= \left[\int \mathcal{D}^2\Delta \exp\left(-\frac{1}{\hbar g}\int d\tau d^3x |\Delta|^2\right)\right]^{-1} \\
&\times \int \mathcal{D}^2\Delta \exp\left[-\frac{1}{\hbar}\int d\tau d^3x \left(g^{-1}|\Delta|^2 + \Delta\psi_\uparrow^\dagger\psi_\downarrow^\dagger + \text{h.c.}\right)\right] .
\end{aligned}
\tag{1.71}
$$

After integrating out the fermion degrees of freedom and discarding the normalization factor, which is not important for the dynamics we seek, the partition function becomes, with certain restrictions on the choice of gauge, that is, the Coulomb gauge with $\nabla \cdot \boldsymbol{A} = 0$,

$$Z = \int \mathcal{D}^2 \Delta \mathcal{D} V \mathcal{D}^3 \boldsymbol{A} \exp\left[-\frac{(S_0 + S_1)}{\hbar}\right] , \tag{1.72}$$

where

$$\frac{S_0(V, \boldsymbol{A}, \Delta)}{\hbar} = \int d\tau d^3x \left(\frac{\boldsymbol{E}^2 + \boldsymbol{B}^2}{8\pi} - ien_i V + \frac{|\Delta|^2}{g}\right) , \tag{1.73}$$

and

$$\frac{S_1}{\hbar} = -\mathrm{Tr} \ln \hat{G}^{-1} \hat{G}_o[0] \; ;$$

$$\hat{G}_o[0] \equiv \hat{G}_o[\Delta = 0] = \begin{pmatrix} \left[\frac{\hbar\partial}{\partial\tau} + \xi(\nabla)\right]^{-1} & 0 \\ 0 & \left[\frac{\hbar\partial}{\partial\tau} - \xi(\nabla)\right]^{-1} \end{pmatrix} . \tag{1.74}$$

$\hat{G}$ denotes the Green's function matrix in Nambu space:

$$\hat{G} = \begin{pmatrix} G & F \\ \bar{F} & \bar{G} \end{pmatrix} , \tag{1.75}$$

while the inverse matrix

$$\begin{aligned} \hat{G}^{-1} &= \begin{pmatrix} \frac{\hbar\partial}{\partial\tau} + ieV + \xi(\nabla + \frac{ie}{\hbar c}\boldsymbol{A}) & \Delta \\ \Delta^* & \frac{\hbar\partial}{\partial\tau} - ieV - \xi(\nabla - \frac{ie}{\hbar c}\boldsymbol{A}) \end{pmatrix} \\ &= \begin{pmatrix} \frac{\hbar\partial}{\partial\tau} + \xi(\nabla) + ie\Phi + \frac{m\boldsymbol{v}_s^2}{2} - \frac{i\hbar}{2}\{\nabla, \boldsymbol{v}_s\} & \Delta \\ \Delta^* & \frac{\hbar\partial}{\partial\tau} - \xi(\nabla) - ie\Phi - \frac{m\boldsymbol{v}_s^2}{2} - \frac{i\hbar}{2}\{\nabla, \boldsymbol{v}_s\} \end{pmatrix} . \end{aligned} \tag{1.76}$$

The trace Tr is taken over the matrix in Nambu space and also over internal coordinates of frequency and momenta, and the curly brackets $\{\ldots, \ldots\}$ denote an anticommutator. In addition, the gauge invariant linear combinations of the electromagnetic fields and the order parameter phase φ are introduced:

$$\Phi = V + \frac{\hbar}{2e}\dot{\varphi} , \quad \boldsymbol{v}_s = \frac{1}{2m}\left(\hbar\nabla\varphi + \frac{2e}{c}\boldsymbol{A}\right) . \tag{1.77}$$

The action may be expanded in terms of these gauge invariant quantities about a saddle point delineated below.

The Green's functions in space and real time are defined as

$$\begin{aligned} G(\boldsymbol{r}_1, t_1; \boldsymbol{r}_2, t_2) &\equiv -i\langle|T\{\psi_\uparrow(\boldsymbol{r}_1, t_1)\psi_\uparrow^\dagger(\boldsymbol{r}_2, t_2)\}\rangle \\ \bar{G}(\boldsymbol{r}_1, t_1; \boldsymbol{r}_2, t_2) &\equiv -i\langle|T\{\psi_\downarrow^\dagger(\boldsymbol{r}_1, t_1)\psi_\downarrow(\boldsymbol{r}_2, t_2)\}\rangle \\ F(\boldsymbol{r}_1, t_1; \boldsymbol{r}_2, t_2) &\equiv -i\langle|T\{\psi_\uparrow(\boldsymbol{r}_1, t_1)\psi_\downarrow(\boldsymbol{r}_2, t_2)\}\rangle \\ \bar{F}(\boldsymbol{r}_1, t_1; \boldsymbol{r}_2, t_2) &\equiv -i\langle|T\{\psi_\downarrow^\dagger(\boldsymbol{r}_1, t_1)\psi_\uparrow^\dagger(\boldsymbol{r}_2, t_2)\}\rangle , \end{aligned} \tag{1.78}$$

where $\psi^{\dagger}_{\sigma}$, and ψ_{σ} are the electron creation and annihilation field operators for spin $\sigma = \{\uparrow, \downarrow\}$, respectively. The imaginary-time operators are obtained by replacing t by $-i\tau$ and t' by $-i\tau'$.

The variation of (1.72) with respect to V, $\boldsymbol{A}$, and Δ, respectively, yields the equations describing the Thomas–Fermi screening, London screening, and the BCS-gap equations.

At this point, to further progress, the term with $\mathrm{Tr}\ln \hat{G}^{-1}$ is expanded up to second order, retaining Gaussian fluctuation terms around the saddle point solution: $\Delta = \Delta_0$, $\Phi = 0$, and $\boldsymbol{A} = 0$, and $V = V_\Delta$ a constant which will absorbed into the chemical potential μ. The gap is written as a uniform term plus a fluctuating term: $\Delta = \Delta_0 + \Delta_1$, with Δ_1 being the fluctuation, and Δ_0 is taken to be real. In addition, the inverse of the Green function $\hat{G}$ in (1.74)–(1.76) is split into an unperturbed part $\hat{G}_0^{-1}$ and $\hat{G}_1^{-1}$ with

$$\hat{G}_0 = \begin{pmatrix} G_0 & F_0 \\ \bar{F}_0 & \bar{G}_0 \end{pmatrix} \;;\quad \bar{F}_0 = F_0 = \Delta_0 \tag{1.79}$$

for the saddle point solution, such that

$$\hat{G}_0^{-1} = \begin{pmatrix} \frac{\hbar\partial}{\partial\tau} + \xi(\nabla) & \Delta_0 \\ \Delta_0 & \frac{\hbar\partial}{\partial\tau} - \xi(\nabla) \end{pmatrix}, \tag{1.80}$$

and

$$\hat{G}_1^{-1} = \begin{pmatrix} \left[\frac{m}{2}\left(\frac{e}{mc}\right)^2 \boldsymbol{A}^2 + ieV\right] - \frac{i}{2}\frac{\hbar e}{mc}\{\nabla, \boldsymbol{A}\} & \Delta_1 \\ \Delta_1^* & -\left[\frac{m}{2}\left(\frac{e}{mc}\right)^2 \boldsymbol{A}^2 + ieV\right] - \frac{i}{2}\frac{\hbar e}{mc}\{\nabla, \boldsymbol{A}\} \end{pmatrix}$$
$$= \begin{pmatrix} +ie\Phi + \frac{mv_s^2}{2} - \frac{i\hbar}{2}\{\nabla, \boldsymbol{v}_s\} & \Delta_1 \\ \Delta_1^* & -ie\Phi - \frac{mv_s^2}{2} - \frac{i\hbar}{2}\{\nabla, \boldsymbol{v}_s\} \end{pmatrix}. \tag{1.81}$$

The trace of the natural log of the inverse Green's function can thus be expanded as

$$\mathrm{Tr}\ln \hat{G}^{-1}\hat{G}_0[0] = \mathrm{Tr}\ln \hat{G}_0^{-1}\hat{G}_0[0] - \mathrm{Tr}\sum_{n=1}^{\infty}\frac{(-1)^n}{n}\left(\hat{G}_0\hat{G}_1^{-1}\right)^n. \tag{1.82}$$

Note that since the gap Δ_0 in the unperturbed matrix G_0 is chosen to be real, we have $\bar{F}_0 = F_0$, $F_0(x_1, x_2) = F_0(x_2, x_1)$, and $G_0(x_1, x_2) = -G_0(x_2, x_1)$, where $x_i \equiv (\boldsymbol{r}_i, t_i)$.

From here on, various treatments diverge based on the assumption of Galilean invariance [46–48], gauge invariance, and so on [49]. In the formulation due to Otterlo *et al.*, by using Ward identities [49, 55] due to gauge invariance and charge and particles number conservation, they recast the effective action into four contributions:

$$S_{\mathrm{eff}} = S_{\mathrm{sc}}[\Delta_{\mathrm{L}}, \Phi, \boldsymbol{v}_s] + S_{\mathrm{n}}[\boldsymbol{E}, \boldsymbol{B}] + S_{\mathrm{p\text{-}h}}[\Delta_{\mathrm{L}}, \Phi, \boldsymbol{v}_s] + S_{\mathrm{em}}[\boldsymbol{E}, \boldsymbol{B}], \tag{1.83}$$

where S_{sc} denotes the superconducting contribution, S_n the normal metallic contribution, $S_{p\text{–}h}$ the contribution from particle–hole symmetry breaking terms, and S_{em} the contribution of the free electromagnetic field. In addition, the gap fluctuation Δ_1 with $\Delta_1 = \Delta_L + i\Delta_T$, where Δ_L and Δ_T are both real, is written in terms of a longitudinal component Δ_L, and a phase-like transverse component Δ_T [49]. Please see (1.94) below regarding longitudinal and transverse projections. In this expansion, all terms linear in the expansion parameters vanish. Only quadratic or bilinear terms survive. Higher order terms are discarded.

To express these action terms explicitly, we make use of the gauge invariant quantities

$$\boldsymbol{v}_s = \frac{1}{2m}\left[\hbar\nabla\varphi + \frac{2e}{c}\boldsymbol{A}\right] \approx \frac{1}{2m}\left[\hbar\nabla\frac{\Delta_T}{\Delta_0} + \frac{2e}{c}\boldsymbol{A}\right], \tag{1.84}$$

$$\Phi = V + \frac{\hbar}{2e}\dot{\varphi} \approx V + \frac{\hbar}{2e}\frac{\dot{\Delta}_T}{\Delta_0}, \tag{1.85}$$

where the approximation holds for small fluctuations, that is, $|\Delta_1| \ll \Delta_0$. The second order expansion leads to to terms of the form $\mathrm{tr}[G\,O\,G'O]$. Here, G and G' symbolize any of the Green functions G, $\bar{G}$, F, or $\bar{F}$, and O is an operator. To evaluate the traces over the internal coordinates, $\mathrm{tr}[G\,O\,G'O']$, we define

$$q \equiv (\boldsymbol{q}, \omega_\mu)\,. \tag{1.86}$$

$$\{B\}_{GG'}(q) = \frac{k_B T}{\hbar}\sum_{\omega_\mu}\int\frac{d^3\boldsymbol{p}}{(2\pi)^3}\,B\,G_0(p+q)G_0'(p)\,, \tag{1.87}$$

where B denotes a function of frequency-momentum p and q. For example,

$$g_0 = \{1\}_{GG}(q)\,,\quad g_1 = \left\{\frac{Q}{\boldsymbol{q}^2}\right\}_{GG}(q)\,,$$
$$g_2 = \left\{\frac{Q^2}{\boldsymbol{q}^4}\right\}_{GG}(q)\,,\quad g_3 = \left\{\frac{(\boldsymbol{p}\times\boldsymbol{q})^2}{2\boldsymbol{q}^4}\right\}_{GG}(q)\,, \tag{1.88}$$

with $Q = \boldsymbol{q}\cdot(\boldsymbol{p}+\boldsymbol{q}/2)$, and likewise for f_i, h_i, and k_i denoting $\{\ldots\}_{FF}$, $\{\ldots\}_{GG^\dagger}$, $\{\ldots\}_{GF}$, respectively. Specializing to the various forms of the operators encountered, we have

$$\mathrm{tr}[G\,O\,G'O'] = \frac{k_B T}{\hbar}\sum_{\omega_\mu}\int\frac{d^3\boldsymbol{q}}{(2\pi)^3}\,O(q)O'(-q)\{1\}_{GG'}\,,$$

$$\mathrm{tr}[G\,O\,G'\{\nabla_a, \boldsymbol{O}'_a\}] = 2i\frac{k_B T}{\hbar}\sum_{\omega_\mu}\int\frac{d^3\boldsymbol{q}}{(2\pi)^3}\,O(q)\boldsymbol{O}'_a(-q)\left\{\left(p+\frac{q}{2}\right)_a\right\}_{GG'}\,,$$

$$\mathrm{tr}[G\{\nabla_a, \boldsymbol{O}_a\}G'\{\nabla_b, \boldsymbol{O}'_b\}] =$$
$$-4\frac{k_B T}{\hbar}\sum_{\omega_\mu}\int\frac{d^3\boldsymbol{q}}{(2\pi)^3}\,\boldsymbol{O}_a(q)\boldsymbol{O}'_b(-q)\left\{\left(p+\frac{q}{2}\right)_a\left(p+\frac{q}{2}\right)_b\right\}_{GG'}\,. \tag{1.89}$$

The four contributions to the effective action can now be written as

$$\begin{aligned}\frac{S_{\text{sc}}[\Delta_{\text{L}}, \Phi, \boldsymbol{v}_{\text{s}}]}{\hbar} &= -\text{Tr}\ln\left(G_0^{-1}[\Delta_0]\hat{G}_0[0]\right) + \frac{1}{k_{\text{B}}T}\text{V}\frac{\Delta_0^2}{g} \\ &+ \frac{k_{\text{B}}T}{2}\sum_{\omega_\mu}\int\frac{d^3\boldsymbol{q}}{(2\pi)^3}\left\{\left[\frac{2}{g} + h_0(q) + h_0(-q) + 2f_0(q)\right]\Delta_{\text{L}}(\boldsymbol{q})\Delta_{\text{L}}(-\boldsymbol{q})\right. \\ &+ \left.\left(\frac{8e^2\Delta_0 k_0(q)}{i\omega_\mu}\right)\Phi(\boldsymbol{q})\Phi(-\boldsymbol{q}) - 8m\Delta_0 k_1(q)\boldsymbol{v}_{\text{s}}(\boldsymbol{q})\boldsymbol{v}_{\text{s}}(-\boldsymbol{q})\right\}, \end{aligned} \tag{1.90}$$

$$\begin{aligned}\frac{S_{\text{n}}[\boldsymbol{E}, \boldsymbol{B}]}{\hbar} &= +\frac{k_{\text{B}}T}{2}\sum_{\omega_\mu}\int\frac{d^3\boldsymbol{q}}{(2\pi)^3} + \left(\frac{2e^2 g_1(q)}{mi\omega_{mu}}\right)\boldsymbol{E}(\boldsymbol{q})\boldsymbol{E}(-\boldsymbol{q}) \\ &- \left(\frac{2e^2}{m^2c^2}\right)[g_2(q) + f_2(q) - g_3(q) - f_3(q)]\boldsymbol{B}(\boldsymbol{q})\boldsymbol{B}(-\boldsymbol{q}) \end{aligned} \tag{1.91}$$

$$\begin{aligned}\frac{S_{\text{p-h}}[\Delta_{\text{L}}, \Phi, \boldsymbol{v}_{\text{s}}]}{\hbar} &= k_{\text{B}}T\sum_{\omega_\mu}\int\frac{d^3\boldsymbol{q}}{(2\pi)^3}\left(\frac{1}{2}[h_0(q) - h_0(-a)]\Delta_{\text{L}}(q)\Delta_{\text{L}}(-q)\right. \\ &+ 2ie[k_0(q) + k_0(-q)]\Phi(q)\Delta_{\text{L}}(-q) \\ &\left.+2[k_q(q) - k_q(-q)]\boldsymbol{q}\cdot\boldsymbol{v}_{\text{s}}(q)\Delta_{\text{L}}(-q)\right), \end{aligned} \tag{1.92}$$

and

$$\frac{S_{\text{em}}[\boldsymbol{E}, \boldsymbol{B}]}{\hbar} = k_{\text{B}}T\sum_{\omega_\mu}\int\frac{d^3\boldsymbol{q}}{(2\pi)^3}\left(\frac{\boldsymbol{E}(\boldsymbol{q})\boldsymbol{E}(-\boldsymbol{q}) + \boldsymbol{B}(\boldsymbol{q})\boldsymbol{B}(-\boldsymbol{q})}{8\pi}\right). \tag{1.93}$$

Here, V is the volume of the system. The subscripts “L” and “T” denote longitudinal and traverse projected components, respectively, based on the project operator associated with a momentum $\boldsymbol{q}$:

$$P_{\text{L}}^{\alpha\beta} \equiv \frac{q^\alpha q^\beta}{\boldsymbol{q}^2}, \quad P_{\text{T}}^{\alpha\beta} = \delta^{\alpha\beta} - P_{\text{L}}^{\alpha\beta}, \tag{1.94}$$

with

$$P_{\text{L}}^2 = P_{\text{L}}, \quad P_{\text{T}}^2 = P_{\text{T}}, \quad P_{\text{L}}P_{\text{T}} = P_{\text{T}}P_{\text{L}} = 0, \quad P_{\text{L}} + P_{\text{T}} = 1. \tag{1.95}$$

For example, electric field $\boldsymbol{E}$ can be projected into

$$\boldsymbol{E}_{\text{L}} = P_{\text{L}}\boldsymbol{E} = (\boldsymbol{E}\cdot\boldsymbol{q})\frac{\boldsymbol{q}}{\boldsymbol{q}^2}, \quad \boldsymbol{E}_{\text{T}} = P_{\text{T}}\boldsymbol{E} = \boldsymbol{E} - \boldsymbol{E}_{\text{L}}. \tag{1.96}$$

In addition, the electric and magnetic fields are expressible in terms of $\boldsymbol{v}_{\text{s}}$ and Φ as

$$\begin{aligned}|\boldsymbol{E}(\boldsymbol{q})|^2 &= \boldsymbol{q}^2|\Phi(\boldsymbol{q})|^2 + \frac{m^2\omega^2}{e^2}|\boldsymbol{v}_{\text{s}}(\boldsymbol{q})|^2 \\ &- \frac{m\omega}{e}\left[\Phi(\boldsymbol{q})\boldsymbol{q}\cdot\boldsymbol{v}_{\text{s}}(-\boldsymbol{q}) + \Phi(-\boldsymbol{q})\boldsymbol{q}\cdot\boldsymbol{v}_{\text{s}}(\boldsymbol{q})\right], \end{aligned} \tag{1.97}$$

and

$$|\boldsymbol{B}(\boldsymbol{q})|^2 = \frac{m^2 c^2}{e^2} \boldsymbol{q}^2 P_{\mathrm{T}}^{\alpha\beta} \boldsymbol{v}_{\mathrm{s}}^{\alpha}(q) \boldsymbol{v}_{\mathrm{s}}^{\beta}(q) \,. \tag{1.98}$$

Note that the various contributions to the effective action are expressed in terms of the *equilibrium, nonperturbed* normal and anomalous Green functions G_0, $\bar{G}_0$, F_0, and $\bar{F}_0$, for $\Delta = \Delta_0$, $V = V_\Delta = \mu$, and $\boldsymbol{A} = 0$ (for the normal contribution, $\Delta_0 = 0$). These standard Green functions are readily computed. See the appendices in references [49, 55], for example.

In the above, by introducing the longitudinal and transverse components, for example, Δ_{L} and Δ_{T} for the gap fluctuation Δ_1 about the BCS equilibrium value Δ_0, the effective action is cast in a form advantageous for illustrating the gauge invariant properties. For the 1D superconducting nanowire system, it is not necessary to do so. By repeated use of the Ward identities associated with gauge-invariance and particle number conservation, the second order expansion can be recast into the following convenient form [50, 51, 55]:

$$S_{\mathrm{eff}} = S_{\mathrm{sc}}[\Delta, \Phi, \boldsymbol{v}_{\mathrm{s}}] + S_{\mathrm{n}}[\Delta, V, \boldsymbol{A}] + S_{\mathrm{em}}[\boldsymbol{E}, \boldsymbol{B}] \,, \tag{1.99}$$

where the particle–hole asymmetry contribution has been left out, as it is usually small. The contribution S_{sc} is given by

$$\begin{aligned}\frac{S_{\mathrm{sc}}[\Delta, \Phi, \boldsymbol{v}_{\mathrm{s}}]}{\hbar} &= -\mathrm{tr}\ln\left(G_0^{-1}[\Delta]\right) + \frac{1}{\hbar}\int_0^\beta\!\!\int d\tau\, d^3x \frac{\Delta^2}{g} + \frac{(2e)^2}{2} \\ &\quad \times \mathrm{tr}[F_0 \Phi F_0 \Phi] - \frac{\hbar^2}{2}\mathrm{tr}[F_0\{\nabla, \boldsymbol{v}_{\mathrm{s}}\} F_0\{\nabla, \boldsymbol{v}_{\mathrm{s}}\}] \,, \end{aligned} \tag{1.100}$$

and the contribution S_{n} by

$$\begin{aligned}\frac{S_{\mathrm{n}}[\boldsymbol{E}, \boldsymbol{B}]}{\hbar} &= \frac{1}{\hbar}\int_0^\beta\!\!\int d\tau\, d^3x \frac{m}{2}\boldsymbol{u}^2 n_{\mathrm{e}}[\Delta] \\ &\quad - \frac{\hbar^2}{4}\mathrm{tr}[G_0\{\nabla, \boldsymbol{u}\} G_0\{\nabla, \boldsymbol{u}\}] + \frac{\hbar^2}{4}\mathrm{tr}[F_0\{\nabla, \boldsymbol{u}\} F_0\{\nabla, \boldsymbol{u}\}] \\ &= \frac{1}{\hbar}\int_0^\beta\!\!\int d\tau\, d^3x \frac{m}{2}\boldsymbol{u}^2 n_{\mathrm{e}}[\Delta] \\ &\quad + \frac{(2e)^2}{4}\mathrm{tr}[G_0 V G_0 V] - \frac{(2e)^2}{4}\mathrm{tr}[F_0 V F_0 V] \,, \end{aligned} \tag{1.101}$$

where n_{e} is the electron density. The velocity $\boldsymbol{u}$ does not depend on the phase of the order parameter φ, and is defined as

$$\boldsymbol{u} = \frac{-e}{m}\left(\int_{-\infty}^{\tau} d\tau' \nabla V(\tau') - \frac{1}{c}\boldsymbol{A}\right) \,. \tag{1.102}$$

The last line for S_n neglects the vector potential contribution, as the magnetic response tends to be small in normal metals. The contribution S_{em} remains as before, and is, in real space representation,

$$\frac{S_{em}[\boldsymbol{E}, \boldsymbol{B}]}{\hbar} = \int_0^\beta \int d\tau d^3x \left(\frac{\boldsymbol{E}^2 + \boldsymbol{B}^2}{8\pi} \right) . \tag{1.103}$$

Going further, and expanding Δ about its BCS value Δ_0, with $\Delta_1(\boldsymbol{r}, \tau) = \Delta(\boldsymbol{r}, \tau) - \Delta_0$, we arrive at the final form for the second order effective action, that is,

$$S_{eff}^{(2)} = S_\Delta + S_J + S_L + S_D + S_{em} . \tag{1.104}$$

The various terms are

$$\frac{S_\Delta}{\hbar} = \int_0^\beta d\tau \int d^3x \frac{\Delta_0^2 + |\Delta_1|^2}{g} + \text{tr}\left[F_0 \Delta_1^* F_0 \Delta_1^* + G_0 \Delta_1 G_0 \Delta_1^* \right] , \tag{1.105}$$

$$\frac{S_J}{\hbar} = \frac{(2e)^2}{2} \text{tr}[F_0 \Phi F_0 \Phi] , \tag{1.106}$$

$$\frac{S_L}{\hbar} = -\frac{\hbar^2}{2} \text{tr}[F_0\{\nabla, \boldsymbol{v}_s\} F_0 \{\nabla, \boldsymbol{v}_s\}] , \tag{1.107}$$

and

$$\frac{S_D}{\hbar} = -\frac{(2e)^2}{4} \text{tr}[G_0 V G_0 V] - \frac{(2e)^2}{4} \text{tr}[F_0 V F_0 V] . \tag{1.108}$$

The equilibrium BCS energy gap as a function of temperature $\Delta_0(T)$ is assumed to obey the BCS gap equation (with $g = \lambda$)

$$\frac{1}{N(0)g} = \int_0^{\omega_D} d\epsilon_n \frac{\tanh\left(\sqrt{\epsilon_n^2 + \Delta_0^2}/(k_B T) \right)}{\sqrt{\epsilon_n^2 + \Delta_0^2}} . \tag{1.109}$$

Physically, the term S_Δ comes from fluctuations in the gap magnitude, S_J from Josephson coupling via the gauge invariant potential Φ, S_L pertains to the London screening of the magnetic field penetrating the superconductor, and the Drude contribution S_D accounts for the Ohmic dissipation of the normal electrons.

Based on this action, and specializing into one dimension, Zaikin, Golubev, *et al.* derived an effective action for 1D superconductors in terms of the Mooij–Schön plasmon mode, which is linearly dispersing, plus a phase-slip core contribution. The phase slips then arise as saddle point solutions (or instantons) of the effective action. Please see Chapter 3.

References

1 Bardeen, J., Cooper, L.N., and Schrieffer, J.R. (1957) Theory of superconductivity. *Physical Review*, **108** (5), 1175–1204.

2 Cooper, L. (1956) Bound electron pairs in a degenerate fermi gas. *Physical Review*, **104** (4), 1189–1190.

3 London, F. and London, H. (1935) The Electromagnetic Equations of the Supraconductor. *Proceedings of the Royal Society A: Mathematical, Physical and Engineering Sciences*, **149** (866), 71–88.

4 de Gennes, P.G. (1966) *Superconductivity of Metals and Alloys*, W.A. Benjamin, New York.

5 Tinkham, M. (1975) *Introduction to Superconductivity*, 2nd edn, McGraw-Hill, New York.

6 Ashcroft, N.W. and Mermin, N.D. (1976) *Solid State Physics*, 1st edn, Brooks Cole.

7 Fetter, A.L. and Walecka, J.D. (2003) Quantum Theory of Many-Particle Systems. New York, Dover.

8 Anderson, P. (1959) Theory of dirty superconductors. *Journal of Physics and Chemistry of Solids*, **11**, 26–30.

9 Bogoliubov, N.N. (1958) A new method in the theory of superconductivity-1. *Soviet Physics JETP*, **7** (1), 41–46.

10 Bogoliubov, N.N. (1958) A new method in the theory of superconductivity-3. *Soviet Physics JETP*, **7** (1), 51–55.

11 de Gennes, P. (1964) Boundary effects in superconductors. *Reviews of Modern Physics*, **36** (1), 225–237.

12 Shanenko, A.A. and Croitoru, M.D. (2006) Shape resonances in the superconducting order parameter of ultrathin nanowires. *Physical Review B*, **73** (1), 012510.

13 Shanenko, A., Croitoru, M., and Peeters, F. (2007) Oscillations of the superconducting temperature induced by quantum well states in thin metallic films: Numerical solution of the Bogoliubov–de Gennes equations. *Physical Review B*, **75** (1), 014519.

14 Croitoru, M.D., Shanenko, A.A., and Peeters, F.M. (2007) Size-resonance effect in cylindrical superconducting nanowires. *Moldavian Journal of the Physical Sciences*, **6** (1), 39–47.

15 Ginzburg, V.L. and Landau, L.D. (1950) *Journal of Experimental and Theoretical Physics*, **20**, 1075.

16 Gorkov, L.P. (1959) Microscopic derivation of the Ginzburg–Landau equations in the theory of superconductivity. *Soviet Physics JETP*, **9** (6), 1364–1367.

17 Little, W. (1967) Decay of persistent currents in small superconductors. *Physical Review*, **156** (2), 396–403.

18 Langer, J.S. and Ambegaokar, V. (1967) Intrinsic resistive transition in narrow superconducting channels. *Physical Review*, **164**, 498–510.

19 McCumber, D.E. and Halperin, B.I. (1970) Time scale of intrinsic resistive fluctuations in thin superconducting wires. *Physical Review B*, **1** (3), 1054–1070.

20 Schmid, A. (1966) A time dependent Ginzburg–Landau equation and its application to the problem of resistivity in the mixed state. *Physik der kondensierten Materie*, **5** (4), 302–317.

21 Abrahams, E. and Tsuneto, T. (1966) Time variation of the Ginzburg–Landau order parameter. *Physical Review*, **152** (1), 416–432.

22 Ivlev, B.I. and Kopnin, N.B. (1984) Theory of current states in narrow superconducting channels. *Soviet Physics Uspekhi*, **142**, 435–471.

23 Dmitrenko, I.M. (1996) On the 90th anniversary of Boris Georgievich Lasarev. *Low Temperature Physics*, **22**, 643–647.

24 Kramer, L. and Watts-Tobin, R.J. (1978) Theory of dissipative current-carrying states in superconducting filaments. *Physical Review Letters*, **40**, 1041–1044.

25 Watts-Tobin, R.J., Krähenbühl, Y., and Kramer, L. (1981) Nonequilibrium theory of dirty, current-carrying superconductors: phase slip oscillators in narrow filaments near T_C. *Journal of Low Temperature Physics*, **42**, 459–501.

26 Vodolazov, D.Y., Peeters, F.M., Piraux, L., Mátéfi-Tempfli, S., and Michotte, S. (2003) Current–voltage characteristics of quasi-one-dimensional superconductors: An S-shaped curve in the constant

voltage regime. *Physical Review Letters*, **91** (15), 157001.

27 Michotte, S., Mátéfi-Tempfli, S., Piraux, L., Vodolazov, D.Y., and Peeters, F.M. (2004) Condition for the occurrence of phase slip centers in superconducting nanowires under applied current or voltage. *Physical Review B*, **69** (9), 094512.

28 Berdiyorov, G.R., Elmurodov, A.K., Peeters, F.M., and Vodolazov, D.Y. (2009) Finite-size effect on the resistive state in a mesoscopic type-II superconducting stripe. *Physical Review B*, **79** (17), 174506.

29 Berdiyorov, G.R., Milošević, M.V., and Peeters, F.M. (2009) Kinematic vortex-antivortex lines in strongly driven superconducting stripes. *Physical Review B*, **79** (18), 184506.

30 Gorkov, L.P. (1958) On the energy spectrum of superconductors. *Soviet Physics JETP*, **7** (3), 505–508.

31 Eilenberger, G. (1968) Transformation of Gorkov's equation for type II superconductors into transport-like equations. *Zeitschrift für Physik*, **214** (2), 195–213.

32 Larkin, A.I. and Ovchinnikov, Y.N. (1969) Quasiclassical method in the theory of superconductivity. *Soviet Journal of Experimental and Theoretical Physics*, **28**, 1200–1204.

33 Usadel, K.D. (1970) Generalized diffusion equation for superconducting alloys. *Physical Review Letters*, **25** (8), 507–509.

34 Eilenberger, G. (1966) General approximation method for the free energy functional of superconducting alloys. *Zeitschrift für Physik*, **190** (2), 142–160.

35 Maki, K. (1968) The critical fluctuation of the order parameter in type-II superconductors. *Progress of Theoretical Physics*, **39**, 897–906,.

36 Maki, K. (1968) Critical fluctuation of the order parameter in a superconductor. I. *Progress of Theoretical Physics*, **40**, 193–200.

37 Maki, K. (1970) Nonlinear responses in type-II superconductors. I. Dirty limit. *Physical Review B*, **2**, 2574–2580.

38 Lüders, G. (1966) Zu de Gennes' Methode der Korrelationsfunktion in der Theorie der Supraleitung. *Zeitschrift für Naturforschung Teil A – Astrophhysik und Physikalische Chemie*, **21**, 680.

39 Lüders, G. (1966) Die Methode der Korrelationsfunktion in der Theorie der Supraleitung. II. Linearisierte Ginzburg–Landau-Gleichung, Diffusionsnäherung. *Zeitschrift für Naturforschung Teil A*, **21**, 1415.

40 Lüders, G. (1966) Die Methode der Korrelationsfunktion in der Theorie der Supraleitung. III. Ableitung der BOLTZMANN-Gleichung. *Zeitschrift für Naturforschung Teil A*, **21**, 1425.

41 Lüders, G. (1966) Die Methode der Korrelationsfunktion in der Theorie der Supraleitung. IV. Paramagnetische Zusätze. *Zeitschrift für Naturforschung Teil A*, **21**, 1842.

42 Schön, G. and Zaikin, A. (1990) Quantum coherent effects, phase transitions, and the dissipative dynamics of ultra small tunnel junctions. *Physics Reports*, **198** (5/6), 237–412.

43 Sondhi, S.L., Girvin, S.M., Carini, J.P., and Shahar, D. (1997) Continuous quantum phase transitions. *Reviews of Modern Physics*, **69** (1), 315–333.

44 Goswami, P. and Chakravarty, S. (2006) Dissipation, topology, and quantum phase transition in a one-dimensional Josephson junction array. *Physical Review B*, **73** (9), 094516.

45 Blatter, G., Feigel'man, M.V., Geshkenbein, V.B., Larkin, A.I., and Vinokur, V.M. (1994) Vortices in high-temperature superconductors. *Reviews of Modern Physics*, **66**, 1125.

46 Greiter, M., Wilczek, F., and Witten, E. (1989) Hydrodynamic relations in superconductivity. *Modern Physics Letters B*, **3** (12), 903–918.

47 Schakel, A.M.J. (1990) On the effective theory of a BCS system at zero temperature. *Modern Physics Letters B*, **4**, 927–934.

48 Aitchison, I., Ao, P., Thouless, D., and Zhu, X. (1995) Effective Lagrangians for BCS superconductors at $T = 0$. *Physical Review B*, **51** (10), 6531–6535.

49 van Otterlo, A., Golubev, D., Zaikin, A., and Blatter, G. (1999) Dynamics and effective actions of BCS superconductors.

The European Physical Journal B, **10** (1), 131–143.

50 Zaikin, A.D., Golubev, D.S., van Otterlo, A., and Zimányi, G.T. (1997) Quantum phase slips and transport in ultrathin superconducting wires. *Physical Review Letters*, **78**, 1552–1555.

51 Golubev, D. and Zaikin, A. (2001) Quantum tunneling of the order parameter in superconducting nanowires. *Physical Review B*, **64** (1), 14504.

52 Khlebnikov, S. and Pryadko, L.P. (2005) Quantum phase slips in the presence of finite-range disorder. *Physical Review Letters*, **95** (10), 107007.

53 Carlson, R. and Goldman, A. (1975) Propagating order-parameter collective modes in superconducting films. *Physical Review Letters*, **34** (1), 11–15.

54 Mooij, J. and Schön, G. (1985) Propagating plasma mode in thin superconducting filaments. *Physical Review Letters*, **55** (1), 114–117.

55 Arutyunov, K., Golubev, D., and Zaikin, A. (2008) Superconductivity in one dimension. *Physics Reports*, **464** (1/2), 1–70.

2
1D Superconductivity: Basic Notions

2.1
Introduction

In this chapter, we explore the basic phenomenon of superconductivity in 1D (one-dimension). As mentioned in the introduction to this book, by 1D, we focus on the situations where the behavior of the order parameter is 1D, or more specifically, when ξ, the superconducting coherence length exceeds, the lateral dimensions. In this limit, the order parameter is nearly constant across the wire, and only variations along the wire need to be considered. On the other hand, since the coherence length ξ far exceeds the Fermi wavelength of the electron system, λ_F, the quasiparticle (hole) fermionic excitations are still bulk-like (i.e., 3D). Typically, the number of transverse quantum channels (or modes) is in the tens to a few thousand range. Of course, for a strictly 1D system, that is, a single-channel limit, one cannot speak of a finite temperature superconductor as fluctuations will destroy the coherence of the Cooper pairs. Variations of the above scenario will also be touched upon. In addition, we will also describe a slightly more relaxed criterion for 1D, that of a wire with length much larger than the transverse dimensions ($\mathsf{L} \gg (\mathsf{w}, \mathsf{h}) \sim \mathsf{d}$, where w is the width, h is the height, and d is the diameter). Some notable predictions of universal scaling behavior of the conductance have been advanced for short wires near a quantum critical point, such as that induced by the presence of magnetic scatters, when the short wires are connected to various configurations of superconducting, normal, and mixed leads, for example, one superconducting, the other normal [1, 2]. In addition, systems of strictly 1D nanowires weakly coupled to each other in the lateral directions, such as ropes of single-walled carbon nanotubes (SWCNTs), or SWCNTs embedded in an nearly insulating matrix, for example, in zeolite, will be discussed [3, 4].

We will explore three topics: (1) size oscillations/shape resonance in a clean nanowire. This is relevant for experiments on single-crystal nanowires. Oscillations of T_c and other properties are expected as the diameter varies atomic layer by layer. (2) Strictly 1D nanowire systems, such as semiconducting nanowires, the coupled single-walled carbon nanotube (SWCNT) system, as well as multiwalled carbon nanotube, will also be discussed. (3) The notion of a phase slip, as first introduced by Langer and Ambegaokar in their analysis of the appearance of a finite

One-Dimensional Superconductivity in Nanowires, First Edition. F. Altomare and A.M. Chang

resistance below T_c in narrow superconducting strips [5, 6]. We will also describe the phase-slip rate estimated by Langer and Ambegaokar [7], and McCumber and Halperin [8], for the thermally activated phase slip (TAP).

The first topic, that is, size oscillations/shape resonances, pertains to the clean limit where the nanowire is sufficiently uniform in its transverse dimensions, and at the same time, free of disorder, such that the electron mean free path exceeds these dimensions. The nanowires relevant in this regime are single-crystal nanowires. In this limit, the quantization of the transverse eigenfunctions are well-defined, forming the so-called quantum well-type subbands.

As a consequence, the superconducting properties oscillate when individual subbands and their associated transverse modes pass sequentially through the chemical potential, as the transverse dimensions are systematically reduced. Such oscillations are now definitively observed in clean, quench-condensed Pb (lead) thin films [9–11], and are expected to occur in 1D single crystal nanowires of sufficient uniformity, as well as in superconducting nanoparticles such as Al nanoparticles of diameter $\sim$ 5 nm [12, 13].

The second topic is of interest, in particular, in view of the many amazing chemical techniques for the growth and fabrication of semiconducting and metallic nanowires, such as InAs or InSb nanowires [14] and carbon nanotubes (CNTs) [15]. Whereas by themselves, InAs nanowires are semiconducting, coupling by proximity to superconducting electrodes can lead to a supercurrent passing through the S-InAsNW-S or S-InSbNW-S system. More intriguingly, the sizable spin–orbit interaction in InAs and InSb, in conjunction with the possibility of reaching the few-channel (or mode) limit by reducing the wire diameter, has led to the prediction and detection of the existence of zero energy, Majorana fermion excitations [16, 17]. Please see Chapter 5 for details and a more complete list of references.

CNTs are notable because they represent a new class of molecular conductors. The possibility of observing superconductivity in molecular systems is intriguing from many perspectives, including potential applications as directed superconducting interconnects. The idea that proximity can induce superconductivity in a (single-walled CNT (SWCNT) is now well-established. Here, we focus on the intrinsic superconductivity within a bundle of metallic SWCNT, weakly coupled to each other through an insulator, or in multiwalled CNTs. The analysis [3] involves a modified version of the BCS theory due to Eliashberg [18, 19], which attempts to account for electron interaction in the strong-coupling limit rather than the weak-coupling limit treated in the original BCS theory.

The third topic is the emergence of the phase slip as the dominant, low energy excitation in 1D. The dissipative mechanism occurring through the generation of phase slips – a topological defect in the superconducting condensate where the phase of the order parameter changes rapidly through 2π over a region the coherence length, ξ – is generally believed to be responsible for the finite linear resistance below T_c [7, 8, 20]. In particular, we will discuss the phase slips as they were first introduced by Langer and Ambegaokar in their analysis of the appearance of a finite resistance below T_c in narrow superconducting strips [7, 21]. We will also describe the phase slip rate estimated in their work, and in the work of McCumber and Halperin [8] for the thermally activated phase slip (TAP). Nonlinear voltages also emerge at high values of current bias below the critical current value.

2.2
Shape Resonances – Oscillations in Superconductivity Properties

In narrow wires, if it were possible to reduce the system to such a size in the transverse dimensions to be comparable to the Fermi wavelength, λ_F, such that only a small number of transverse quantum channels are occupied below the Fermi energy, one would anticipate that the dramatic oscillations in the electronic density of states would have a noticeable effect on the physical properties, including superconductivity. The density of states oscillations are associated with the van Hove singularities as each channel crosses the Fermi level, affected by gradually reducing the diameter of the nanowire. Moreover, it would be necessary for the occupied quantum channels to number more than a few; or else the fluctuations will rapidly drive the phase transition toward $T = 0$. In the strictly 1D limit, only a $T = 0$ transition into the superconducting state is possible.

In the homogeneous 3D case treated in the BCS theory, the electronic density of states (DOS) at the Fermi level, $N(0)$, enters into the expression of the energy gap, and hence the transition temperature in an essential way is

$$\Delta_0 = 1.76 T_c = 1.14 \hbar \omega_D e^{-\frac{1}{N(0)g}} . \tag{2.1}$$

Although this relationship between Δ_0, T_c, and $N(0)$ no longer holds exactly in reduced dimensions, when quantum size effects become important, the basic trend of the dependence is preserved. Thus, even slight oscillations in the DOS can be translated into sizable oscillations in Δ_0 and in T_c.

Ideally, one would want the nanowires to be extremely uniform in diameter, as well as straight and devoid of disorder, to minimize localization effects, and to render the transverse channels to be well-defined. Such a scenario may be achievable in materials systems, such as metallic carbon multiwalled carbon nanotubes, or a bundle of single-walled carbon nanotubes which are coupled with each other. However, even in metallic systems, such as Al, Pb, Sn, and so on, size quantization effects may be observable, despite the typically larger number of occupied transverse channels, in the range of 1000 for a diameter of $\sim$ 10 nm. This is because of the larger Fermi energy in elemental metals, of order 10 eV. The typical level spacing is still on the order of several meV, exceeding the superconducting gap, which is of order 1 meV or smaller. Thus, unless the disorder is excessive, the quantum channels in these disorder wires may still be well-defined. For instance, in aluminum nanowires, the transport mean free path is of order 10 nm, compared to a Fermi wavelength of 0.3 nm. Thus, the level-broadening from scattering is of order 16 meV, compared to the Fermi energy of 11.7 eV. Even in this limit, quantum channels may still be well-defined, although the spacing of the subband levels may reflect the randomness, that is, displaying a Wigner–Dyson type statistics rather than Poisson statistics [22–26].

In a related system of ultrathin two-dimensional (2D) superconducting (SC) films, for which the technique of low-temperature quenched evaporation offers control down to submonolayer thicknesses, size oscillations in T_c are clearly seen in experiment [9–11]. This situation in metallic nanowires in the size range of 20 nm, which are thought to be single crystalline, but in isolated arrays rather than individual nanowires [27], offers hints of such oscillations. The reduction in T_c in individual aluminum nanowires as the diameter decreases from 10 nm down to $\sim$ 5.4 nm may also be related to such quantization effects [28]. This reduction contrasts with typical trends in thin Al films and wires, where T_c tends to increase with a decrease in film thickness or wire diameter.

Other related systems include coupled nanowire systems, more notably coupled SWCNTs, as well as superconducting nanograins of diameter $\sim$ 5 nm. In the former situation, one may imagine incorporating identical carbon nanotubes within a matrix, but of smaller and smaller diameters. As the diameter is reduced, the number of channels decreases accordingly. As each successive channel crosses above the Fermi level, the electronic density of states drops abruptly, leading to an abrupt change in the physical properties. Similarly, in a clean superconducting grain, or quantum dot, the discrete energy level spectrum may exhibit size resonances. Here, as in the case of individual 1D metallic nanowires (where $\xi \gg (\mathrm{w}, \mathrm{h}) \sim \mathrm{d}$), the precise shape of the grain (of the cross section in 1D nanowires) can give rise to regular (integrable) or chaotic underlying dynamics. The behavior of the electronic density of states can be dramatically different in the two cases. In the integrable case, sharp spikes in the density of states arise, with a periodicity determined by the Fermi wavelength λ_F. In contrast, in the chaotic case, such sharp features tend to be smoothed out, leading to a smooth variation of the superconducting properties with the diameter.

To study these systems with reduced dimensionality, the Bogoliubov–de Gennes equations introduced in Section 1.3 provide the most convenient and widely used formalism. The solution must, in addition, satisfy the self-consistency equation for the energy gap Δ, and must account for the Coulomb potential between electrons and with the positive background of ions as well as impurity ions. At the same time, the boundary condition at the surfaces dictates that the wavefunction vanishes there. This type of inhomogeneous system is precisely what the BdG equations are particularly suited to handle [29–31].

In a clean 2D film, the occurrence of size oscillations is particularly straight forward to envision. The thickness, when small, gives rise to well-defined quantized states in this thickness direction. Such states are termed subbands, or quantum well states [32, 33]. The presence of such discrete energy levels leads to oscillations in the electronic density of states at the Fermi level. Since in the simplest scenario, the energy depends exponentially on the inverse of this DOS, even small modulations in it can have a dramatic effect on the gap Δ, and all physical quantities, which depend on Δ. Other notable oscillatory behaviors include position-dependent oscillations of the gap as a function of position in the thickness z-direction, a difference in the behavior of films containing an even versus an odd number of monolayers, as well as the need to introduce an oscillatory modulation of the electron–phonon interaction.

2.2.1 Early Treatments of Shape Resonances in 2D Films

The study of such size oscillations in clean 2D films commenced with the work of Blatt and Thompson in the early 1960s [34, 35]. They referred to these as shape resonances. Coming before the BdG equations became widely employed, they based their analysis on Anderson's idea of pairing time-reversed electronic eigenfunctions. Blatt and Thompson computed the energy gap at zero temperature $T = 0$, which, in this spatially inhomogeneous system with boundaries in the direction perpendicular to the film, is no longer strictly independent of the quantum numbers, which index the electronic eigenfunctions in the normal state. However, it turns out that any dependence is noticeable only within a narrow transition region, when an additional quantum channel (mode) passes through the Fermi energy as the thickness is reduced.

It is instructive to follow Blatt and Thompson's treatment to gain insight into the problem of conventional SC in the presence of size quantization. In the clean limit, the 2D film system is a translational invariant in the plane of the film. The wavefunction in these two directions is thus of the planewave form. In the absence of the phonon-mediated, attractive electron–electron interaction responsible for Cooper pairing, the Hamiltonian is

$$H = \frac{\hbar^2}{2m}\left(p_x^2 + p_y^2 + p_z^2\right) + U_{\text{ion}}(z) + U_{\text{e–e}}(z) - \mu \,, \tag{2.2}$$

where U_{ion} represents the attractive potential of the uniform background position ions, $U_{\text{e–e}}$ the self-consistent repulsive potential from all other electrons, and μ is the chemical potential indicating the position of the Fermi level. From the x- and y-translational invariance of the system, these potentials only depend on the z-coordinate. The eigenfunctions are labeled by n in the z-direction, and k_x and k_y in the x- and y-directions, respectively. The eigenfunctions, ψ_{n,k_x,k_y} are written as

$$\psi_{n,k_x,k_y}(x, y, z) = \frac{1}{\mathsf{L}} e^{ik_x x} e^{ik_y y} w_n(z) \,, \tag{2.3}$$

where L is the size of the system in the x- and y-directions, and a periodic boundary condition is imposed. In the z-direction, the thickness is h and the wavefunction vanishes at the boundaries: $w_n(z = 0) = w_n(z = h) = 0$. Writing the eigenenergy, which is measured from the Fermi level at μ as

$$\epsilon_{n,k_x,k_y} = e_{\text{n}} + \frac{\hbar^2}{2m}\left(k_x^2 + k_y^2\right) \,, \tag{2.4}$$

w_n satisfies the Schrödinger equation

$$-\frac{\hbar^2}{2m}\frac{d^2 w_n}{dz^2} + [U_{\text{ion}}(z) + U_{\text{e–e}}(z)]w_n = (e_{\text{n}} + \mu)w_n \,. \tag{2.5}$$

Note that the energies are referenced to the chemical potential μ. For the superconducting state, a BCS model is assumed, with a contact potential – see Sections 1.2 and 1.3:

$$V(\boldsymbol{r}_i, \boldsymbol{r}_j) = -V\mathsf{V}\delta(\boldsymbol{r}_i - \boldsymbol{r}_j) = -g\delta(\boldsymbol{r}_i - \boldsymbol{r}_j) \,, \tag{2.6}$$

with the volume $\mathsf{V} = \mathsf{L}^2\mathsf{h}$. Whereas in the homogeneous 3D case, the scattering matrix element from paired state $(\boldsymbol{k}, \uparrow; -\boldsymbol{k}, \downarrow)$ to $(\boldsymbol{k}', \uparrow; -\boldsymbol{k}', \downarrow)$ is independent of $\boldsymbol{k}$ and $\boldsymbol{k}'$, provided both states have energy within $\hbar\omega_{\mathrm{D}}$ of the Fermi surface. Here, the position-dependent values of $w_n(z)$ lead to a matrix element, which depends on the w_n eigenfunctions

$$\begin{aligned} V_{nn';\boldsymbol{k},\boldsymbol{k}',-\boldsymbol{k}',-\boldsymbol{k}} &= -\frac{g}{\mathsf{L}^2}\int_0^h |w_n(z)w_{n'}(z)|^2 dz \\ &\quad \text{if} \quad (|\epsilon_{n,k_x,k_y}|, |\epsilon_{n',k'_x,k'_y}|) < \hbar\omega_{\mathrm{D}} \,, \\ &= 0 \quad \text{otherwise} \,. \end{aligned} \tag{2.7}$$

The self-consistent equation for the gap, $\Delta_{n,k_x,k_y} \equiv \Delta_{n\boldsymbol{k}}$, for $n\boldsymbol{k} : |\epsilon_{n\boldsymbol{k}}| < \hbar\omega_{\mathrm{D}}$, now reads

$$\begin{aligned} \Delta_{n\boldsymbol{k}} &= \frac{1}{2}\sum_{n'\boldsymbol{k}':|\epsilon_{n'\boldsymbol{k}'}|<\hbar\omega_{\mathrm{D}}} \frac{V_{n\boldsymbol{k},n'\boldsymbol{k}'}\Delta_{n'\boldsymbol{k}'}}{\sqrt{\epsilon^2_{n'\boldsymbol{k}'} + \Delta^2_{n'\boldsymbol{k}'}}} \\ &= \frac{\pi\rho}{2k_{\mathrm{F}}}\sum_{n'} \Delta_{n'}A_{n'}\int_0^h [w_n(z)w_{n'}(z)]^2 dz \,. \end{aligned} \tag{2.8}$$

Here, $\sqrt{\epsilon^2_{n'\boldsymbol{k}'} + \Delta^2_{n'\boldsymbol{k}'}}$ is the quasiparticle energy $E_{n\boldsymbol{k}}$. Also, k_{F} denotes the Fermi wavenumber, and ρ is the dimensionless coupling $\rho = gD(E_{\mathrm{F}} = \mu)$, with $D(E_{\mathrm{F}}) \equiv N(0)$ being the normal electron density of states at the Fermi energy for both spins.

In the second expression, the subscript $\boldsymbol{k}$ in the gap Δ'_n has been dropped, as the translational invariance in the x–y plane ensures that the gap is independent of k'_x and k'_y, provided the condition $n'\boldsymbol{k}' : |\epsilon_{n'\boldsymbol{k}'}| < \hbar\omega_{\mathrm{D}}$ is satisfied. The A_ns arise after the usual conversion of the sum over the k'_x and k'_y states into an integral. The remaining sum over n' is estimated by converting to and performing the integral of the type

$$\frac{1}{2}\int_{-\hbar\omega}^{\hbar\omega} d\epsilon \frac{D(E_{\mathrm{F}})}{\sqrt{\epsilon^2 + \Delta^2}} = \int_0^{\hbar\omega} d\epsilon \frac{D(E_{\mathrm{F}})}{\sqrt{\epsilon^2 + \Delta^2}} = D(E_{\mathrm{F}})\sinh^{-1}\left(\frac{\hbar\omega}{\Delta}\right) , \tag{2.9}$$

which leads to

$$\begin{aligned} A_n &= \sinh^{-1}\left(\frac{\hbar\omega_{\mathrm{D}}}{\Delta_n}\right) \qquad e_{\mathrm{n}} < -\hbar\omega_{\mathrm{D}} \\ &= \frac{1}{2}\sinh^{-1}\left(\frac{\hbar\omega_{\mathrm{D}}}{\Delta_n}\right) - \frac{1}{2}\sinh^{-1}\left(\frac{e_{\mathrm{n}}}{\Delta_n}\right) \qquad |e_{\mathrm{n}}| < \hbar\omega_{\mathrm{D}} \\ &= 0 \quad e_{\mathrm{n}} > \hbar\omega_{\mathrm{D}} \,. \end{aligned} \tag{2.10}$$

To determine the chemical potential μ (at $T = 0$) given the electron number density N/V, we make use of the relation of the total number of electrons N in the volume $\mathsf{V} = \mathsf{L}^2\mathsf{h}$, that is,

$$\frac{N}{\mathsf{V}} = \frac{2}{\mathsf{V}} \int dx\,dy\,dz \sum_{n'\boldsymbol{k}':|\epsilon_{n'k'}|<\hbar\omega_{\mathrm{D}}} \left| \frac{w_{n'}(z) e^{i(k'_x x + k'_y y)}}{\sqrt{\mathsf{L}}} \right|^2$$
$$= 2 \sum_{n':|\epsilon_{n'k'}|<\hbar\omega_{\mathrm{D}}} \int_0^{e_{n'}} d\epsilon_{n'} \frac{1}{2} \left[1 + \frac{\epsilon_{n'k'_x k'_y}}{\sqrt{\Delta_{n'}^2 + \epsilon_{n'k'_x k'_y}^2}} D_{2\mathrm{D}}(\epsilon_{n'k'_x k'_y}) \right] ,$$
$$= \frac{m}{2\pi\hbar^2 \mathsf{h}} \sum_n H_n , \tag{2.11}$$

where the second equality holds after integration over the volume $\mathsf{V} = \mathsf{L}^2\mathsf{h}$ and converting the sums over the k_x and k_y states into an energy sum, with the 2D density of state per spin in the normal state $D_{2\mathrm{D}}(\epsilon)$:

$$D_{2\mathrm{D}}(\epsilon) = \frac{m}{2\pi\hbar^2} . \tag{2.12}$$

We have also made use of the relation (1.6)

$$\left| \frac{w_{n'}(z) e^{i(k'_x x + k'_y y)}}{\sqrt{\mathsf{L}}} \right|^2 = \frac{1}{2} \left(1 + \frac{\epsilon_{n'k'_x k'_y}}{E_{n'k'_x k'_y}} \right) = \frac{1}{2} \left(1 + \frac{\epsilon_{n'k'_x k'_y}}{\sqrt{\Delta_{n'}^2 + \epsilon_{n'k'_x k'_y}^2}} \right) . \tag{2.13}$$

The quantity H_n, computed below, is defined via the last equality, yielding

$$H_n = -2e_{\mathrm{n}} , \quad e_{\mathrm{n}} < -\hbar\omega_{\mathrm{D}} ,$$
$$= \hbar\omega_{\mathrm{D}} - \sqrt{\Delta_n^2 + \hbar^2\omega_{\mathrm{D}}^2} - e_{\mathrm{n}} + \sqrt{\Delta_n^2 + e_{\mathrm{n}}^2} , \quad |e_{\mathrm{n}}| < \hbar\omega_{\mathrm{D}} ,$$
$$= 0 , \quad e_{\mathrm{n}} > \hbar\omega_{\mathrm{D}} . \tag{2.14}$$

This relationship between μ and N/V, together with the gap (2.9) form a close set of equations, which can be solved numerically. Here, the values on the right-hand side of the second equal sign are obtained using the relations between the u's and v's and ϵ_{nk}s and E_{nk}s. The numerical solution results in a behavior of the gap shown in Figure 2.1. The gap is strongly dependent on the thickness h. As each subband crosses the Fermi level, a sharp drop is observed for the value of the gap. The gap has a slight dependence on the index, n, for the parameters shown. The scale of the cut-off introduced by the Debye phonon energy determines the sharpness of the drop, as the thickness h is reduced, as can be seen in the expanded view in Figure 2.2.

To gain further insight, Blatt and Thompson made the assumption of zero average charge through out most of the thickness of the film, with a small region of

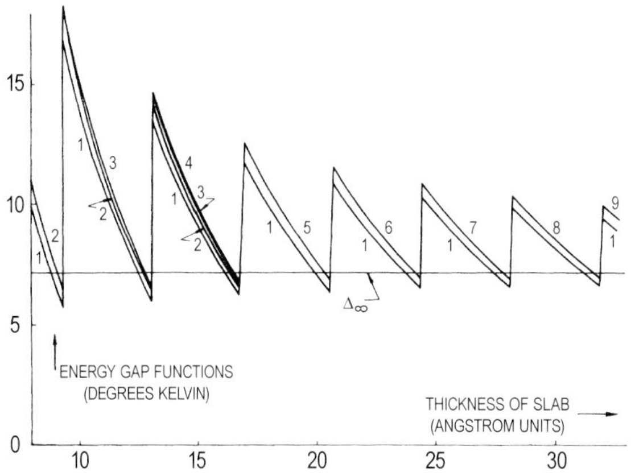

Figure 2.1 Shape resonances–oscillations of the superconducting energy gap parameter as a function of the thickness of the film. At each resonance, a new value of *n* starts to contribute. The horizontal line denotes the bulk value Δ_∞. Note that the oscillations are not symmetric about the bulk value. The enhancement of the peaks is larger than the depression in the troughs. When the Fermi level crosses a new quantum level, a sharp and narrow transition region is observed [34].

the order of 0.2 nm at the top and bottom film surfaces, within which this exact balance between the electronic density and the positively-charged background ions is lost. Neglecting these small regions, this enables one to set $U_{\text{ion}} + U_{\text{e–e}}$ in (2.5) to zero, thus allowing to solve for an approximate form of $w_n(z)$. This yields

$$w_n(z) = \sqrt{\frac{2}{\mathrm{h}}} \sin\left(\frac{n\pi z}{\mathrm{h}}\right) , \tag{2.15}$$

with eigenenergy

$$e_{\mathrm{n}} = \frac{\hbar^2}{2m}\frac{n^2\pi^2}{\mathrm{h}^2} - \mu . \tag{2.16}$$

The self-consistent equation for the gap Δ_n can now be written as

$$\left(1 - \frac{A_n}{2}\frac{k_{\mathrm{F}}}{\pi\rho\mathrm{h}}\right)\Delta_n = \frac{k_{\mathrm{F}}}{\pi\rho\mathrm{h}}\sum_{n'}^{n_{\max}} \Delta_{n'} A_{n'} . \tag{2.17}$$

$n_{\max}$ is determined from the condition

$$|\epsilon_{n'\boldsymbol{k}'}| < \hbar\omega_{\mathrm{D}} , \tag{2.18}$$

and can be computed as

$$n_{\max} = \text{fixed}\left(\frac{\mathrm{h}}{\pi}\frac{2m}{\hbar^2}\sqrt{\mu + \hbar\omega_{\mathrm{D}}}\right) . \tag{2.19}$$

Here, "fixed" denotes the integer part of the resultant number.

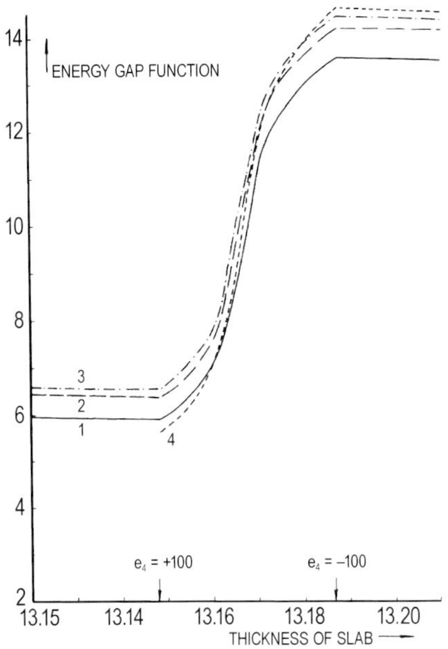

Figure 2.2 Expanded via Figure 2.1, showing the sharp and narrow transition regions. From [34].

When A_n only depends on Δ_n, occurring for most values of the film thickness h, with the exception of small regions for which $|e_n| < \hbar\omega_D$, the gap Δ_n is independent of the index n:

$$\Delta_n = \hbar\omega_D \left[\sinh\left(\frac{k_F h}{\pi\rho} \frac{1}{n_{max} + \frac{1}{2}} \right) \right]^{-1}, \quad n \leq n_{max}$$
$$= 0\,, \quad n > n_{max}\,. \tag{2.20}$$

With this result, the chemical potential μ can be found in closed form, that is,

$$\mu = \frac{\pi h \hbar^2}{n_{max} m} \left[\frac{N}{V} + \frac{\pi}{6h^3} n_{max} \left(n_{max} + \frac{1}{2} \right) (n_{max} + 1) \right]. \tag{2.21}$$

This follows from the relation between the volume density N/V and the chemical potential μ.

The behavior of the chemical potential as a function of h is illustrated in Figure 2.3. As the thickness h is decreased, μ shows a general trend to increase. This is a well-known effect in metals, whether superconducting or otherwise [35].

For a fixed density, while varying the thickness h, for a given resonance with a maximum n value, n_{max}, the corresponding range of h may be estimated as

$$(h_{n_{max}} + \Delta h_{t,n_{max}}) \leq h \leq (h_{n_{max}} - \Delta h_{t,n_{max}+1})\,, \tag{2.22}$$

where $\Delta h_{t,n_{max}}$ denotes the halfwidth in h characterizing the small transition region as the n_{max}th subband crosses the chemical potential (Fermi energy), as indicated

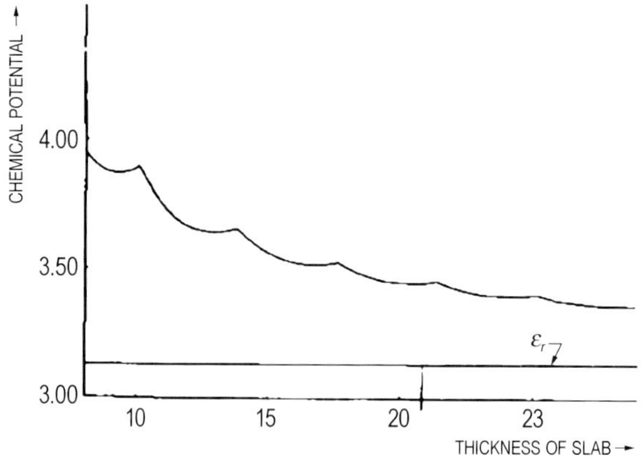

Figure 2.3 Oscillations of the chemical potential μ as a function of the thickness of the film. From [35].

in Figure 2.2. The threshold thickness $\mathsf{h}_{n_{\max}}$ corresponding to a maximum n ($n_{\max}$) may be estimated by neglecting the small energy $\hbar\omega_{\mathrm{D}}$ compared to μ in (2.19). Writing

$$\frac{N}{\mathsf{V}} = \frac{m}{\pi\hbar \mathsf{h}} \sum_{n'=1}^{n_{\max}} -e_{\mathrm{n}} \,, \tag{2.23}$$

with $e_{\mathrm{n}} = \frac{\hbar^2}{2m}\frac{\pi^2 n'^2}{\mathsf{h}^2} - \mu$, we find

$$\mathsf{h}^3_{n_{\max}} = \frac{\pi \mathsf{V}}{2N}\left[n^3_{\max} - \frac{1}{3}n_{\max}\left(n_{\max} + \frac{1}{2}\right)(n_{\max} + 1)\right] . \tag{2.24}$$

2.2.2
Bogoliubov–de Gennes Equations, Finite Temperature, and Parabolic-Band Approximation for Realistic Materials

To go beyond the simplified model considered by Blatt and Thompson, and to make contact with experimental measurements, for example, those performed on the Pb thin film system [9–11], Shanenko, Croitoru, and Peeters [32, 33] solved the self-consistent BdG equations, supplemented by a parabolic band approximation for the band structure of real materials such as Al, Pb, Sn, Zn, and so on. The parabolic band approximation amounts to replacing the $\boldsymbol{k} \cdot \boldsymbol{p}$ and pseudopotential, full band calculations of the band structure, with a simple, effective mass approximation

$$H_{\mathrm{band}} \approx H_{\mathrm{PB}} = \frac{\hbar^2 \boldsymbol{k}^2}{2m^*} \,, \tag{2.25}$$

where m^* is the effective band mass. Moreover, the electron–phonon coupling V may be adjusted depending on the thickness, as surface effects become increasingly important with reduced thickness. The parabolic band approximation leads to the notion of an effective chemical potential μ_{eff}, which differs substantially from the actual chemical potential.

The full calculation based on the BdG equations (see Section 3), solved numerically, indicates substantially more finer structure than found by Blatt and Thompson in their simplified analytic calculation. For instance, the energy gap shows oscillations in the thickness (z) direction arising from Friedel oscillations introduced by the top and bottom boundaries. The formalism for the numerical computations is as follows. As in the Blatt–Thompson model, periodic boundary conditions are imposed in the plane of the film, for the x and y directions. Furthermore, as before, the gap can only have spatial dependence in the thickness z-direction: $\Delta(\boldsymbol{r}) = \Delta(z)$. The same holds true for the self-consistent single particle potential: $U_{\text{sc}}(\boldsymbol{r}) = U_{\text{sc}}(z)$. The u's and v's thus are given by

$$\begin{aligned} \mathrm{u}_{nk_xk_y}(\boldsymbol{r}) &= \frac{e^{ik_xx}}{\sqrt{\mathrm{L}_x}}\frac{e^{ik_yy}}{\sqrt{\mathrm{L}_y}}u_{nk_xk_y}(z)\,, \\ \mathrm{v}_{nk_xk_y}(\boldsymbol{r}) &= \frac{e^{ik_xx}}{\sqrt{\mathrm{L}_x}}\frac{e^{ik_yy}}{\sqrt{\mathrm{L}_y}}v_{nk_xk_y}(z)\,, \end{aligned} \tag{2.26}$$

with k_x, k_y, n indexed as before. The normalization for u and v reads

$$\int dz\left(|u_{nk_xk_y}|^2 + |v_{nk_xk_y}|^2\right) = 1\,, \tag{2.27}$$

while the boundary conditions in the thickness z-direction are

$$u_n(0) = u_n(z = \mathrm{h}) = 0\,; \quad v_n(0) = v_n(z = \mathrm{h}) = 0\,. \tag{2.28}$$

Such boundary conditions lend themselves naturally to a basis for expansion in terms of the usual sine functions associated with an infinite square well potential in the thickness z-direction, that is,

$$\phi_l(z) = \sqrt{\frac{2}{\mathrm{h}}}\sin\left(\frac{l\pi z}{\mathrm{h}}\right)\,. \tag{2.29}$$

The projection of the u_n and v_n onto the ϕ_l basis yields the coefficients

$$u_n^l = \int_0^{\mathrm{h}} dz\,\phi_l^*(z)u_n(z)\,, \quad v_n^l = \int_0^{\mathrm{h}} dz\,\phi_l^*(z)v_n(z)\,. \tag{2.30}$$

The BdG equations in terms of these coefficients are written as (with $k_l = l\pi/\mathrm{h}$)

$$\begin{aligned} &\left[\frac{\hbar^2}{2m}\left(k_l^2 + k_x^2 + k_y^2\right) - \mu\right]u_n^l + \sum_{l'}\left(U_{\text{sc},ll'}u_n^{l'} + \Delta_{ll'}v_{nk_xk_y}^{l'}\right) \\ &\quad = E_n u_{nk_xk_y}^l\,, \end{aligned} \tag{2.31}$$

$$\begin{aligned} &-\left[\frac{\hbar^2}{2m}\left(k_l^2 + k_x^2 + k_y^2\right) - \mu\right]v_n^l - \sum_{l'}\left(U_{\text{sc},ll'}v_{nk_xk_y}^{l'} - \Delta_{ll'}u_{nk_xk_y}^{l'}\right) \\ &\quad = E_{nk_xk_y}v_{nk_xk_y}^l\,, \end{aligned} \tag{2.32}$$

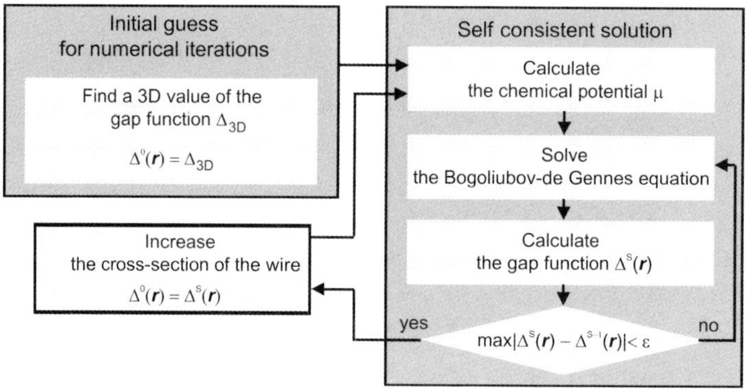

Figure 2.4 A schematic flow diagram of the numerical iteration procedure to solve the self-consistent BdG equations. From [39].

with $k_n = (n\pi/\mathrm{h})$. The quantities $\Delta_{ll'}$ and $U_{\mathrm{sc},ll'}$ are given by

$$\Delta_{ll'} = \int_0^{\mathrm{h}} dz \phi_l^*(z)\Delta(z)\phi_{l'}(z) , \quad U_{\mathrm{sc},ll'} = \int_0^{\mathrm{h}} dz \phi_l^*(z) U(z)\phi_{l'}(z) . \tag{2.33}$$

In Figure 2.4, a flow chart diagram of the numerical, self-consistent calculation procedure is shown.

2.2.3
Numerical Solutions and Thin Film Shape Resonances

In Section 2.2.1, we saw that a simplified model calculation produced oscillations in the superconducting gap and in the chemical potential as the thickness of a clean 2D film is reduced. To make direct contact with realistic experimental measurements on elemental metal, such as Pb [9–11], Shanenko, Croitoru, and Peeters [32] solved the BdG equations using the formulation presented in the previous Section 2.2.2, in which the band structure is approximated by a parabolic band (PB). Such an approximation produces many of the relevant behaviors, such as the oscillation of the superconducting gap, Δ, with thickness, oscillations in the ratio of the transition temperature to the bulk transition $T_c(\mathrm{h})/T_{\mathrm{c,bulk}}$, where h is the film thickness.

As may be expected, the oscillations occur with a period in the thickness h determined by the Fermi wavelength λ_F. The introduction of the PB approximation turns out to require the accompanying use of an effective chemical potential, μ_{eff}, which is often an order of magnitude below the actual μ, in order to obtain agreement with experimental data. To obtain sufficient accuracy in the numerical computation, the dimensions in the 2D-plane, L_x and L_y, exceeded 500 nm, to approximate an infinite film.

In Figure 2.5a, we show the oscillations of the transition temperature, relative to the bulk temperature, $T_c/T_{\mathrm{c,bulk}}$, as a function of film thickness, for cadmium, alu-

Table 2.1 Parameters used in the numerical calculations. From [32].

Metal	$\hbar\omega_D/k_B$ (K)	$gN_{bulk}(0)$	Fermi level (μ_{bulk}) (eV)
Cd	164	0.18	7.47
Al	375	0.18	11.7
Sn	195	0.25	10.2
Pb	96	0.39	9.47

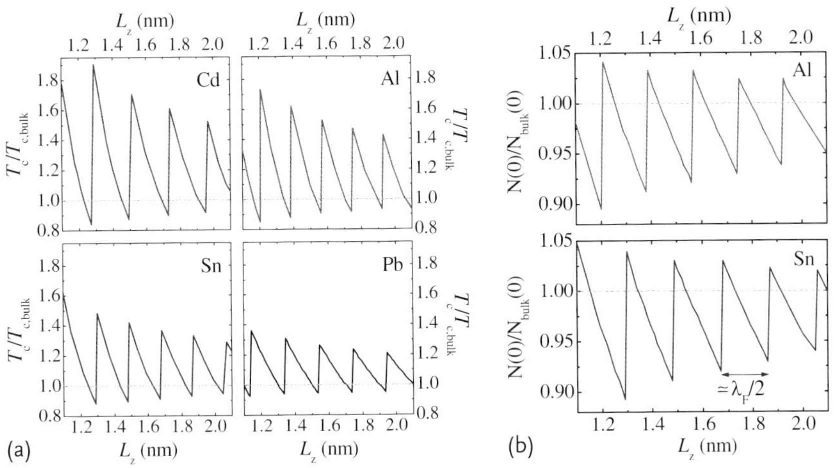

Figure 2.5 (a) Shape resonances–oscillations of the superconducting transition temperature divided by the bulk transition temperature: $T_c/T_{c,bulk}$ versus the thickness measured in monolayers of atoms. The panels correspond to numerical results for cadmium (Cd), aluminum (Al), tin (Sn), and lead (Pb). The parameters used are the bulk values presented in Table 2.1. (b) For comparison to the curves in (a), the single-particle density of state at the chemical potential, divided by the bulk value: $N(0)/N_{bulk}(0)$, is plotted versus the thickness for Al and Sn films. From [32].

minum, tin, and lead. These are contained in Figure 2.5a. For reference, the single-electron density of states at the Fermi level relative to bulk values $N(0)/N_{bulk}(0)$ are plotted for aluminum and tin in Figure 2.5b. The parameters used in the calculations are presented in Table 2.1. The shape of the oscillations are more complex than the simple expectation of a Friedel-type oscillation $\sim \cos(2k_F h)/h$, found experimentally for the surface energy [36, 37]. In agreement with the model calculations by Blatt and Thompson (Section 2.2.1), the enhancements at the peaks are more pronounced than the depressions at the troughs. The size of these deviations from bulk values are sensitive to the quantities ω_D and $gN_{bulk}(0)$, where g denotes the electron–phonon coupling strength. The film density of states $N(0)$ was estimated by dividing the number of states within the Debye window divided by the energy scale $2\hbar\omega_D$, and by the volume $V = L_x L_y h$. As one may expect from the behavior of T_c, the depressions in $N(0)$ are more pronounced than the enhancements.

Because of the presence of Friedel-like oscillations arising from the hard boundaries at the top and bottom surfaces of the thin film, the superconducting gap $\Delta(z)$ exhibits oscillatory modulations. These are exhibited in Figure 2.6. The oscillation periods in Figures 2.5 and 2.6 occur very close to the thickness change of $\Delta h \approx \lambda_F/2$ and the position change of $\Delta z \approx \lambda_F/2$, respectively. However, the numerical values of the period $\sim$ 0.15 nm ($\sim$ 1/2 ML), for instance, in lead (Pb(111)), is substantially smaller than what is found in experiment, which is approximately 2 ML ($\sim$ 0.6 nm) [9–11]! This difference is attributed to the PB approximation which is used to simplify the numerical calculations, rather than the full band structure. Introduction of the effective chemical potential is necessary to achieve agreement with experiment (Section 7.6.2). The value for $\mu_{\text{eff}} \approx 1$ eV. This is reduced by an order of magnitude from the bulk value of $\mu = 9.47$ eV for Pb. Despite the need for a reduced value of μ_{eff} compared to the bulk value, it is necessary to maintain the same $gN(0)$ product as in the bulk, to achieve good agreement of the numerical calculations with experiment.

Experiments in Pb thin films of thickness $\sim$ 20 ML ($\sim$ 5.8 nm) deposited on silicon (111) crystal indicate an oscillatory dependence on even and odd numbers of layers. In Figure 2.7, such oscillations obtained in numerical calculations are

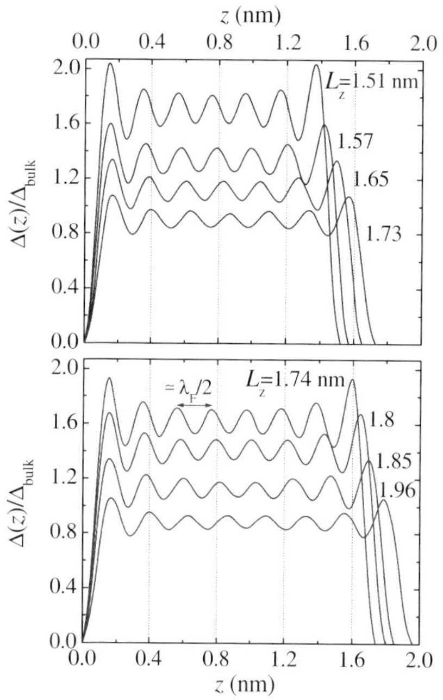

Figure 2.6 Distribution of the relative order parameter: $\Delta(z)/\Delta_{\text{bulk}}(z)$ for different thicknesses $L_z = h$. Friedel oscillations are clearly visible. The order parameter vanishes at the boundaries as required by the hard-wall boundary conditions. From [32].

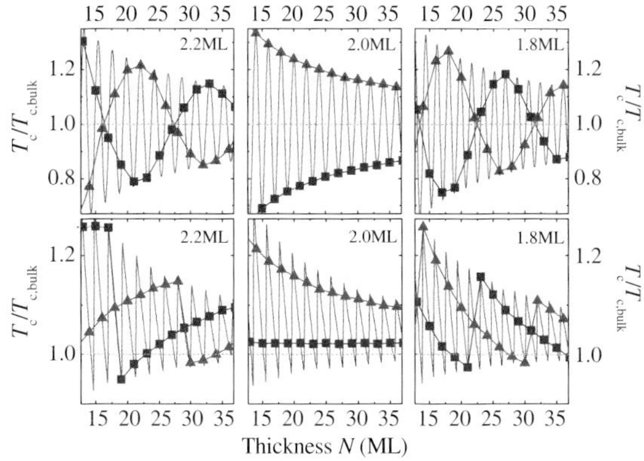

Figure 2.7 The dependence of the relative transition temperature: $T_c/T_{c,\mathrm{bulk}}$ for odd-layered (squares) and even-layered (triangles). These films are Pb, deposited on the (111) crystallographic plane of silicon. The quantum well states (subbands) periods are 2.2, 2.0, and 1.8 ML. The upper curves show approximate results using the simplified Friedel model. The lower curves are the solutions to the BdG equations. From [32].

shown. A fuller discussion and comparison with experimental data will be presented in Chapter 7, Section 7.6.2.

Furthermore, experimental findings demonstrated that the Eliashberg mass-enhancement factor for the electron–phonon coupling, λ, is reduced from the bulk value, due to the influence of the silicon substrate, on which the films are deposited. It is thus necessary to introduce a thickness dependence of the coupling constant g, which scales inversely as the number of monolayers N_{ML}:

$$g = g_{\mathrm{bulk}} - \frac{g_1 + \cos(2k_{\mathrm{F}} a_{\mathrm{e,o}} N_{\mathrm{ML}})}{N_{\mathrm{ML}}}, \tag{2.34}$$

where $a_{\mathrm{e}} = 2\pi a$ for even-layered films and $a_{\mathrm{o}} = \pi a$ for odd-layered films with a the lattice constant, and g_1 is an adjustment constant.

2.2.4 1D Nanowires – Shape Resonances and Size Oscillations

The same formalism used to study shape resonances and size oscillations in clean thin film can be applied to clean nanowires. Here, the shape of the cross section will determine the energy eigenstates and spectrum for the transverse directions (y- and z-direction) perpendicular to the length of the wire (x-direction). For a cross section which yields integrable dynamics, the level spacings are expected to obey Poisson statistics, whereas for a shape which is nonintegrable or chaotic, level repulsion familiar in the nuclear context, should lead to a Wigner–Dyson-type statistics for the level spacings [22–26].

As a starting point, we will describe the case of a circular cross section, as was done by Shanenko and Croitoru [38], and by Croitoru, Shanenko, and Peeters [33, 39]. The circle has a radius R. This shape of the cross section leads to an integrable dynamics. For a cylindrical wire, the coordinates are ρ, θ, and x. The gap is dependent on the coordinate ρ only, and the periodic boundary condition is applied in the length x-direction. The us and vs are written as

$$u_{nmk_x}(\rho\theta x) = u_{nmk_x}(\rho)\frac{e^{im\theta}}{\sqrt{2\pi}}\frac{e^{ik_x x}}{\sqrt{L_x}} ,$$
$$v_{nmk_x}(\rho\theta x) = v_{nmk_x}(\rho)\frac{e^{im\theta}}{\sqrt{2\pi}}\frac{e^{ik_x x}}{\sqrt{L_x}} . \tag{2.35}$$

The boundary conditions at the surface of the cylinder are

$$u_n(R) = 0 = v_n(R) , \tag{2.36}$$

$$\left[-\frac{\hbar^2}{2m^*}\left(\frac{\partial^2}{\partial\rho^2} + \frac{1}{\rho}\frac{\partial}{\partial\rho} - \frac{m^2}{\rho^2} - k_x^2\right) + U_{\text{sc}}(\rho) - \mu\right]$$
$$\times u_{nmk_x}(\rho) + \Delta(\rho)v_{nmk_x}(\rho) = E_{nmk_x}u_{nmk_x}(\rho) ,$$
$$-\left[-\frac{\hbar^2}{2m^*}\left(\frac{\partial^2}{\partial\rho^2} + \frac{1}{\rho}\frac{\partial}{\partial\rho} - \frac{m^2}{\rho^2} - k_x^2\right) U_{\text{sc}}(\rho) - \mu\right]$$
$$\times v_{nmk_x}(\rho) + \Delta(\rho)u_{nmk_x}(\rho) = E_{nmk_x}v_{nmk_x}(\rho) . \tag{2.37}$$

To perform the numerical calculations, as in the 2D thin film case, the wavefunctions are expanded in terms of the following basis states, namely,

$$\frac{e^{ik_x x}}{\sqrt{L_x}} , \quad \frac{e^{im\theta}}{2\pi} , \quad \frac{\sqrt{2}}{R J_{n+1}(\alpha_{mn})} J_n\left(\frac{\alpha_{mn}\rho}{R}\right) , \tag{2.38}$$

with

$$u^l_{nmk_x} = \int \rho d\rho \frac{\sqrt{2}}{R J_{l+1}(\alpha_{ml})} J_l\left(\frac{\alpha_{ml}\rho}{R}\right) u_{nmk_x}(\rho) ,$$
$$v^l_{nmk_x} = \int \rho d\rho \frac{\sqrt{2}}{R J_{l+1}(\alpha_{ml})} J_l\left(\frac{\alpha_{ml}\rho}{R}\right) v_{nmk_x}(\rho) . \tag{2.39}$$

The coupled equations for the expansion coefficients $u^l_{nmk_x}$

$$\frac{\hbar^2}{2m}\left(\frac{\alpha^2_{ml}}{R^2} + k_x^2 - \mu\right) u^l_{nmk_x} + \sum_{l'} U_{\text{sc},ll'} u^{l'}_{nmk_x} + \sum_{l'} \Delta_{ll'} v^{l'}_{nmk_x}$$
$$= E_{nmk_x} u^l_{nmk_x} , \tag{2.40}$$

$$-\frac{\hbar^2}{2m}\left(\frac{\alpha^2_{ml}}{R^2} + k_x^2 - \mu\right) v^l_{nmk_x} - \sum_{l'} U_{\text{sc},ll'} v^{l'}_{nmk_x} + \sum_{l'} \Delta_{ll'} u^{l'}_{nmk_x}$$
$$= E_{nmk_x} v^l_{nmk_x} . \tag{2.41}$$

The quantities $\Delta_{ll'}$ and $U_{ll'}$ are given by

$$\Delta_{ll'} = \int_0^R \rho d\rho \phi_l^*(\rho)\Delta(\rho)\phi_{l'}(\rho) , \quad U_{sc,ll'} = \int_0^R \rho d\rho \phi_l^*(\rho)\, U_{sc}(\rho)\phi_{l'}(\rho) , \tag{2.42}$$

with

$$\phi_l(\rho) = \frac{\sqrt{2}}{R J_{l+1}(\alpha_{ml})} J_l\left(\frac{\alpha_{ml}\rho}{R}\right) , \tag{2.43}$$

where the J_ls are the Bessel functions of order l.

In Figure 2.8, we show the relative gap $\Delta(R)/\Delta_{bulk}$ and relative chemical potential $\mu(R)/\mu_{bulk}$ versus the nanowire radius R. The numerical data are plotted as discrete points, while the continuous curve represents a spline interpolation of the data points. Two different electron densities are used for the Al nanowires. In Figure 2.9, the position dependence of the gap is shown for aluminum and tin. Friedel-type oscillations are clearly visible.

What is striking in these figures is the unusually large amplitude of the oscillations in the gap, as the radius is reduced. Even compared to the 2D films, for example, in Figure 2.5, the peak value of the gap relative to the bulk here is large, and can exceed 30 [38, 40]! Unfortunately, to date, such large amplitude size oscil-

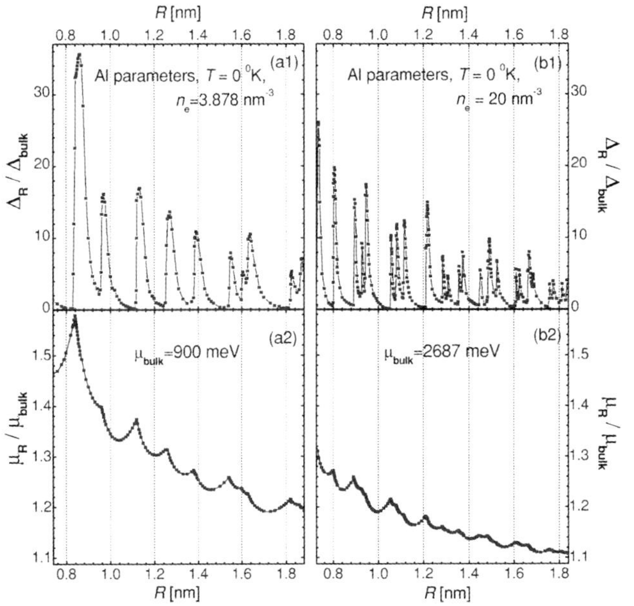

Figure 2.8 Shape resonances in the superconducting gap and chemical potential relative to the bulk values: $\Delta(R)/\Delta_{bulk}$ and $\mu(R)/\mu_{bulk}$, shown in the upper and lower panels, respectively. Two different electron densities are used for the aluminum nanowires. From [38].

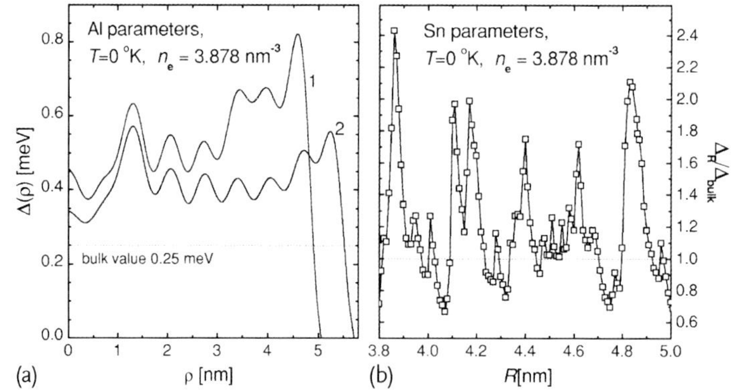

Figure 2.9 (a) The superconducting gap $\Delta(\rho)$ versus the radial position ρ for R = 5.06 nm – (1), and R = 5.75 nm – (2) for aluminum. (b) The relative gap: $\Delta(R)/\Delta_{bulk}$ for tin. From [38].

lations have yet to be confirmed in experiments. This may be a consequence of the difficulty in achieving truly uniform, single crystalline nanowires. Nevertheless, because in the best nanowires, for example, Al nanowires [28, 41], the electron mean free path can be comparable, or just slightly less than the diameter, there could be some hints that size resonance may be starting to affect the measured superconducting properties. This issue will be discussed in the experimental Section 7.6.2.

2.3
Superconductivity in Carbon Nanotubes – Single-Walled Bundles and Individual Multiwalled Nanotubes

It is well-known that intercalated graphite, for example KC_8, and doped Bucky ball films, for example Cs_3C_{60}, undergo a superconducting transition, the former at ~ 0.8 K, and the latter ~ 40 K. The strong enhancement of T_c in the Buckminsterfullerene compound is believed not to be caused by any difference in the electronic density of states at the Fermi level, but rather, by the curvature-enhanced electron–phonon coupling. The curvature opens up scattering channels which were not accessible in a flat geometry. Benedict *et al.* [3] theoretically analyzed the normal state conductivity, as well as superconductivity in small-diameter carbon nanotubes, and found that indeed, the electron–phonon scattering matrix elements, which in a flat geometry would be zero by parity symmetry considerations in the π^* states of sp^2 hybridized orbitals, would now become nonzero from an admixture of σ^* character.

As a consequence, the pair potential has an extra term, that is,

$$U_{NT} \approx U_{graphite} + \frac{1}{4} U_{curve} \left(\frac{R_0}{R} \right)^2 , \qquad (2.44)$$

where R is the radius of the nanotube, and R_0 is a reference radius.

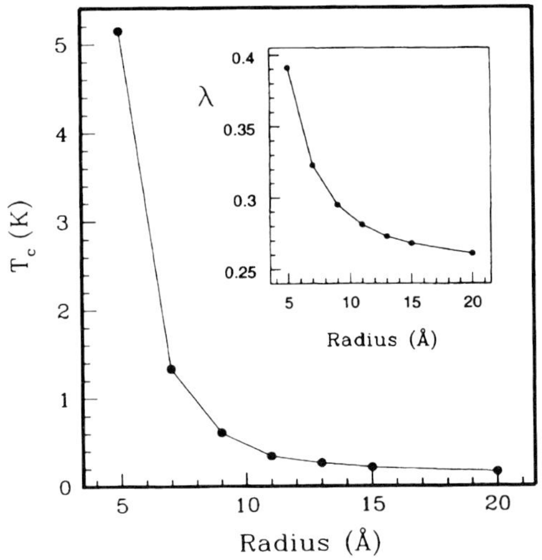

Figure 2.10 Variation of the superconducting transition temperature T_c and the electron–phonon coupling λ, with a carbon nanotube radius. From [3].

Considering an array of parallel NT at regular spacing in a lattice, and assuming an intertube electronic coupling $\sim$ 0.5 eV, which is an order of magnitude smaller than the axial bandwidth of $\sim$ 5 eV, the authors argued that the electronic density of states should be a weak function of the intertube spacing. Utilizing theoretical estimates of the screened Coulomb repulsion parameter μ^* for a strong-coupling superconductor [18, 19, 42], the transition temperature is estimated to be in the few kelvin range, for NT radius in the 0.5–2 nm range. These estimates are for single-walled nanotubes. The estimated behavior of T_c vs. R is presented in Figure 2.10. In fact, such superconducting behavior in arrays of SWCNTs encased in zeolite has now been observed, with a $T_c \sim 10$ K [4]. Because the geometry is that of an array of lines, the finite temperature transition is of the Kosterlitz–Thouless–Berezinskii type [43].

Instead of a bundle of SWCNTs, it is also possible to consider the occurrence of superconductivity in multiwalled CNTs. The multiwalls introduce different 1D quantum channels (modes), which are best described as Luttinger liquids. The different modes corresponding to the different shells of the multiwall structure can scatter into each other via phonons, giving rise to a coherent state with superconducting correlations, which is lower in energy than the multichannel Luttinger liquid [44]. The large number of quantum channels in a multiwalled tube also help screen the Coulomb repulsion, thus reducing the degree of suppression of superconductivity arising from strong-coupling effects.

Normally, the different chirality of neighboring NT shells prevents electrons from tunneling between them. However, for a Cooper pair with zero total momentum, this mismatch does not arise, and Cooper pairs are believed to be able to tunnel without being forbidden by symmetry considerations. The electron–electron

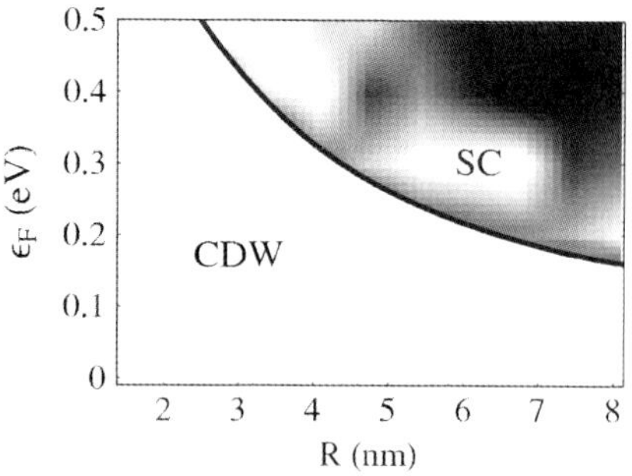

Figure 2.11 Phase diagram of doped MWCNTs in terms of the shells and the doping level represented by the shift in the Fermi energy relative to its value in the undoped nanotubes. The upper grayscale region corresponds to a superconducting phase with p-wave symmetry. The grayscale is a density plot for the logarithm of the transition gap energy relative to bandwidth. From [44].

interaction causes each 1D quantum channel to behave as a Luttinger liquid. Consequently, the interaction vertices depend logarithmically on the energy. To account for the tunneling between different Luttinger modes, a set of scaling equations were obtained that naturally account for such logarithmic energy dependence. A superconducting instability appears as the electron–phonon coupling becomes sufficiently large, and overcoming the repulsive Coulomb interaction, lead to a phase transition from a charge-density-wave (CDW) ground state to a superconducting ground state, as shown in Figure 2.11. Alternatively, superconductivity in individual ropes of SWCNT, which have a character that is a combination of the two cases, are also expected to occur, as discussed in experimental Section 9.2.

2.4 Phase Slips

The possibility of phase slips in the order parameter as a mechanism for dissipation in a narrow superconductor wire was first explored by Little in 1967 [20], in an attempt to address the question of whether thermodynamic fluctuations of the order parameter could provide a means for the supercurrent to decay in a ring geometry. In the same year, Langer and Ambegaokar (LA) [7] put forth a quantitative analysis of the lowest free energy barrier in order for a narrow wire to undergo a transition from one local free energy minimum to a next local minimum in either direction while under a current-bias. Langer and Ambegaokar sought to address experimental indications that the measured "supercurrent" departs from the mean field predictions [45], and the possibility of a finite width in the resistive transition through the critical temperature T_c. In other words, even in the superconducting state below T_c, a finite resistance can be measured. The term supercurrent is thus put in parenthesis since in many situations, the resistance is finite, albeit small. The mean field supercurrent is determined by the depairing limit which occurs when the velocity of the Cooper pair becomes large due to the large current, and the threshold for scattering to break apart a Cooper pair into unpaired quasiparticles is exceeded.

To address these issues, fluctuations in the order parameter, and specifically, the phase of the order parameter (with a concomitant local suppression of the magnitude), was proposed as the focus of the fluctuation mechanism. Once a lowest free energy barrier is available, the superconducting state can transition between adjacent minima, which differ by a phase of 2π along the entire length – see below – via fluctuations, either thermal fluctuations at higher temperatures close to, but below T_c, or by quantum-mechanical tunneling at low temperatures [46–49] (see Chapter 3).

To identify the lowest free energy barrier, LA used the GL free energy functional for a superconductor as their starting point, but specialized into one dimension. The criterion is the lateral dimensions, width (w) and height (h), are smaller than the superconducting coherence length ξ: $(\mathrm{w}, \mathrm{h}) < \xi$. Under this condition, the order parameter is approximated as constant across the narrow wire. Based on this model, LA computed the saddle point solution, which gives the lowest free energy barrier ΔF between neighboring, metastable minimum. The metastability arises due to the nonequilibrium situation occurring under a nonzero current drive.

2.4.1 Finite Voltage in a Superconducting Wire and Phase Slip

Before embarking on the details of the calculation of the saddle point free energy barrier, let us motivate how a phase slip can provide a means to generate a nonzero resistance within a superconducting state. If a voltage is applied across the two ends (lengthwise) of a narrow superconducting wire, with a voltage of $V(x)$ along the length $0 \leq x \leq L$, where $V(0) = 0$ and $V(L) = V$, then at a position x, the order parameter phase will increase linearly in time according to the Josephson relation

$$\frac{d\varphi(x)}{dt} = \frac{(-2e)}{\hbar} V(x) . \tag{2.45}$$

Since in the uniform case, where the magnitude of the order parameter is independent of position x, the gradient of the phase $\nabla\varphi$ gives the current density, this increase in the phase with time implies a corresponding increase in the current. To balance out such an implied increase, thus ensuring a steady state current, the increase in phase with time due to the applied voltage must be counter-balanced by a corresponding slippage of the order parameter phase at a rate on average dictated by the Josephson relation above (2.45).

An intuitive picture of what transpires during a spatial-temporal phase-slip event in a narrow wire can be obtained by appealing to the connection between the passage of a superconducting quantized vortex across the wire; in both situations, the phase changes by 2π, and within the core of a vortex, just as in the case of a phase-slip event, the order parameter magnitude goes to zero at least one spatial point [7, 8, 20, 21, 50]. Within this scenario, it is instructive to analyze the phase-slip event in a ring geometry. Here, one can define a winding number associated with a particular current carrying state, that is,

$$\text{w.n.} \equiv \frac{1}{2\pi}\int_0^{2\pi} \nabla\varphi \, d\theta \ , \tag{2.46}$$

where the angle θ denotes the azimuthal angle around the circular ring. The phase-slip event, which increases or decreases the phase by 2π, and thus correspondingly the winding number increases or decreases by unity, depends on the direction of the phase-slip event. The passage of a vortex either from the outside, across the narrow wire, to inside the ring or vice versa, does exactly this. For a detailed discussion, please see Halperin, Refael, and Demler [21] for a particularly lucid description.

The superconducting state is a macroscopic coherent state in which all electrons, bound into Cooper pairs, behave in a coherent manner. When spatially uniform, all Cooper pairs behave identically, with a "wavefunction" which, in the Ginzburg–Landau theory, is given by the complex, superconducting order parameter, Ψ (see Section 1.4).

The term phase slip immediately conjures up the spatial or temporal change of the phase associated with the order parameter. In the context of the Josephson junction, or a superconducting ring in the form of an annulus, the phase must change by an integer multiple of 2π. The most accessible slippage process, energetically speaking, is where the change is by one unit of 2π. This, in principle, can go either way, that is, be positive or negative. In equilibrium, the processes of either polarity are equally likely. In a nonequilibrium situation under the passage of a current, a situation commonly termed current-biased condition, the free energy barrier in one direction of polarity is lower, thus favoring one of the two directions depending on the direction of the current.

2.4.2
Phase Slip in a Josephson Junction

The most intuitive and analogous system in which a phase slip is of importance is the Josephson junction. Here, a thin insulator is placed between two superconductors. Under appropriate conditions, a supercurrent can pass through the insulator from one superconductor to the other. The situation in a SC nanowire is similar in many respects. However, there are important differences which will be discussed below. To pave the way to understanding the phase-slip process, we begin by considering the Josephson junction.

To understand the dynamics of a Josephson junction, the simplest model is the RCSJ (resistively and capacitively-shunted junction) model. The Josephson junction is characterized by the Josephson energy-phase relationship

$$U_J(\phi) = -E_J \cos\varphi \ , \tag{2.47}$$

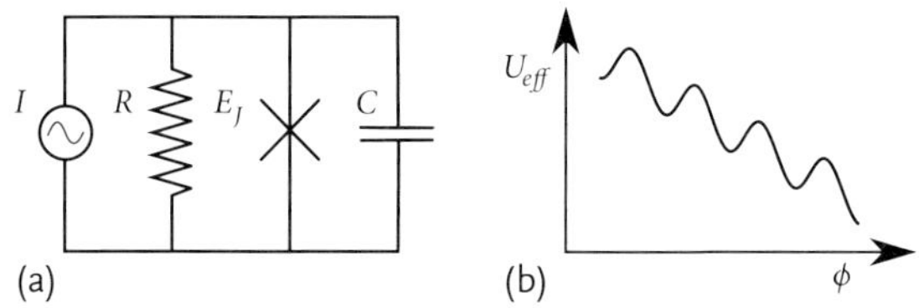

Figure 2.12 (a) An RCSJ model of a Josephson junction. (b) The effective potential energy under a constant current-bias, showing the tilted washboard potential.

where E_J is the Josephson energy at zero phase difference, by a capacitance C across the junction associated with the insulator dielectric between the superconducting metal electrodes, and by the quasiparticle normal resistance R_N, plus the resistance of external circuitry, all of which run in parallel, as seen in Figure 2.12a. The junction capacitance stores an energy

$$U_C = \frac{q^2}{2C} \tag{2.48}$$

as charges build up on the SC electrodes, leading to a Hamiltonian for the systems

$$H_{JJ} = U_C + U_J = \frac{q^2}{2C} - E_J \cos\varphi \,, \tag{2.49}$$

where, for the moment, we have neglected the shunting resistance, which is dominated by the external circuit resistive element at low temperatures.

The Josephson energy, $U_J(\phi)$, is related to the Josephson supercurrent across the junction by

$$I_J = \left(\frac{-2e}{\hbar}\right)\frac{\partial U_J}{\partial\varphi} = \frac{E_J}{\hbar/(-2e)}\sin\varphi \,. \tag{2.50}$$

In the presence of an externally imposed current, that is, under current-bias, I, the Hamiltonian becomes

$$H_{\text{IbiasJJ}} = H_{JJ} - \frac{\hbar}{2e} I\varphi \,. \tag{2.51}$$

With the commutation relation between the Cooper pair number N (related to the charge q) and the phase φ

$$[N, \varphi] = 2i \,, \tag{2.52}$$

the charge can then be interpreted as a "momentum," conjugate to the phase φ. The energy associated with the junction capacitance U_C can be thought of as a kinetic energy, and the system lives in a tilted washboard potential in the phase-coordinate space (φ space), as shown in Figure 2.12b.

The voltage across the junction is related to the phase φ through the Josephson relationship

$$V = \frac{\hbar}{2e^*}\frac{d\varphi}{dt} = \frac{\hbar}{-2e}\frac{d\varphi}{dt} \,, \tag{2.53}$$

with $2e^* = -2e$. Thus, temporal fluctuations or variations in the phase lead to a nonzero voltage across the junction, and can lead to dissipation if the system can move downhill in the potential landscape, at this fixed current-bias I. On the other hand, if the system is trapped in a local minimum of the tilted washboard potential, perturbations about the minimum will lead to oscillations, which will average to zero over time. Thus, it is necessary to pass through the energy barrier on the low energy side, either by acquiring sufficient thermal energy to surmount the barrier, or via quantum-mechanical tunneling through the barrier. In the former case, the system also has the possibility of moving uphill energetically, to the local minimum on the left.

The energy barrier is readily computed, and is given by

$$\Delta F^{\pm} \approx 2E_{\mathrm{J}} \pm \frac{\pi\hbar}{2e}|I| \,, \tag{2.54}$$

for $|I|/I_{\mathrm{c}} \ll 1$, where $I_{\mathrm{c}} = 2eE_{\mathrm{J}}/\hbar$. At a sufficiently large bias-current $I = I_{\mathrm{c}}$, the energy barrier on the low side disappears. This current I_{c}, at which the barrier first reaches zero, thus represents the critical current of the Josephson junction. The $-$ sign denotes the barrier to the lower energy (downhill) minimum, while the $+$ sign represents the higher energy (uphill) neighboring minimum.

The rate at which the Josephson junction escapes from its local minimum to the neighboring minima to the downhill and uphill side as a result of thermal agitation is, respectively, given by

$$\Gamma^{\pm} = \Omega^{\pm} \exp\left(-\frac{\Delta F^{\pm}}{k_{\mathrm{B}}T}\right) \,, \tag{2.55}$$

with $\Omega^{\pm}$ being the barrier attempt frequency.

The attempt frequency is approximately given by the Josephson plasma frequency [51]:

$$\Omega^{\pm} \approx \frac{1}{2\pi\hbar}\sqrt{\frac{(2e)^2 E_{\mathrm{J}}}{C}} = \frac{1}{2\pi\hbar}\sqrt{2E_{\mathrm{C}}E_{\mathrm{J}}} \,, \tag{2.56}$$

where $E_{\mathrm{C}} = (2e)^2/(2C)$ is the Coulomb charging energy.

The voltage is given by the difference in the forward and backward processes, that is,

$$\begin{aligned} V &= \frac{2\pi\hbar}{-e}\Omega^{\pm}\left[\exp\left(-\frac{\Delta F^{-}}{k_{\mathrm{B}}T}\right) - \exp\left(-\frac{\Delta F^{+}}{k_{\mathrm{B}}T}\right)\right] \\ &= \frac{\hbar}{-2e}\sqrt{2E_{\mathrm{C}}E_{\mathrm{J}}}\, e^{-2E_{\mathrm{J}}/(k_{\mathrm{B}}T)} \sinh\left(\frac{I}{2I_{\mathrm{T}}}\right) \,, \end{aligned} \tag{2.57}$$

where the thermal current $I_{\mathrm{T}} = 2e/[2\pi\hbar/(k_{\mathrm{B}}T)]$.

2.4.3
Langer–Ambegaokar Free Energy Minima in the Ginzburg–Landau Approximation

The starting point in Langer and Ambegaokar's analysis is the Ginzburg–Landau free energy functional in one dimension (see (1.29)) [7] under a constant current-bias,

$$F - F_n = \int_{-L/2}^{L/2} \mathsf{A} dx \left[\alpha|\Psi|^2 + \frac{\beta}{2}|\Psi|^4 + \frac{1}{2M}\left|\left(\frac{\hbar}{i}\nabla + \frac{2e}{c}A\right)\Psi\right|^2 \right] , \quad (2.58)$$

where $\Psi = \Psi(x)$ is the complex order parameter, and $M = 2m$ is the Cooper pair mass. The assumption is that the width and height (or the diameter) of the wire are sufficiently small compared to the superconducting coherence length ξ, such that the order parameter is uniform across the cross section A of the narrow wire. Thus, the y (width-direction) and z (height-direction) integrals are simple results in the cross-sectional area, and the remaining integral is strictly along the x, length-direction.

To find the stationary solutions, the variation of the free energy functional with respect to $\Psi(x)$, while holding the ends fixed, yields the condition for a stationary solution (1.33). In the absence of an applied magnetic field, and neglecting the small self field generated by the bias current and induced supercurrent, this condition is given by

$$-\frac{\hbar^2}{2M}\frac{d^2\Psi}{dx^2} - \alpha\Psi + \beta|\Psi|^2\psi = 0 . \quad (2.59)$$

With the imposition of the periodic boundary condition along the length

$$\Psi\left(x = -\frac{\mathsf{L}}{2}\right) = \Psi\left(x = +\frac{\mathsf{L}}{2}\right) , \quad (2.60)$$

the solutions for the ground state, which is spatially uniform in $|\Psi|^2$, are well-known and given by

$$\Psi_k = f_k e^{ikx} , \; f_k^2 = \frac{\left(\alpha - \frac{\hbar^2 k^2}{2M}\right)}{\beta} , \quad (2.61)$$

with $k = 2n\pi/\mathsf{L}$, where n is an integer. The electrical current density is (1.37)

$$J = \frac{-2e\hbar}{2Mi}\left(\Psi^*\frac{d\Psi}{dx} - \Psi\frac{d\Psi^*}{dx}\right) = \frac{-2e\hbar k}{M} f_k^2 = \frac{-2e\hbar k}{M}\frac{\left(\alpha - \frac{\hbar^2 k^2}{2M}\right)}{\beta} . \quad (2.62)$$

Beyond a critical current density J_c, corresponding to $k < k_c = (1/\hbar)\sqrt{(2M\alpha)/3}$, where

$$J_c = (-4e)\sqrt{\frac{2}{M}}\frac{\alpha^{3/2}}{3\sqrt{3}\beta} \propto (T_c - T)^{3/2} , \quad (2.63)$$

the solution in (2.61) no longer represents a local minimum, as discussed below. This critical electrical current density is the mean-field critical density.

Following McCumber and Halperin, it is convenient to recast this set of equations in terms of a normalized order parameter [8]:

$$\psi(x) \equiv \frac{\Psi}{\sqrt{|\alpha(T)|/\beta}} \,. \tag{2.64}$$

By further scaling the length by the Ginzburg–Landau coherence length (1.39),

$$\xi(T) = \sqrt{\frac{\hbar^2}{2M|\alpha(T)|}} = \frac{\xi(0)}{\sqrt{1 - T/T_c}}\,, \quad \xi(0) = \sqrt{\frac{\hbar^2}{2M|\alpha(T=0)|}}\,, \tag{2.65}$$

and an energy expressed in terms of the thermodynamic critical field,

$$H_c(T) = \sqrt{\frac{4\pi}{\beta}}\alpha(T)\,, \tag{2.66}$$

(2.58) above now reads

$$F - F_n = \mathrm{AL}\frac{H_c^2(T)}{4\pi}\int_{-\mathrm{L}/2}^{\mathrm{L}/2}\frac{dx}{\mathrm{L}} \times \left\{-|\psi|^2 + \frac{1}{2}|\psi|^4 + \left|\left[\xi(T)\frac{d}{dx} + i\xi(T)\frac{2e}{\hbar c}A_x\right]\psi\right|^2\right\} \,. \tag{2.67}$$

With a rescaling of the momentum k as $\kappa = k\xi(T)$, the uniform stationary solutions take the form

$$\psi_\kappa = \frac{f_k}{\sqrt{|\alpha(T)|/\beta}}e^{ikx} = g_\kappa e^{i\kappa x/\xi(T)}\,, \quad g_\kappa^2 = (1-\kappa^2)\,, \tag{2.68}$$

with $\kappa = 2\pi\xi(T)/\mathrm{L}$, where n is an integer. The expression for the free energy, referenced to the normal free energy, is thus

$$F\{\psi_\kappa\} - F_n = \mathrm{AL}\frac{H_c^2(T)}{4\pi}\left[-(1-\kappa^2)g_\kappa^2 + \frac{1}{2}g_\kappa^4\right] = -\mathrm{AL}\frac{H_c^2(T)}{8\pi}(1-\kappa^2)^2\,. \tag{2.69}$$

The wavefunction can be represented pictorially in a 3D plot, where the z-axis represents the real part of ψ, and the y-axis represents the imaginary part of ψ, as shown in Figure 2.13.

The electrical current density can be converted into a dimensionless density:

$$j \equiv \frac{J}{J_c} = \frac{3\sqrt{3}}{2}\kappa(1-\kappa^2)\,; \quad \kappa \le \sqrt{\frac{1}{3}} = \kappa_c\,. \tag{2.70}$$

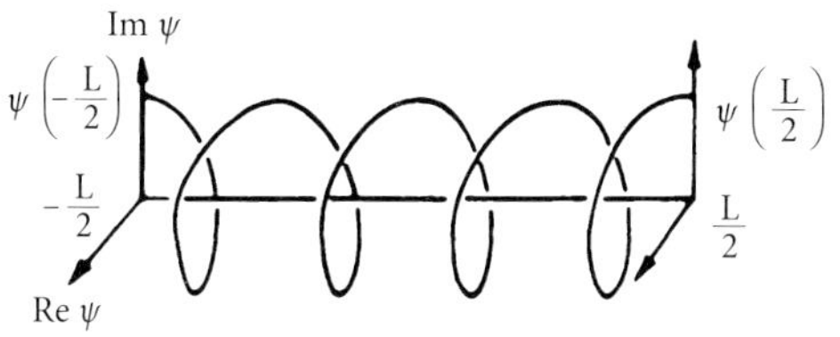

Figure 2.13 The order parameter Φ for a uniform current-carrying state. The real and imaginary parts are plotted versus the distance x along the 1D wire. From [7].

The stationary equation for the rescaled order parameter ψ is

$$(1 - |\psi|^2)\psi + \left[\xi(T)\frac{d}{dx} + i\xi(T)\frac{2e}{\hbar c}A_x\right]^2 \psi = 0 . \quad (2.71)$$

This equation can be cast into two coupled equations for the magnitude and phase. Writing, with $A_x = 0$ in the absence of a magnetic field

$$\psi(x) = f(x)\exp\left[i\varphi(x)\right] , \quad (2.72)$$

the coupled equations are

$$\xi(T)^2\frac{d^2 f}{dx^2} = -f + f^3 + \left(\frac{4}{27}\right)\frac{j^2}{f^3} ,$$
$$\frac{dj}{dx} = 0 , \quad j = \frac{3\sqrt{3}}{2} f^2 \xi(T)\frac{d\varphi}{dx} . \quad (2.73)$$

Here, j, the dimensionless current density, is independent of x. Since the Ginzburg–Landau free energy is only valid for temperature T near T_c, these expressions should NOT be applied to situations far below T_c.

It is instructive to examine the free energy functional to demonstrate that for $\kappa < \sqrt{1/3}$, the solution represents a local minimum in the functional space of $\psi(x)$. To do so, it is necessary to expand to second order in the deviations from the stationary states. In order for a stationary state to be a minimum, it is necessary that the matrix characterizing coefficients in the quadratic terms have eigenvalues which are strictly positive. The eigenvalues represent the curvature of the free energy about the stationary points in a normal coordinate system. (Please see Appendix C in Langer and Ambegaokar [7]). Writing the wavefunction in the form of a deviation from the stationary solutions

$$\psi(x) = \psi_\kappa(x) + e^{i\kappa x/\xi(T)}s(x) , \quad (2.74)$$

the free energy then takes the form

$$F\{\psi(x)\} - F_n = F_\kappa\psi(x) + F_2 s(x) + \text{higher order terms} . \quad (2.75)$$

The second order contribution F_2 is given by

$$F_2\{s(x)\} = \mathsf{AL}\frac{H_c^2(T)}{4\pi}\int_{-\mathsf{L}/2}^{\mathsf{L}/2}\frac{1}{\mathsf{L}}dx$$
$$\times\left[\left|\xi(T)\frac{d(e^{i\kappa x/\xi(T)}s)}{dx}\right|^2 - \kappa^2|s|^2 + \frac{1}{2}(1-\kappa^2)(2|s|^2 + s^2 + s^{*2})\right]$$
$$= \mathsf{AL}\frac{H_c^2(T)}{4\pi}\int_{-\mathsf{L}/2}^{\mathsf{L}/2}\frac{\xi(T)}{\mathsf{L}}dx\left[-\xi(T)^2 e^{-i\kappa x/\xi(T)}s^*\frac{d^2(e^{ikx}s)}{dx^2} - \kappa^2|s|^2\right.$$
$$\left. + \frac{1}{2}(1-\kappa^2)(2|s|^2 + s^2 + s^{*2})\right], \tag{2.76}$$

where the second expression follows after an integration by parts with the condition of fixed boundary conditions at the ends. To ensure that the integral F_2 is positive (or zero) for arbitrary deviations, we seek solutions to the eigenequation

$$-\xi(T)^2 e^{-i\kappa x/\xi(T)}\frac{d^2(e^{i\kappa x/\xi(T)}s)}{dx^2} - \kappa^2 s + (1-\kappa^2)(s+s^*) = \lambda s\,. \tag{2.77}$$

If the eigenvalue λ can be shown to be nonnegative for any solution, then the integral of the left-hand side, after multiplying by s^*, equals the integral in F_2 above, and must be positive semidefinite since the integrand is then equal to $\lambda|s|^2$.

Writing s in terms of its real and imaginary parts: $s = s_\mathrm{R} + i s_\mathrm{I}$, the eigenequation yields two-coupled real equations, that is,

$$-\xi(T)^2\frac{d^2 s_\mathrm{R}}{dx^2} + 2\kappa\xi(T)\frac{ds_\mathrm{I}}{dx} + 2(1-\kappa^2)s_\mathrm{R} = \lambda s_\mathrm{R}\,, \tag{2.78}$$

$$-\xi(T)^2\frac{d^2 s_\mathrm{I}}{dx^2} - 2\kappa\xi(T)\frac{ds_\mathrm{R}}{dx} = \lambda s_\mathrm{I}\,. \tag{2.79}$$

The solutions are the plane waves e^{iqx}, and thus form a complete set:

$$s_{\mathrm{R},q} = \mathrm{Re}\left(c_1 e^{iqx}\right)\;;\quad s_{\mathrm{I},q} = \mathrm{Re}\left(c_2 e^{iqx}\right)\,. \tag{2.80}$$

These lead to a secular determinant for the eigenvalues λ_q:

$$\det\begin{pmatrix}[\xi(T)^2q^2 + 2(1-\kappa^2) - \lambda_q] & [2i\xi(T)q\kappa] \\ [-2i\xi(T)q\kappa] & [\xi(T)^2q^2 - \lambda_q]\end{pmatrix}$$
$$= \lambda_q^2 - 2\lambda_q[\xi(T)^2q^2 + (1-\kappa^2)] + \xi(T)^2q^2[\xi(T)^2q^2 + 2(1-\kappa^2)]$$
$$- 4\xi(T)^2q^2\kappa^2 = 0\,. \tag{2.81}$$

The eigenvalues are

$$\lambda_q = \xi(T)^2q^2 + (1-\kappa^2) \pm \sqrt{(1-\kappa^2)^2 + 4\xi(T)^2q^2\kappa^2}\,. \tag{2.82}$$

Aside from the trivial case of $q = 0$ for which $\lambda_0 = 0$, as long as $\kappa < \sqrt{1/3}$ so that the current density is below the critical value, the eigenvalue is positive for all other $q \neq 0$. For $\kappa \geq \sqrt{1/3}$, the "–" solution crosses zero at $q = \pm 1/\xi(T)\sqrt{2(3\kappa^2 - 1)}$, and turns negative for $|q|$ smaller than this value $1/\xi(T)\sqrt{2(3\kappa^2 - 1)}$ for which F_2 is no longer positive, and hence the extremum no longer a minimum.

2.4.4 Transition Rate and Free Energy Barrier

In the above, the free energy is obtained for a metastable state which is uniform the magnitude of the order parameter $|\psi|$ along the entire length of the wire, and which carries a current density given by (2.70). The state is indexed by the dimensionless wavenumber $\kappa = k\xi(T) = (2\pi n/\mathsf{L})\xi(T)$. In order to deduce the rate of a phase slip, which involves changing the momentum by one unit of $2\pi/\mathsf{L}$, it is necessary to find the lowest energy free energy barrier, $\Delta F^{\pm}$, where the "+" and "–" superscripts denote the energetically uphill $(n-1) \to n$ and downhill $n \to (n-1)$ processes, respectively. To be more precise, there is a corresponding change in k and κ, respectively, of

$$\begin{aligned} \delta k &= \pm\left[\frac{2\pi n}{\mathsf{L}} - \frac{2\pi(n-1)}{\mathsf{L}}\right] \\ &= \pm\frac{2\pi}{\mathsf{L}}\,, \\ \delta\kappa &= \pm\frac{2\pi\xi(T)}{\mathsf{L}}\,. \end{aligned} \tag{2.83}$$

The free energy difference between the n and $n-1$ local minima is

$$\Delta F_{n,n-1} = -\mathsf{AL}\frac{1}{\beta}\left[\left(\alpha - k_n^2\right)^2 - \left(\alpha - k_{n-1}^2\right)^2\right] = \frac{2\pi\hbar}{|2e|}\mathsf{A}J = \frac{I\phi_0}{c}\,, \tag{2.84}$$

where $\phi_0 = 2\pi\hbar c/(2e) = hc/2e$ is the flux quantum ($\phi_0 = h/2e$ in SI units), and $I = \mathsf{A}J$ is the electrical current through the wire. Here, we have kept the leading terms in $2\pi/L$. Note that although the term $I\phi_0/c$ ($I\phi_0$ in SI units) in the expression for $\Delta F_{n,n-1}$ appears identical to that for the Josephson junction discussed previously in Section 2.4.2, there is an important difference: For the Josephson junction case, the current I is a fixed quantity, applied through an external circuit through the junction, and is the same for all the local minima in the tilted washboard potential shown in Figure 2.12b. Here, for each given stationary state corresponding to local minimum index by n, the current density is given by (2.62), (2.63), and (2.70), and is different for each minimum, increasing as n increases (but with J below J_c). Therefore, care must be taken when considering a transition at a fixed current imposed by an external circuit.

Conceptually, if ΔF_0 can be found for the lowest barrier at zero current $I = 0$, then the corresponding free energy barrier for the uphill ("+") and downhill processes are

$$\Delta F^{\pm} = \Delta F_0 \pm \frac{1}{2}\Delta F_{n,n-1} = \Delta F_0 \pm \frac{1}{2}\frac{I\phi_0}{c} \, . \tag{2.85}$$

This form will be shown to be correct below. Once the free energy barrier is determined, then, just as in the case of the Josephson junction, the thermally activated rates for the uphill and downhill transitions between $n \leftrightarrow (n-1)$ states are

$$\Gamma^{\pm} = \Omega^{\pm} \exp\left(-\frac{\Delta F^{\pm}}{k_{\mathrm{B}} T}\right) . \tag{2.86}$$

The corresponding voltage along the length of the nanowire is then

$$V = \frac{2\pi\hbar}{(-2e)}[\Gamma^{+} - \Gamma^{-}] = \frac{4\pi\hbar}{2e}\exp\left(-\frac{\Delta_0}{k_{\mathrm{B}} T}\right)\sinh\left(\frac{I\phi_0}{2ck_{\mathrm{B}} T}\right) . \tag{2.87}$$

The expression for the sinh can be recast in the form

$$\sinh\left(\frac{I\phi_0}{2ck_{\mathrm{B}} T}\right) = \sinh\left(\frac{I}{2I_{\mathrm{T}}}\right) , \quad I_{\mathrm{T}} = \frac{2e}{2\pi\hbar/(k_{\mathrm{B}} T)} , \tag{2.88}$$

where, as before, I_{T} is a thermal current scale given by the Cooper pair charge (magnitude) divided by the thermal time $2\pi\hbar/(k_{\mathrm{B}} T)$.

The task at hand is thus to determine the free energy barrier $\Delta F^{\pm}$, including its dependence on the current density J, and the attempt frequency $\Omega^{\pm}$.

2.4.5
Free Energy Barrier for a Phase Slip in the Ginzburg–Landau Theory

Langer and Ambegaokar sought a saddle point solution for the Ginzburg–Landau free energy. The criteria are that the solution must be insensitive to the boundary, and thus is a bulk intrinsic effect, and that it must be a saddle point close to the uniform solution corresponding to the local minimum at hand, such that this saddle point is a minimum in all but one directions. In the remaining direction, it must be an inflection point. The Ginzburg–Landau equations for a stationary state enables an analytic solution to be obtained. We will employ the expressions for the normalized order parameter ψ given in (2.64) and (2.72), namely,

$$\psi(x) = f(x)\exp(i\varphi(x)) \, , \tag{2.89}$$

and thus the coupled equations are (2.73)

$$\xi(T)^2 \frac{d^2 f}{dx^2} = -f + f^3 + \left(\frac{4}{27}\right)\frac{j^2}{f^3} \, ,$$
$$j = \frac{3\sqrt{3}}{2} f^2 \xi(T)\frac{d\varphi}{dx} = \text{constant} \, . \tag{2.90}$$

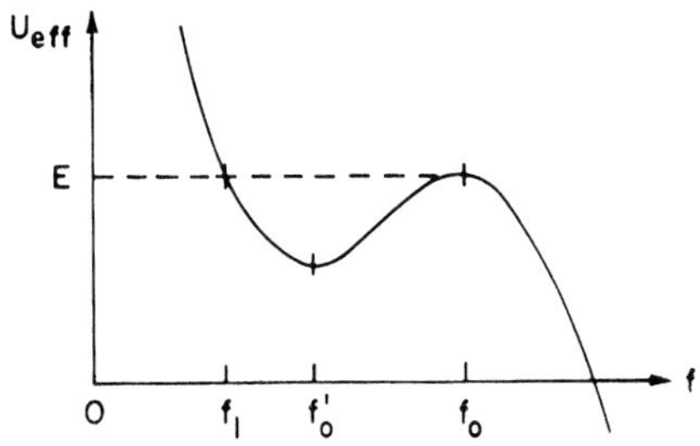

Figure 2.14 The effective potential U_{eff} for $J < J_c$, or equivalently, $j < 1$. From [7].

These equations, expressed in terms of the magnitude $f(x)$, and the phase $\varphi(x)$, can be solved by appealing to the analogy of a particle moving under the influence of a potential. We see that the right-hand side of the first equation in (2.90) can be recast as a force due to an effective potential U_{eff}

$$-f + f^3 + \left(\frac{4}{27}\right)\frac{j^2}{f^3} = -\frac{dU_{\text{eff}}}{df}\,, \quad U_{\text{eff}} = \frac{1}{2}f^2 - \frac{1}{4}f^4 + \frac{2j^2}{27f^2}\,. \tag{2.91}$$

The term containing $2j^2/27$ acts much like a centrifugal barrier, as shown in Figure 2.14. By exploiting a conservation law in the "energy" E

$$\frac{dE}{dx} = 0\,, \quad E = \frac{\xi(T)^2}{2}\left(\frac{df}{dx}\right)^2 + U_{\text{eff}}(f)\,, \tag{2.92}$$

the solutions were found by examining the turning points [7]. The complete solution, expressed in closed form, is [7, 8]

$$\psi_s(x) = \left[\sqrt{1-3\kappa^2}\tanh\sqrt{\frac{1-3\kappa^2}{2}}\frac{x-x_0}{\xi(T)} - i\sqrt{2}\kappa\right]e^{i\kappa(x-x_0)/\xi(T)}\,. \tag{2.93}$$

Here, x_0 represents an arbitrary position for the center of the saddle point solution, along the wire length: $-L/2 \le x_0 \le L/2$. The expression for $f_s^2(x)$ is

$$f_s^2(x) = -2\kappa^2 + (1-3\kappa^2) - (1-3\kappa^2)\text{sech}^2\left[\sqrt{\frac{1-3\kappa^2}{2}}\frac{x-x_0}{\xi(T)}\right]\,. \tag{2.94}$$

For a given current j, the phase $\varphi_s(x)$ can be found, up to a constant, from

$$\varphi_s(x) - \text{constant} = \int_{-L/2}^{x} dx'\frac{d\varphi_s(x')}{dx'} = \int_{L/2}^{x} dx' 3\frac{2}{3\sqrt{3}\xi(T)}\frac{j}{f_s^2(x')}\,. \tag{2.95}$$

In Figure 2.15, we show a graphical representation of this saddle point order parameter wavefunction. Notice that the magnitude is reduced in the core of the phase slip. This saddle point solution represents the lowest energy configuration of the order parameter, which is close to a uniform state. In order for a phase slip to take place, the 1D superconducting wire system must either pass through this

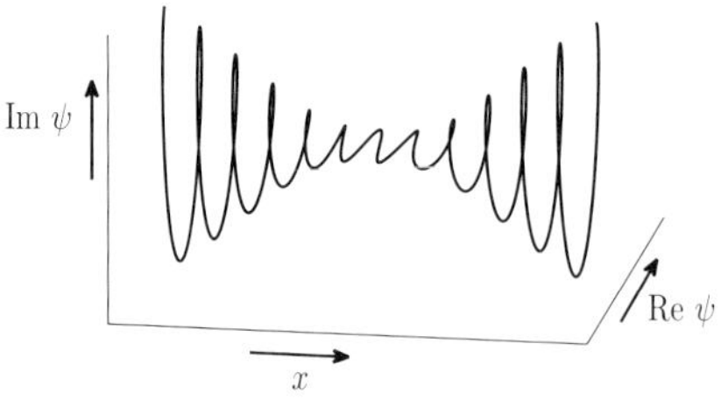

Figure 2.15 The dimensionless order parameter ψ, as a function of distance x along the 1D narrow wire. Notice that the magnitude given by $|\psi| = \sqrt{\mathrm{Re}(\psi)^2 + \mathrm{Im}(\psi)^2}$ is reduced in the core region. The entire region of reduced $|\psi|$ is of order ξ, the superconducting coherence length. Courtesy of P. Li.

configuration on the way to the adjacent local minimum, or else tunnel through the associated one with this configuration. Since the Ginzburg–Landau theory is only valid close to T_c, the quantum treatment will be discussed in the following Chapter 3, to treat the low temperature regime. Here, we are only concerned with the thermally activated process.

This solution is uniform, except in a region of order $\xi(T)$ about its center x_0. Within this region, which represents the core region of the saddle point solution at the heart of the phase-slip process, the magnitude $f(x)$ is depressed, while the phase changes rapidly across it. This phase change can be found by evaluating (2.90) above:

$$\Delta\varphi = \int_{-L/2}^{L/2} \frac{2}{3\sqrt{3}} \frac{dx}{\xi(T)} \frac{j}{f_s^2} = kL + 2\tan^{-1}\left(\sqrt{\frac{1-3\kappa^2/2}{\kappa}}\right) . \tag{2.96}$$

This is in contrast to the metastable uniform state index by n, for which

$$\Delta\varphi = kL = 2n\frac{\pi}{L} . \tag{2.97}$$

Because of the depression of the order parameter magnitude within the core region, it lies energetically higher than the local minima to its side. From this solution for the saddle point, the free energy can be computed from the expression

$$F_s - F_n = -\mathsf{AL}\frac{H_c^2(T)}{8\pi} \int_{-L/2}^{L/2} dx\, f^4 . \tag{2.98}$$

This form is obtained with the aid of the stationary condition in (2.59), and after an integration by part on the gradient term in the free energy in (2.58).

The barriers are found to be

$$\Delta F^-(\kappa, T) = \mathsf{A}\frac{H_c^2(T)}{\pi}\xi(T) \times \left[\frac{\sqrt{2}}{3}\sqrt{1-3\kappa^2} - \kappa(1-\kappa^2)\tan^{-1}\left(\frac{\sqrt{1-3\kappa^2}}{\sqrt{2}\kappa}\right)\right] , \tag{2.99}$$

$$\Delta F^{+}(\kappa, T) = \mathsf{A}\frac{H_c^2(T)}{\pi}\xi(T)$$
$$\times \left\{\frac{\sqrt{2}}{3}\sqrt{1-3\kappa^2} + \kappa(1-\kappa^2)\left[\pi - \tan^{-1}\left(\frac{\sqrt{1-3\kappa^2}}{\sqrt{2}\kappa}\right)\right]\right\} . \tag{2.100}$$

Using the relationship

$$j = \frac{J}{J_c} = \frac{3\sqrt{3}}{2}\kappa(1-\kappa^2) , \tag{2.101}$$

we can recast the above free energy barrier heights in the form

$$\Delta F^{-}(\kappa, T) = \frac{\sqrt{2}}{3\pi}\mathsf{A}H_c^2(T)\xi(T)$$
$$\times \left[\sqrt{1-3\kappa^2} - \sqrt{\frac{2}{3}}\, j \tan^{-1}\left(\frac{\sqrt{1-3\kappa^2}}{\sqrt{2}\kappa}\right)\right] , \tag{2.102}$$
$$\Delta F^{+}(\kappa, T) = \frac{\sqrt{2}}{3\pi}\mathsf{A}H_c^2(T)\xi(T)$$
$$\times \left\{\sqrt{1-3\kappa^2} + \sqrt{\frac{2}{3}}\, j \left[\pi - \tan^{-1}\left(\frac{\sqrt{1-3\kappa^2}}{\sqrt{2}\kappa}\right)\right]\right\} . \tag{2.103}$$

Taking their average, we arrive at the quantities ΔF_0 defined above in the expressions for $\Delta F^{\pm}$ (2.85):

$$\Delta F_0(\kappa, T) = \frac{\sqrt{2}}{3\pi}\mathsf{A}H_c^2(T)\xi(T)$$
$$\times \left\{\sqrt{1-3\kappa^2} + \sqrt{\frac{2}{3}}\, j \left[\frac{1}{2}\pi - \tan^{-1}\left(\frac{\sqrt{1-3\kappa^2}}{\sqrt{2}\kappa}\right)\right]\right\} . \tag{2.104}$$

We thus arrive at the form introduced above, that is,

$$\Delta F^{\pm}(\kappa, T) = \Delta F_0(\kappa, T) \pm \frac{1}{2} I\phi_0 , \tag{2.105}$$

with $I = J\mathsf{A}$. The expression for ΔF^{-} for the downhill free energy barrier has the limiting forms

$$\Delta F^{\pm} = \frac{\sqrt{2}}{3\pi}\mathsf{A}H_c^2(T)\xi(T)\left(\sqrt{1-3\kappa^2} - \frac{1}{2}\frac{I\phi_0}{c}\right) , \quad I \ll I_c(T)$$
$$\propto \left(1 - \frac{I}{I_c}\right)^{5/4} , \quad I \to I_c(T). \tag{2.106}$$

It is possible to mathematically prove that the solution is truly a saddle point by analyzing the free energy functional in the presence of small perturbations, as was done for the stationary solutions above. The explicit calculations can be found in the 1967 paper by Langer and Ambegaokar [7], and will not be discussed here. However, see Section 2.4.7 on the McCumber–Halperin estimate of the attempt frequency $\Omega^{\pm}$, which analyses the general case of an extremum.

2.4.6 Physical Scenario of a Thermally-Activated Phase Slip

Having deduced the free energy barrier heights and the saddle point solutions of the lowest energy, which are accessible to the local minimum at hand, we are now able to discuss the physical process of a phase slip. We envision a system originally in a local minimum corresponding to the stationary solution indexed by n, with a corresponding $k_n = 2n\pi/\mathsf{L}$ ($\kappa_n = 2n\pi\xi(T)/\mathsf{L}$). Due to thermal agitation, the system oscillates with a small amplitude, with an oscillation frequency determined by the curvature at the bottom of the local minimum and the spectrum of thermal noise. This agitation, along with the natural oscillation frequency, and in conjunction with the curvatures at the saddle point, determine an attempt frequency. This attempt frequency was computed by McCumber and Halperin [8] within a time-dependent Ginzburg–Landau theory framework, but carried out using the Fokker–Planck stochastic equations for the probability density of configurations rather than by adding a stochastic noise term in the Langevin form to the TDGL equation. Through direct comparison to experiments, the resulting estimate has been proven to be much more accurate than the somewhat simplistic form of the attempt frequency utilized by Langer and Ambegaokar. Setting aside the value for the moment, the rate at which the saddle point is accessed by the system is given by an equation of the form presented in (2.86). The free energy barrier leads to an exponential suppression of the rate with an exponent of $-\Delta F^{\pm}/(k_{\mathrm{B}}T)$. This, of course, is multiplied by the attempt frequency $\Omega^{\pm}$, which turns out to be a significant factor, as pointed out by McCumber and Halperin.

The system starts in a uniform state, with a fixed k_n, and electrical current $J_n = -2e\hbar k_n f_{k_n}^2/M$ (2.62). It meanders and by chance acquires sufficient energy to reach a saddle point configuration. Since the index n is a topological quantity characterizing the system, it cannot change unless the system goes through a point of zero energy. In other words, as long as there is a gap in the excitation through the length of the narrow wire, this topological invariant must stay unchanged. This invariant is, in fact, equal to the total phase change along the entire wire in units of 2π, which is an integer under the imposition of a periodic boundary condition (i.e., basically a ring geometry): $n = k_n\mathsf{L}/2\pi$. On the other hand, according to (2.96) for the total phase of a saddle point solution, the phase has an extra piece when compared to the uniform case (2.97). Since this total phase must remain unaltered until the phase slip event takes place. Given that the saddle solution itself does NOT contain any position in x where the order parameter actually vanishes, a sudden change in the phase is prohibited! What must happen is that the k_s associated with the saddle must be slightly different to ensure that the total phase remains the same. This yields a slight reduction in the k_s value, relative to k_n:

$$\delta k = k_s - k_n = -\left(\frac{2}{\mathsf{L}}\right)\tan^{-1}\left[\sqrt{\frac{3(1-3\kappa^2)/2}{\kappa}}\right]. \qquad (2.107)$$

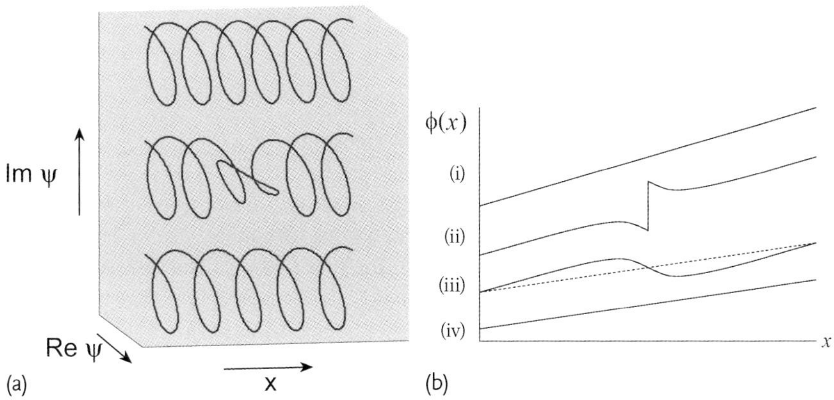

Figure 2.16 (a) Top to bottom shows the sequence of the behavior of the order parameter ψ as it undergoes a phase slip. Note that the top curve has one more winding than the bottom. (b) A plot of the phase $\phi(x)$ vs. distance x along the nanowire. (i) Corresponds to the top in (a), where there is an extra 2π phase (slope of line is steeper than (iv)). After the phase slip to reduce the winding by one, or equivalent, the phase by 2π, the state is now in (iv), and the slope is less steep. Curve (ii) depicts the situation just before the phase slip takes place. At the position of the kink, the magnitude of the order parameter must be zero. Otherwise, the gradient term in the free energy would diverge. Courtesy of A. Del Maestro (a) and S. Khlebnikov (b))!

Far from the phase-slip core, the saddle point solution is uniform. Thus, the current of the saddle point configuration must differ slightly from that of the initial state k_n.

After surmounting the barrier, assuming there is an efficient channel to dissipate the excess energy, the system moves downhill energetically. At some point, the system must pass through a state of lower energy which has a zero in the order parameter at a position near the core of that saddle point the system has just passed through. At that instant, the system makes the jump in phase by 2π, corresponding to a change in the winding number by one unit. A pictorial representation of the process is shown in Figure 2.16. Because of the compression of the spatial region over which the excess 2π phase shift resides as the phase-slip process is rapidly approached, the position at which the phase slip is to take place must have its order parameter amplitude rapidly approach zero. Otherwise, the gradient term in the free energy function would become undesirably large.

Thus, the entire picture is as follows: The system wanders toward a saddle point as a consequence of a rare fluctuation. As this saddle configuration is approached, the current changes momentarily, causing a momentary build up of charge. Thus, there could be additional electrostatic energy associated with the build up, but since the phase slip the process is fast, the amount of build up could be small. Subsequent to the phase slip, the excess energy acquired in order to surmount the free energy barrier may be dissipated efficiently in the electron channel via two possible mechanisms: (a) through the normal core via the quasiparticles, and (b) by sending out 1D plasmons – linearly dispersing excitations of a 1D Luttinger liquid – away from the phase-slip core.

2.4.7
McCumber–Halperin Estimate of the Attempt Frequency

Having described the physical picture of the phase-slip process, what remains is to present the quantitative estimate of the attempt frequency $\Omega^{\pm}$. McCumber and Halperin based their calculation using the formalism of the time-dependent Ginzburg–Laudau (TDGL) equations. However, it is well-known that the TDGL equations are only accurate under very restrictive conditions [52, 53]. The actual computation was performed using the Fokker–Planck equations, which is used to determine the probability density functional, rather than solving the TDGL equation containing a Langevin noise term with white noise Gaussian correlations.

Langer and Ambegaokar argued that the attempt frequency should be estimated as the ratio of the number of electrons N_e divided by a microscopic relaxation time τ_e in the normal state of order picoseconds. In place of this, McCumber and Halperin stressed that the attempt frequency should instead be the product of three contributions: (1) the number of statistically independent units for a phase-slip process $\sim \mathsf{L}/\xi(T)$, and (2) a phase space factor associated with the number of energy cells in phase space. In an activated process, the cell size is of order $k_B T$. The energy scale is the free energy barrier $\Delta F^{\pm}$. As a random process, the contributing factor is thus $(\Delta F^{\pm}/(k_B T))^{1/2}$, and (3) a characteristic diffusion time for exploring the free energy extrema (minimum and saddle), given by the Ginzburg–Landau relaxation time, $\tau_{GL} = \pi\hbar/[8k_B(T_c - T)]$. Additional factors of order unity associated with entropy factors also appear.

Roughly speaking, the resultant attempt frequency is given by

$$\Omega(T) \sim \frac{\mathsf{L}}{\xi(T)}\left(\frac{\Delta F^{\pm}}{k_B T}\right)^{1/2} \times \frac{\text{entropy factor}}{\tau_{GL}} . \tag{2.108}$$

The numerical value for the typical experimental parameters encountered is as much as 10 orders of magnitude smaller than the initial estimate of N_e/τ_e. To date, the McCumber–Halperin estimate has proven to be adequate in providing reasonable fittings to a large body of experimental data on the resistive transition in narrow 1D superconducting wires, and has consequently become the generally accepted estimate.

The mathematical analysis that yielded the above estimate, with the various contributing factors, is based on the TDGL equation. Despite its many flaws and inaccuracies, it is argued that near T_c, the TDGL formalism should give an adequate estimate, at the least, order of magnitude-wise. Since the phase-slip rate is a product of this attempt frequency and an exponential term in the ratio of the free energy barrier to the thermal energy $k_B T$, small changes in the barrier height will absorb any inaccuracies in the attempt frequency at the order of magnitude level.

According to (1.40), the relaxation to equilibrium in the TDGL formulation is given by [52, 54]

$$\left(i\hbar\frac{\partial}{\partial t} - 2\mu\right)\Psi = -\frac{i}{\tau_{\rm GL}}\frac{\hbar}{|\alpha(T)|}\frac{\partial f}{\partial \Psi^*}$$
$$= -\frac{i}{\tau_{\rm GL}}\frac{\hbar}{|\alpha(T)|}\left[\alpha(T) + \beta(T)|\Psi|^2 - \frac{\hbar^2}{2M}\left(\nabla + \frac{2ie}{\hbar c}A\right)^2\right]\Psi\,. \qquad (2.109)$$

Here, f is the free energy density. The term $\mu = (-eV)$ is an electrochemical potential, which may vary spatially and temporally. In the simplest case, we are interested in the situation where no magnetic field, or a small magnetic field is present. The self-magnetic field generated by the current is also small $B \leq 10^{-5}$ T, estimated from the typical maximum supercurrent sustainable in very narrow nanowires (diameter $\sim$ 10 nm) in the microampere range. We thus set the vector potential equal to zero. The electrochemical term will also be neglected. McCumber and Halperin justified this choice by appealing to the requirement of local charge neutrality on the timescale of the fluctuations.

Near $T_{\rm c}$, there is a sizable normal component of the current due to the thermal activation across the superconducting gap. The total current through the wire at a point and time is (with $\psi = \sqrt{\beta/\alpha}\Psi$)

$$I(x,t) = -\left(\frac{\mathsf{A}H_{\rm c}^2\xi^2 c}{\phi_0}\right)|\psi(x,t)|^2\nabla\varphi(x,t) + \mathsf{L}G_n\nabla\frac{\mu}{-e}\,. \qquad (2.110)$$

The first term is the supercurrent, while the second reflects conduction via the quasiparticles, with a conductance approximated by the DC normal state conductance. Charge neutrality requires that any local fluctuation in the supercurrent is counter-balanced by a fluctuation in the normal contribution. This requirement yields $\nabla I(x,t) = 0$. Using the Josephson relation $\partial\nabla\varphi/\partial t = (-2e/\hbar)\nabla V = 2\nabla\mu/\hbar$, we find, with processes dominated by phase fluctuations so that $|\psi|^2 \approx 1$,

$$\frac{\partial\nabla^2\mu}{\partial t} = -\nabla^2\frac{\mu}{\tau_\mu}\,. \qquad (2.111)$$

The relaxation time τ_μ is

$$\tau_\mu = \frac{LG_n\phi_0^2}{2\pi\mathsf{A}H_{\rm c}^2\xi^2c^2} \propto \tau_{\rm GL}\,. \qquad (2.112)$$

Although the ratio $\tau_\mu/\tau_{\rm GL}$ is generally not small, so that the effect of spatial-temporal fluctuations in μ is nonnegligible near the regions of the Hilbert space of wavefunctions relevant to the phase-slip process, in particular at the saddle point, the supercurrent term in ∇I proportional to $\nabla|\psi|^2\nabla\varphi$ is identically zero since the current is spatially uniform. Thus, the effect of the chemical fluctuations is argued to be negligible at this level of accuracy.

With the vector potential A and the electrochemical potential μ set to zero, the TDGL equation now reads

$$\frac{\partial\psi}{\partial t} = -\frac{4\pi}{\mathsf{A}H_c^2(T)\xi(T)\tau_{\mathrm{GL}}(T)}\left[(1-|\psi|^2)\psi + \xi(T)^2\frac{\partial^2\psi}{\partial x^2}\right] . \tag{2.113}$$

To address thermal fluctuations, this equation should be modified by adding a Langevin, stochastic noise term.

To perform the actual computation, the equivalent Fokker–Planck formulation based on the determination of the probability-density functional was employed [55–57]. The density $\rho[\psi(x), t]$ must relax to the Boltzmann distribution $\rho \propto \exp\{-F[\psi(x), t]/(k_{\mathrm{B}}T)\}$ in a steady state. In such formulations, the attempt frequency depends on the properties of the eigenmodes for excursions about the extrema (minima and saddle points) of the free energy function in the Hilbert space. The equation for the probability density ρ is written as

$$\tau_{\mathrm{GL}}(T)\frac{\partial\rho}{\partial t} = \sum_n\left[e_n\frac{\partial}{\partial q_n}q_n + \frac{2\pi k_{\mathrm{B}}T}{\mathsf{A}H_c^2(T)\xi(T)}\frac{\partial^2}{\partial q_n^2}\right]\rho . \tag{2.114}$$

Here, q_n denotes the normal mode coordinates of the eigenmodes, and e_n denotes the corresponding eigenvalues.

Thus, as we did before to illustrate that the stationary solutions to the Ginzburg–Landau equation with uniform current density, put forth by Langer and Ambegaokar (2.61), represent local minima, it is necessary to expand the Ginzburg–Landau expression to second order about the stationary solutions. Writing the order parameter in the form

$$\psi(x) = \left[f_{\mathrm{R}}(x) + i f_{\mathrm{I}}(x) + s_{\mathrm{R}}(x) + i s_{\mathrm{I}}(x)\right]e^{i\kappa x/\xi(T)} , \tag{2.115}$$

where the stationary solution is $\psi_{st} = f e^{i\kappa x/\xi(T)}$ with $f = f_{\mathrm{R}} + i f_{\mathrm{I}}$, and the excursion $\Delta\psi = s e^{i\kappa x/\xi(T)}$ with $s = s_{\mathrm{R}} + i s_{\mathrm{I}}$, the resulting expansion, referenced to the respective free energy of stationary solutions, can be cast in matrix form as

$$F_2\{s(x)\} = \mathsf{AL}\frac{H_c^2(T)}{4\pi}\int_{-\mathsf{L}/2}^{\mathsf{L}/2}\frac{1}{\mathsf{L}}dx \quad \boldsymbol{S}^{\mathrm{T}}(x)\boldsymbol{M}\boldsymbol{S}(x) , \tag{2.116}$$

where the vector $\boldsymbol{S}(x)$ is a two-component vector

$$\boldsymbol{S}(x) = \begin{pmatrix} s_{\mathrm{R}}(x) \\ s_{\mathrm{I}}(x) \end{pmatrix} \tag{2.117}$$

and the matrix $\boldsymbol{M}(x)$ is

$$\boldsymbol{M}(x) = \begin{pmatrix} \left[-\xi(T)^2\frac{d^2}{dx^2}-(1-\kappa^2)+3f_{\mathrm{R}}^2+f_{\mathrm{I}}^2\right] & \left[2f_{\mathrm{R}}f_{\mathrm{I}}+2\kappa\xi(T)\frac{d}{dx}\right] \\ \left[2f_{\mathrm{R}}f_{\mathrm{I}}-2\kappa\xi(T)\frac{d}{dx}\right] & \left[-\xi(T)^2\frac{d^2}{dx^2}-(1-\kappa^2)+f_{\mathrm{R}}^2+3f_{\mathrm{I}}^2\right] \end{pmatrix}. \tag{2.118}$$

The eigenvalues λ_n and eigenvectors $\mathcal{Q}_n(x)$ of the matrix operator $\boldsymbol{M}(x)$ are solutions to the equation

$$\boldsymbol{M}(x)\mathcal{Q}_n(x) = \lambda_n \mathcal{Q}_n(x)\ . \tag{2.119}$$

The eigenvectors are normalized:

$$\int_{-L/2}^{L/2} \frac{dx}{L} \mathcal{Q}_m^{\mathrm{T}}(x)\mathcal{Q}_n(x) = \delta_{mn}\ , \tag{2.120}$$

and form a complete set that spans the Hilbert functional space.

The second order free energy can be recast in the form

$$F_2\{s(x)\} = \mathrm{A}\xi(T)\frac{H_{\mathrm{c}}^2(T)}{4\pi}\sum_{n=1}^{\infty}\lambda_n q_n^2\ , \tag{2.121}$$

where the q_n represent the degree of excursion in the direction of the nth eigenvector.

The actual computation of these eigenvalues and the eigenvectors is tedious, and will not be presented here. Interested readers are referred to the original work by McCumber and Halperin [8]. We will simply summarize what was found. For the minima, there is one zero-mode for which the eigenvalue is zero ($\lambda_{m1} = 0$). This was seen before in the initial analysis of the quadratic departure, and is associated with and arbitrary choice of the reference value for the phase (see Section 2.4.3). All other values are positive. In contrast, for the saddle point solution, there is one negative eigenvalue ($\lambda_{\mathrm{sp},1} < 0$) associated with the direction of descent, and two zero eigenvalues ($\lambda_{\mathrm{sp},2} = \lambda_{\mathrm{sp},3} = 0$). All others are positive. One zero eigenvalue is associated with a uniform shift in the phase as before, while the second is with translation along the wire.

The Fokker–Planck formalism expresses the transition rates in the form

$$\Gamma^{\pm}(\kappa, T) = \Omega^{\pm}(\kappa, T)\exp\left(-\frac{\Delta F^{\pm}(\kappa, T)}{k_{\mathrm{B}}T}\right)\ . \tag{2.122}$$

The attempt frequencies $\Omega^{\pm}$ are given by

$$\Omega^{\pm}(\kappa, T) = N\left(\frac{\Delta F^{\pm}}{k_{\mathrm{B}}T}\right)^{1/2}|\lambda_{\mathrm{sp}1}| \times\left[\frac{\lambda_{m2}(\kappa_m)\lambda_{m3}(\kappa_m)}{|\lambda_{\mathrm{sp},1}|}\prod_{i=4}^{\infty}\frac{\lambda_{m,i}(\kappa_m)}{\lambda_{\mathrm{sp},i}(\kappa_s)}\right]^{1/2}\frac{1}{\tau_{\mathrm{GL}}}\ . \tag{2.123}$$

Here, the subscript m denotes the local minimum in which the system originally resides, and sp denotes the saddle point. This is in the form discussed at the beginning of this section.

Summarizing all the mathematical analyses, and combining the free energy barrier height obtained by Langer and Ambegaokar (2.99) and (2.100), and the attempt frequency estimated by McCumber and Halperin (2.108), the thermally activated phase-slip (TAP) rate is given by

$$\Gamma^{\pm} = \frac{L}{\xi(T)} \left(\frac{\Delta F^{\pm}}{k_B T} \right)^{1/2} \frac{1}{\tau_{GL}} \exp\left(-\frac{\Delta F^{\pm}}{k_B T} \right), \quad \tau_{GL} = \frac{\pi}{8} \frac{\hbar}{T_c - T}. \tag{2.124}$$

The linear resistance R_{TAP} due to TAPs at small current I is given by

$$R_{TAP} = R_q \frac{L}{\xi(T)} \left(\frac{\Delta F_0}{k_B T} \right)^{1/2} \frac{\hbar}{\tau_{GL} k_B T} \exp\left(-\frac{\Delta F_0}{k_B T} \right), \tag{2.125}$$

where $R_q = h/(4e^2) \approx 6.453\,\mathrm{k\Omega}$ is the quantum resistance, and ΔF_0 is given in (2.104). These expressions are valid close to, but below T_c, as they have been derived from the Ginzburg–Landau equations. They have provided excellent fits to experimental data, as will be seen in Chapter 7.

At temperature significantly below T_c, however, it is expected to become increasingly inaccurate. As Giordano first noted [46], at low temperatures, the thermal activation TAP process must cease to be effective, and a quantum tunneling process, specifically the macroscopic quantum tunneling (MQT) of phase slips must take over below some characteristic temperature. As a first approximation, Giordano proposed to replace the thermal energy scale $k_B T$ in the above expressions by the Ginzburg–Landau energy scale $E_{GL} = (1/a)\hbar/\tau_{GL}$, where a is a proportionality constant of order unity. This results in the tunneling rate and resistance given by

$$\Gamma^{\pm} = \frac{L}{\xi(T)} \left(\frac{\Delta F^{\pm}}{E_{GL}} \right)^{1/2} \frac{1}{\tau_{GL}} \exp\left(-\frac{\Delta F^{\pm}}{E_{GL}} \right) \tag{2.126}$$

and

$$R_{QPS} = R_q \frac{L}{\xi(T)} \left(\frac{\Delta F_0}{E_{GL}} \right)^{1/2} \exp\left(-\frac{\Delta F_0}{E_{GL}} \right), \tag{2.127}$$

respectively. Of course, since the Ginzburg–Landau theory itself is only meaningful near T_c, such an expression is, strictly speaking, only valid not too far below T_c. Nevertheless, it has been widely used to estimate the tunneling rate in the analysis of experimental data.

Nonlinear current–voltage (I–V) relationships for TAP and for QPS can readily be obtained from theses expressions (2.124)–(2.127), respectively, in conjunction with (2.86)–(2.88).

References

1 Sachdev, S., Werner, P., and Troyer, M. (2004) Universal conductance of nanowires near the superconductor-metal quantum transition. *Physical Review Letters*, **92** (23), 237003.

2 Del Maestro, A., Rosenow, B., Müller, M., and Sachdev, S. (2008) Infinite randomness fixed point of the superconductor–metal quantum phase transition. *Physical Review Letters*, **101** (3), 035701.

3 Benedict, L., Crespi, V., Louie, S., and Cohen, M. (1995) Static conductivity and superconductivity of carbon nanotubes: Relations between tubes and sheets. *Physical Review B*, **52** (20), 14935–14940.

4 Tang, Z.K., Zhang, L., Wang, N., Zhang, X.X., Wen, G.H., Li, G.D., Wang, J.N., Chan, C.T., and Sheng, P. (2001) Superconductivity in 4 angstrom single-walled carbon nanotubes. *Science (New York, N.Y.)*, **292** (5526), 2462–2465.

5 Parks, R.D. and Groff, R.P. (1967) Evidence for thermodynamic fluctuations in a superconductor. *Physical Review Letters*, **18**, 342–345.

6 Hunt, T.K. and Mercereau, J.E. (1967) Quantum phase correlation in small superconductors. *Physical Review Letters*, **18**, 551–553.

7 Langer, J.S. and Ambegaokar, V. (1967) Intrinsic resistive transition in narrow superconducting channels. *Physical Review*, **164**, 498–510.

8 McCumber, D.E. and Halperin, B.I. (1970) Time scale of intrinsic resistive fluctuations in thin superconducting wires. *Physical Review B*, **1** (3), 1054–1070.

9 Guo, Y., Zhang, Y.F., Bao, X.Y., Han, T.Z., Tang, Z., Zhang, L.X., Zhu, W.G., Wang, E.G., Niu, Q., Qiu, Z.Q., Jia, J.F., Zhao, Z.X., and Xue, Q.K. (2004) Superconductivity modulated by quantum size effects. *Science (New York, N.Y.)*, **306** (5703), 1915–1917.

10 Zhang, Y.F., Jia, J.F., Han, T.Z., Tang, Z., Shen, Q.T., Guo, Y., Qiu, Z., and Xue, Q.K. (2005) Band structure and oscillatory electron–phonon coupling of Pb thin films determined by atomic-layer-resolved quantum-well states. *Physical Review Letters*, **95** (9), 096802.

11 Bao, X.Y., Zhang, Y.F., Wang, Y.P., Jia, J.F., Xue, Q.K., Xie, X.C., and Zhao, Z.X. (2005) Quantum size effects on the perpendicular upper critical field in ultrathin lead films. *Physical Review Letters*, **95** (24), 247005.

12 Ralph, D.C., Black, C.T., and Tinkham, M. (1995) Spectroscopic measurements of discrete electronic states in single metal particles. *Physical Review Letters*, **74** (16), 3241–3244.

13 Black, C., Ralph, D., and Tinkham, M. (1996) Spectroscopy of the superconducting gap in individual nanometer-scale aluminum particles. *Physical Review Letters*, **76** (4), 688–691.

14 Morales, A.M. (1998) A laser ablation method for the synthesis of crystalline semiconductor nanowires. *Science*, **279** (5348), 208–211.

15 Iijima, S. (1991) Helical microtubules of graphitic carbon. *Nature*, **354** (6348), 56–58.

16 Fu, L. and Kane, C. (2009) Probing neutral Majorana fermion edge modes with charge transport. *Physical Review Letters*, **102** (21), 216403.

17 Nadj-Perge, S., Pribiag, V.S., van den Berg, J.W.G., Zuo, K., Plissard, S.R., Bakkers, E.P.A.M., Frolov, S.M., and Kouwenhoven, L.P. (2012) Spectroscopy of spin–orbit quantum bits in indium antimonide nanowires. *Physical Review Letters*, **108** (16), 166801.

18 Eliashberg, G.M. (1961) Temperature Greens function for electrons is a superconductor. *Soviet Physics JETP*, **12** (5), 1000–1002.

19 McMillan, W. (1968) Transition temperature of strong-coupled superconductors. *Physical Review*, **167** (2), 331–344.

20 Little, W.A. (1967) Decay of persistent currents in small superconductors. *Physical Review*, **156**, 396–403.

21 Halperin, B.I., Refael, G., and Demler, E. (2010) Resistance in superconductors. *International Journal of Modern Physics B*, **24** (20/21), 4039–4080.

22 Wigner, E.P. (1955) Characteristic vectors of bordered matrices with infinite dimensions. *Annals of Mathematics*, **63**, 548–564.
23 Dyson, F.J. (1962) Statistical theory of the energy levels of complex systems. I. *Journal of Mathematical Physics*, **3** (1), 140.
24 Dyson, F.J. (1962) Statistical theory of the energy levels of complex systems. II. *Journal of Mathematical Physics*, **3** (1), 157.
25 Dyson, F.J. (1962) Statistical theory of the energy levels of complex systems. III. *Journal of Mathematical Physics*, **3** (1), 166.
26 Dyson, F.J. and Mehta, M.L. (1963) Statistical theory of the energy levels of complex systems. IV. *Journal of Mathematical Physics*, **4** (5).
27 Tian, M., Kumar, N., Xu, S., Wang, J., Kurtz, J., and Chan, M. (2005) Suppression of superconductivity in zinc nanowires by bulk superconductors. *Physical Review Letters*, **95** (7), 76802.
28 Li, P., Wu, P., Bomze, Y., Borzenets, I., Finkelstein, G., and Chang, A. (2011) Switching currents limited by single phase slips in one-dimensional superconducting Al nanowires. *Physical Review Letters*, **107** (13), 137004.
29 Bogoliubov, N.N. (1958) A new method in the theory of superconductivity-1. *Soviet Physics JETP*, **7** (1), 41–46.
30 Bogoliubov, N.N. (1958) A new method in the theory of superconductivity-3. *Soviet Physics JETP*, **7** (1), 51–55.
31 DE GENNES, P. (1964) Boundary effects in superconductors. *Reviews of Modern Physics*, **36** (1), 225–237.
32 Shanenko, A., Croitoru, M., and Peeters, F. (2007) Oscillations of the superconducting temperature induced by quantum well states in thin metallic films: Numerical solution of the Bogoliubov–de Gennes equations. *Physical Review B*, **75** (1), 014519.
33 Peeters, F.M., Shanenko, A., and Croitoru, F.M. (2010) Nanoscale superconductivity, in *Handbook of Nanophysics: Principles and Methods* (ed. K.D. Sattler), 1st edn, CRC Press.
34 Blatt, J. and Thompson, C. (1963) Shape resonances in superconducting thin films. *Physical Review Letters*, **10** (8), 332–334.
35 Thompson, C.J. and Blatt, J.M. (1963) Shape resonances in superconductors – II Simplified theory. *Physics Letters*, **5** (1), 6–9.
36 Upton, M.H., Wei, C.M., Chou, M.Y., Miller, T., and Chiang, T.C. (2004) Thermal stability and electronic structure of atomically uniform Pb films on Si(111). *Physical Review Letters*, **93** (2), 026802.
37 Czoschke, P., Hong, H., Basile, L., and Chiang, T.C. (2004) Quantum Beating Patterns Observed in the Energetics of Pb Film Nanostructures. *Physical Review Letters*, **93** (3), 036103.
38 Shanenko, A.A. and Croitoru, M.D. (2006) Shape resonances in the superconducting order parameter of ultrathin nanowires. *Physical Review B*, **73** (1), 012510.
39 Croitoru, M.D., Shanenko, A.A., and Peeters, F.M. (2007) Size-resonance effect in cylindrical superconducting nanowires. *Moldavian Journal of the Physical Sciences*, **6** (1), 39–47.
40 Thompson, C.J. (1964) Shape resonances in superconducting cylinders. *Journal of Physics and Chemistry of Solids*, **26**, 1053–1060.
41 Altomare, F., Chang, A.M., Melloch, M.R., Hong, Y., and Tu, C.W. (2006) Evidence for macroscopic quantum tunneling of phase slips in long one-dimensional superconducting Al wires. *Physical Review Letters*, **97** (1), 17001.
42 Allen, P.B. (1975) Transition temperature of strong-coupled superconductors reanalyzed. *Physical Review B*, **12** (3), 905–922.
43 Wang, Z., Shi, W., Lortz, R., and Sheng, P. (2012) Superconductivity in 4-angstrom carbon nanotubes – a short review. *Nanoscale*, **4** (1), 21–41.
44 Perfetto, E. and González, J. (2006) Theory of superconductivity in multiwalled carbon nanotubes. *Physical Review B*, **74** (20), 201403.
45 Bardeen, J. (1962) Critical fields and currents in superconductors. *Reviews of Modern Physics*, **34** (4), 667–681.

46 Giordano, N. (1994) Superconducting fluctuations in one-dimension. *Physica B*, **203** (3/4), 460–466.

47 Zaikin, A., Golubev, D., van Otterlo, A., and Zimányi, G. (1997) Quantum phase slips and transport in ultrathin superconducting wires. *Physical Review Letters*, **78** (8), 1552–1555.

48 Golubev, D. and Zaikin, A. (2001) Quantum tunneling of the order parameter in superconducting nanowires. *Physical Review B*, **64** (1), 14504.

49 Khlebnikov, S. and Pryadko, L.P. (2005) Quantum phase slips in the presence of finite-range disorder. *Physical Review Letters*, **95** (10), 107007.

50 Abrikosov, A.A. (2004) Type II superconductors and the vortex lattice. *Reviews of Modern Physics*, **76**, 975–979.

51 Ivanchenko, Y.M. and Zilberman, I.A. (1968) Destruction of Josephson current by fluctuations. *JETP Letters-USSR*, **8** (4), 113.

52 Schmid, A. (1966) A time dependent Ginzburg–Landau equation and its application to the problem of resistivity in the mixed state. *Physik der kondensierten Materie*, **5** (4), 302–317.

53 Arutyunov, K., Golubev, D., and Zaikin, A. (2008) Superconductivity in one dimension. *Physics Reports*, **464** (1/2), 1–70.

54 Abrahams, E. and Tsuneto, T. (1966) Time variation of the Ginzburg–Landau order parameter. *Physical Review*, **152** (1), 416–432.

55 Landauer, R. and Swanson, J. (1961) Frequency factors in the thermally activated process. *Physical Review*, **121** (6), 1668–1674.

56 Langer, J. (1968) Theory of nucleation rates. *Physical Review Letters*, **21** (14), 973–976.

57 Chandrasekhar, S. (1943) Stochastic problems in physics and astronomy. *Reviews of Modern Physics*, **15**, 1–89.

3
Quantum Phase Slips and Quantum Phase Transitions

3.1
Introduction

In this chapter, we will explore the quantum theories of superconductivity in nanowires. The topics that will be covered include effective low energy theories of superconductivity in one dimension, macroscopic quantum tunneling of the order-parameter phase, so-called quantum phase slips (QPS), quantum phase transitions in short and long wires, and quantum critical theories motivated by the presence of pair-breaking processes such as magnetic impurities, which suppress superconductivity.

The development of these theories was necessitated by the need to provide a description of superconductivity in nanowires, as the temperature approaches absolute zero; such theories are expected to be relevant at finite temperatures significantly below the superconducting transition temperature T_c, when quantum mechanical processes dominate over classical thermal activation. The goal is to pave the way toward understanding the limiting behavior of superconductivity as the diameter of a nanowire shrinks toward zero. Issues of interest to be addressed include the amount of residual wire resistance in the $T \to 0$ limit, that is, whether the resistance goes to zero, the diameter size-scale where quantum processes such as the QPS become important, and the amount of supercurrent a nanowire is capable of carrying. These are of central importance to applications using nanowire-based devices.

From these theories, novel behaviors are predicted. Particularly noteworthy is the emergence behaviors and phenomena in QPS-based devices made from short wires, of lengths shorter than the thermal wavelength of the Mooij–Schön plasmon in the superconducting state. For instance, nanowire-based qubits, dual Josephson relationship and Shapiro steps in the current are envisioned. Such predictions are predicated on a duality between a Josephson junction and a short superconducting nanowire connected to large superconducting leads, in which the role of the Cooper pair number operator and the phase of the order parameter are interchanged. These predictions will be discussed in the following Chapter 4.

In this chapter, the specific topics to be discussed include the derivation of effective low-energy theories, the computation of the action for a QPS, including the

One-Dimensional Superconductivity in Nanowires, First Edition. F. Altomare and A.M. Chang

core and the hydrodynamic contributions, interaction between phase slips, and the quantum phase transitions between a superconductor and a metallic state (SMT), as well as short-wire behavior, for which a superconducting–insulator transition (SIT) has been observed experimentally instead. Understanding these topics in turn provides the basis for computing the QPS macroscopic quantum tunneling rate, and scaling of the residual resistance with temperature below T_c, as well as the nonlinear voltage–current characteristics. The important roles played by disorder, dissipation, and nanowire diameter in driving the quantum transitions and in setting the tunneling rate emerge from the analysis. An entirely different type of theory, one in which both amplitude and phase fluctuation are important due to the presence of a quantum critical point, will also be summarized. Here, universal scalings of the transport properties are predicted with the scaling form dependent on the boundary condition set by the large electrical leads connected to a short nanowire, that is, whether the leads are superconducting, normal or mixed.

In the process of developing such theory, many nontrivial concepts will be explored. These include phase-only effective theories exhibiting behaviors similar to Luttinger liquids via the central role played by the linearly dispersing plasmon mode, momentum conservation, and energy relaxation issues in the macroscopic quantum-tunneling process of the phase, Kosterlitz–Thouless–Berezinskii (KTB) quantum phase transitions, duality, and quantum criticality.

Many of these find common ground with other many-body systems, for example, dissipation driven KTB transitions and quantum criticality, and with concepts and methodologies widely utilized in the high-energy, quantum field theory arena. For example, Khlebnikov has highlighted the concepts of duality, quark confinement-deconfinement transition in relation to the short nanowire SIT, and even gravity-gauge field duality in the context of understanding the maximum current sustained in the superconducting state uncovered in experimental studies. He has also introduced highly sophisticated instanton methods for computing macroscopic quantum tunneling rates, and to evaluate the path integrals.

Due to the highly technical nature of much of the discussion, the sections in this chapter are structured such that many technical details may be by-passed without loss of continuity. Each section will start with a summary of the issues and main results, as well as a layout of presentation.

Before embarking on the discourse on these intriguing theories, many of which have already been found to be of direct relevance to experiments, we wish to emphasize the central role of disorder and dissipation in these theories. Disorder modifies key parameters. For example, it reduces the superconducting coherence length ξ. The coherence length sets the size and volume of the QPS core and consequently the amount of loss in condensation energy, which in turn determines the phase-slip rate. The amount of disorder also set the resistance of the normal electron channel. This resistance parameterizes the dissipation in the normal fluid running parallel to the superfluid channel, as well as the dissipation in the core of the phase slip by the normal electrons. Large dissipation, corresponding to a small resistance, can help stabilize superconductivity by reducing fluctuations. Furthermore, the presence of disorder is essential in helping to supply or absorb the sizable momentum change produced in the phase-slip process.

The quantum theories for the dynamics and thermodynamics of narrow superconducting wires discussed in this chapter may be classified into two classes – (1) phase-only theories, and (2) quantum critical theories. In the phase-only theories, long wires undergo a superconductor-metal transition (SMT) tuned by the wire diameter, which sets the propagation impedance of the nanowire for the sound-like plasmon mode associated with phase fluctuations. In addition, dissipation in the normal electron channel could also help stabilize superconductivity. The scale for length comparison is set by the thermal plasmon wavelength. In short wires, the amount of dissipation tunes the nanowire through a superconductor–insulator transition (SIT). In the quantum critical theories, the presence of an SMT does not require disorder. However, disorder not only modifies the parameters of the theory, but remarkably, changes the critical point to an exotic variety, in which the dynamics exhibit unusual exponential dependence on the correlating length scale, rather than a power law.

The first class of theory emphasizes the role of the phase slip, specifically the quantum tunneling of phase slips. These phase-only theories assume that outside of the QPS core, only fluctuations of the phase of the superconducting order parameter are important at low temperatures, while amplitude fluctuations are small. Such phase fluctuations are intimately related to the Mooij–Schön plasmon mode, which is linearly dispersing in a one-dimensional nanowire [1].

Disorder increases the normal state resistance of a nanowire, as well as the impedance on the scale of superconducting coherence length ξ – a combined effect of the normal resistance on the scale ξ, acting in parallel with the plasmon propagation impedance. Depending on the length of the wire, when one of these two quantities exceeds the quantum resistance $R_Q = h/(2e)^2 = h/(4e^2)$, fluctuations destroy superconducting order. Strictly speaking, such QPTs occur at $T = 0$ limit, as the parameters in the Hamiltonian are varied continuously. In phase-only theories, notably the work of Zaikin, Golubev, Otterlo, and Zimanyi [2], Golubev and Zaikin [3], Refael, Demler, Oreg, and Fisher [4], Meidan, Oreg, Refael [5], Khlebnikov and Pryadko [6], Khlebnikov [7, 8], Giamarchi [9], and Büchler, Geshkenbein, and Blatter [10], excess disorder in combination with the extremely small wire diameter leads to the unbinding of QPS and anti-QPS pairs. Such pairs interact logarithmically as a function of their separation in either space and/or time with an attractive potential, while the QPS–QPS or anti-QPS–anti-QPS interactions are logarithmically repulsive. Such an interaction occurs at space-time distances outside of the core of the phase slip, which is of order ξ, the superconducting coherence length in the dirty limit, relevant for most, if not all, experimental realizations to date. In an imaginary-time formulation in terms of path integrals, this time is the imaginary time.

The presence of the logarithmic interaction is analogous to what happens in 2D continuous *XY*-spin models and related models, for example, superconductivity in 2D films, in which a Kosterlitz–Thouless–Berezinskii transition occurs at a finite temperature T_0 [11, 12]. Here, the transition was originally found for a classical spin system, and takes place at a temperature at which topological defects of opposite windings, for example, a superconducting vortex and antivortex in the case of 2D superconducting films, unbind into free entities. The logarithmic interaction leads to power law dependence in the correlation functions.

In the case of the 1D superconducting nanowires, such power law dependence extends to the transport properties. Power law relations are thus expected in both linear and nonlinear transport, the latter occurring under a finite source-drain bias. This is the case for long wires. In short wires, boundary effect can further reduce the quantum phase-slip rate, and therefore, the amount of residual resistance in the superconducting state. In all cases, the linear resistance at low current drives approaches zero in the $T = 0$ limit.

In the second class of theory, the quantum phase transition is driven by quantum critical fluctuations, both in the amplitude and in the phase, due to the presence of pair-breaking mechanisms. The pair-breaking mechanism is envisioned to arise from magnetic impurities on the surface of a nanowire, or can be effected by the application of a magnetic field parallel to the nanowire. The starting point is based on the analysis of the 2D Josephson junction array in the presence of dissipation via coupling to a normal metal plane. There, the long-range repulsive Coulomb coupling, when sufficiently strong compared to the Josephson coupling, enhances fluctuations, and leads to a destruction of the superconducting correlation between cells of Josephson junctions. The critical fluctuations give rise to a superconductor-insulation (SIT) quantum phase transition. Extending these ideas to 1D for narrow short wires, where $(\mathsf{w}, \mathsf{h}) \ll \mathsf{L}$, Sachdev, Werner, and Troyer [13], and Del Maestro *et al.* [14] argued for the existence of a superconductor-normal metal transition (SMT). The phase slips produce operators, which, in the renormalization group sense, are irrelevant, and thus do not control the behavior at or near the critical point. This class of theory is termed "pair-breaking". The quantum critical behavior obeys certain hyperscaling rules, and the conductance of a short 1D wire, depending on the boundary conditions applied, that is, whether the ends are connected to large leads of the normal, superconducting, or mixed varieties, is argued to exhibit universal scaling of differing forms. Here, 1D simply means that the length greatly exceeds the width and is thus often less stringent than the requirement that the width/height dimensions be smaller than the superconducting coherence length, utilized in the LAMH-type theories discussed in Chapter 2. In principle, the SMT here does not require any disorder; it can occur in a clean or a dirty superconducting system. However, in the presence of strong disorder which gives rise to spatial fluctuations of the physical parameters such as the strength of the pair-breaking rate, and concomitant large range disorder-induced interaction in the imaginary-time direction, Del Maestro, Rosenow, Sachdev *et al.* [14] and Del Maestro, Rosenow, Hoyos, Vojta [15] uncovered a correspondence to the infinite-randomness fixed point (IRFP) [16], which now characterizes the quantum critical point. Such an exotic fixed point was originally derived for the random-transverse-field Ising model (RTFIM) by Fisher [17, 18]. Such a fixed point exhibits unusual energy scales for temporal fluctuations, which scales exponentially with a power law of the correlation length, instead of the usual power law scaling.

3.2 Zaikin–Golubev Theory

The theory of Zaikin and Golubev and collaborators [2, 3] is based on an impurity averaged, low energy effective theory of superconductivity in one dimension. In this chapter, we will summarize the derivation of the impurity-averaged 1D theory, starting from the imaginary-time, path integral formulation introduced in Section 1.6. What emerges is an effective theory which contains contributions both from fluctuations of the amplitude, and from fluctuations of the phase. To compute the saddle point action (instanton action) due to a phase slip, fluctuations in both quantities must be taken into account. The phase slip is characterized by a core size, of the order of the coherence length ξ, and a timescale, of order and exceeding the inverse gap $\sim \hbar/\Delta(T)$. Within the spatial extent of the core, the amplitude of the order parameter is significantly depressed from the nearly uniform value outside. Thus, the effective action must be able to account for this depression, which costs condensation energy. Outside of the core, however, only phase fluctuations are important. Such fluctuations give rise to the hydrodynamic linearly-dispersing Mooij–Schön plasmon mode [1]. This mode leads to a hydrodynamic contribution to the QPS action, in addition to the core action.

On the scale of space-imaginary time distances that are large compared to the core size, the Mooij–Schön mode also mediates an interaction between phase slips with a logarithmic dependence on their separation. Such an interaction form is appropriate in long wires of sufficient length. Even if the wire ends are attached to large superconducting leads, the order parameter fluctuations are not affected by lead-imposed boundary conditions. The presence of this long-range, logarithmic interaction is responsible for the existence of the KTB-type quantum phase transitions in the 1D nanowire system. Within this scenario, when the interaction is sufficiently strong to bind QPS and anti-QPS pairs, the nanowire is superconducting. Otherwise, unbinding becomes prevalent and free QPSs proliferate, destabilizing superconductivity and resulting in a normal metallic state.

Zaikin, Golubev, Otterlo, and Zimanyi emphasized the importance of disorder in the phase-slip process [2]. In the absence of disorder, that is, in a clean superconducting nanowire, the free energy barrier for the creation of a QPS event is given by

$$\Delta F_{\mathrm{cln}} \sim \frac{1}{4} N(0)[\Delta(T)]^2 \mathsf{A}\xi_{\mathrm{cln}} \,,$$

with a coherence length given by the BCS value, $\xi_{\mathrm{cln}} \sim \xi_{\mathrm{BCS}}$, with $\Delta(T)$ the superconducting gap, and $N(0)$ the normal state density of states for both spins at the Fermi level. This expression yields a barrier, which is prohibitively large to allow quantum tunneling even for nanowires with a diameter $\sim$ 10 nm, as was pointed out by Duan [19]. Zaikin *et al.* recognized that in the nanowires investigated in experiment, superconductivity takes place in the dirty limit, where the transport mean free path $l_{\mathrm{mfp}} \sim$ 5 nm is much smaller than the BCS coherence length $l_{\mathrm{mfp}} \ll \xi_{\mathrm{BCS}} \sim \xi_{\mathrm{cln}} \sim 0.1$–$2\ \mu$m. Thus, the clean limit ξ_{cln} must be replaced by the dirty limit coherence length, $\xi \approx 0.85\sqrt{l_{\rm mfp}\xi_{\rm BCS}} \ll \xi_{\rm cln}$, and a corresponding reduction in the free energy barrier takes place, allowing sizable quantum tunneling rates in nanowires of diameter 10 nm and below [2, 3, 20].

This section is organized as follows: Section 3.2.1 contains the rather technical discussion of the derivation of the 1D effective low energy action for a 1D superconducting nanowire. The action of a QPS (instanton) is presented in Sections 3.2.2 and 3.2.3 for the QPS core and hydrodynamic contributions, respectively. These are followed by discussions of the QPS interaction (Section 3.2.4), QPS rate and quantum phase transition (Section3.2.5), and finally, the power law scaling forms of the transport quantities (Section 3.2.6).

3.2.1 Derivation of the Low Energy Effective Action

The Zaikin–Golubev quantum theory of phase slips was formulated to describe the dynamics in 1D superconducting nanowires of narrow transverse dimensions $\mathsf{d} \sim (\mathsf{w}, \mathsf{h}) < \xi$ and length $\mathsf{L} > \xi$ [2, 3, 20]. The theory starts from the path integral formulation of the partition function for a superconductor in the imaginary-time formulation with a Euclidean action [21, 22]. Alternatively, an equivalent formulation in real time enables the study of real-time quantum dynamics [20, 21]. In Section 1.6, we introduced the path integral formulation of the partition function in the Keldysh formalism [23] for a specific realization of the background potential (including impurities) and in three dimensions.

To arrive at the quantum theory for 1D superconducting nanowires, the first task is to perform an impurity-average. The averaging accounts for the diffusive motion of the electrons in the disorder potential. Subsequent to the averaging process, the specialization to the 1D limit yields the 1D effective action. Such an action displays the Mooij–Schön plasmon mode, which arises out of the electro-dynamics reflecting the long-range Coulomb interaction, and is a central feature of 1D systems. The Mooij–Schön mode is linearly dispersing and is central in carrying away the energy dissipated in a phase-slip process. The presence of such a mode is familiar in interacting 1D systems, which exhibit Luttinger liquid physics [24, 25].

3.2.1.1 Disorder Averaging

In real time, the Keldysh formalism requires the introduction of unitary evolution along a closed contour in time containing forward and backward branches [23, 26–28]. The imaginary-time formulation is obtained by passing into imaginary time, $t \to -i\tau$, with τ the imaginary time. The Keldysh formulation enables the treatment of nonequilibrium situations, for example, under external drive such as electromagnetic radiation, or under finite current or voltage drive, and at the same time, allows for the averaging of quenched disorder in a straightforward manner. This latter property of the Keldysh formalism, where the impurity-averaging for quenched disorder can be performed on the partition function, rather than on the much more cumbersome free energy which involves a logarithm ($\propto \ln Z$), is a consequence of the return back to $t = -\infty$ in the Keldysh contour.

The partition function can be written as

$$Z = \frac{\mathrm{Tr}\{\hat{U}_{\mathrm{C}}\hat{\rho}\}}{\mathrm{Tr}\{\hat{\rho}\}} \tag{3.1}$$

where the denominator is given by

$$\mathrm{Tr}\{\hat{\rho}(t = -\infty)\} . \tag{3.2}$$

Here, $\mathrm{Tr}\{\ldots\}$ denotes the trace, $\hat{\rho}$ is the density matrix, and $\hat{U}_{\mathrm{C}}$ is the time evolution operator on the closed Keldysh contour [26–28]. For a Hamiltonian which has a source term coupled to an observable, where this coupling is, by design, asymmetric on the forward and backward branches of the Keldysh contour rendering $\hat{U}_{\mathrm{C}}$ different from unity, a derivative of the free energy with respect to the source variable, evaluated as the source goes to zero, yields the expected value of the observable [26].

At the same time, in the limit of zero source, $\hat{U}_{\mathrm{C}} \to 1$ since the Hamiltonian is now symmetric on the two branches, and the starting and end points of the unitary evolution are identically at $t = -\infty$. Therefore, $Z[0]$, the zero-source value, always equals unity, and may be omitted from the denominator in computing the derivative of the free energy (proportional to the derivative of $\ln Z$) to calculate the expected value of the observable! Moreover, the denominator, $\mathrm{Tr}\{\hat{\rho}(t = -\infty)\}$, equals $\mathrm{Tr}\{\hat{\rho_0}\}$ because at $t = -\infty$, no disorder is present, and this is true for any realization of the disorder ensemble as the disorder is adiabatically turned on some time after $t = -\infty$. In the end, averaging the partition function over disorder is equivalent to averaging over the free energy.

The derivation of the effective 1D action was contained in the original works of Zaikin and Golubev and coworkers [2, 3, 20]. The detailed derivation is highly technical, and here, we will summarize the most salient aspects.

We start with the imaginary-time action given by (1.103)–(1.108) and perform the disorder averaging to account for the short l_{mfp} [20, 22, 27, 28]. The equilibrium, nonperturbed normal and anomalous Green functions G_0, $\bar{G}_0$, F_0, and $\bar{F}_0$, for $\Delta = \Delta_0$, $V = V_\Delta = \mu$, and $\boldsymbol{A} = 0$, are first expressed in terms of the electron eigenwavefunctions of the single-particle Hamiltonian, ϕ_n:

$$H_0\phi_n(\boldsymbol{r}) = \epsilon_n\phi_n(\boldsymbol{r}) ,$$
$$H_0 = \xi\left(\nabla + \frac{ie}{\hbar c}\boldsymbol{A}\right) + U , \tag{3.3}$$

with U the single-particle potential, which includes background charge, impurity, and boundary contributions. The eigenenergies ϵ_n represent the electron energies in the normal, nonsuperconducting state. The Green functions in space and real time are defined as

$$G_0(\boldsymbol{r}, t; \boldsymbol{r}'; t') \equiv -i\langle|T\{\psi_\uparrow\psi_\uparrow^\dagger\}\rangle$$
$$\bar{G}_0(\boldsymbol{r}, t; \boldsymbol{r}'; t') \equiv -i\langle|T\{\psi_\downarrow^\dagger\psi_\downarrow\}\rangle$$
$$F_0(\boldsymbol{r}, t; \boldsymbol{r}'; t') \equiv -i\langle|T\{\psi_\uparrow\psi_\downarrow\}\rangle$$
$$\bar{F}_0(\boldsymbol{r}, t; \boldsymbol{r}'; t') \equiv -i\langle|T\{\psi_\downarrow^\dagger\psi_\uparrow^\dagger\}\rangle , \tag{3.4}$$

where $\psi_\sigma^\dagger$, and ψ_σ are the electron creation and annihilation field operators for spin σ, respectively. The imaginary-time operators are obtained by replacing t and t' by $-i\tau$ and $-i\tau'$, respectively. Making use of the BCS coherence factors (see (1.6)–(1.8)),

$$u_n = \frac{1}{\sqrt{2}}\sqrt{1+\frac{\epsilon_n}{E_n}} \tag{3.5}$$

and

$$v_n = \frac{1}{\sqrt{2}}\sqrt{1-\frac{\epsilon_n}{E_n}}\,, \tag{3.6}$$

with

$$E_n = \sqrt{\epsilon_n^2 + \Delta_n^2}\,, \tag{3.7}$$

where we have replaced the eigenfunction index $\boldsymbol{k}$ for a translationally invariant system, with n, for the more general case; the Green functions are expressed in terms of the single-particle eigenfunctions as (Lehmann representation)

$$\begin{aligned}
G_0(\boldsymbol{r}_1, t_1; \boldsymbol{r}_2, t_2) &= -i\sum_n \phi_n(\boldsymbol{r}_1)\phi_n(\boldsymbol{r}_2)\left\{\left[\Theta(t) - \frac{1}{1+e^{E_n/(k_\mathrm{B}T)}}\right]u_n^2 e^{-iE_n t}\right.\\
&\quad\left. + \left[\frac{1}{1+e^{E_n/(k_\mathrm{B}T)}} - \Theta(-t)\right]v_n^2 e^{+iE_n t}\right\},\\
\bar{G}_0(\boldsymbol{r}_1, t_1; \boldsymbol{r}_2, t_2) &= -i\sum_n \phi_n(\boldsymbol{r})\phi_n(\boldsymbol{r}_2)\left\{\left[\Theta(t) - \frac{1}{1+e^{E_n/(k_\mathrm{B}T)}}\right]v_n^2 e^{-iE_n t}\right.\\
&\quad\left. + \left[\frac{1}{1+e^{E_n/(k_\mathrm{B}T)}} - \Theta(-t)\right]u_n^2 e^{+iE_n t}\right\},\\
F_0(\boldsymbol{r}_1, t_1; \boldsymbol{r}_2, t_2) &= \bar{F}_0(\boldsymbol{r}_1, t_1; \boldsymbol{r}_2, t_2)\\
&= -i\sum_n \phi_n(\boldsymbol{r})\phi_n(\boldsymbol{r}')u_n v_n\left\{\left[\Theta(t) - \frac{1}{1+e^{E_n/(k_\mathrm{B}T)}}\right]e^{-iE_n t}\right.\\
&\quad\left. - \left[\frac{1}{1+e^{E_n/(k_\mathrm{B}T)}} - \Theta(-t)\right]e^{+iE_n t}\right\},
\end{aligned} \tag{3.8}$$

with $t = t_1 - t_2$. Here, the eigenfunctions are chosen to be real, due to the time-reversal symmetry in the Hamiltonian H_0, in the absence of time-reversal breaking terms.

For example, in a 3D system in the clean limit, the frequency-momentum representation of the imaginary-time Green functions in Nambu space are given by

$$\begin{aligned}
\hat{G}_0(\omega, \boldsymbol{k}) &= \begin{pmatrix} G_0(\omega, \boldsymbol{k}) & F_0(\omega, \boldsymbol{k}) \\ \bar{F}_0(\omega, \boldsymbol{k}) & \bar{G}_0(\omega, \boldsymbol{k}) \end{pmatrix}\\
&= \frac{1}{\omega^2 + \epsilon_{\boldsymbol{k}}^2 + \Delta_0^2}\begin{pmatrix} -i\omega + \epsilon_{\boldsymbol{k}} & \Delta_0 \\ \Delta_0^* & -i\omega - \epsilon_{\boldsymbol{k}} \end{pmatrix}.
\end{aligned} \tag{3.9}$$

The averaging over disorder follows the standard diffusion approximation. In the second order expansion about the saddle point, the equilibrium Green functions appear twice for each term. Thus, the averaging requires the correlators in the product of four wavefunctions. The correlators needed are

$$\sum_{mn}\left\langle\delta\left(\frac{\epsilon_n}{\hbar}-\omega_1\right)\delta\left(\frac{\epsilon_m}{\hbar}-\omega_2\right)\phi_n(\boldsymbol{r})\phi_n(\boldsymbol{r}')\phi_m(\boldsymbol{r}')\phi_m(\boldsymbol{r})\right\rangle$$
$$=\frac{D(E_\mathrm{F})}{2\pi}\mathrm{Re}\mathcal{D}\left[(\omega_1-\omega_2);\boldsymbol{r},\boldsymbol{r}'\right],$$
$$\sum_{mn}\left\langle\delta\left(\frac{\epsilon_n}{\hbar}-\omega_1\right)\delta\left(\frac{\epsilon_m}{\hbar}-\omega_2\right)\left(\nabla^\alpha_{r_1}-\nabla^\alpha_{r_4}\right)\left(\nabla^\beta_{r_3}-\nabla^\beta_{r_2}\right)\right.$$
$$\left.\times\,\phi_n(\boldsymbol{r}_1)\phi_n(\boldsymbol{r}_2)\phi_m(\boldsymbol{r}_3)\phi_m(\boldsymbol{r}_4)\right\rangle\Big|_{\boldsymbol{r}_4=\boldsymbol{r}_1=\boldsymbol{r},\boldsymbol{r}_3=\boldsymbol{r}_2=\boldsymbol{r}'}$$
$$=\frac{4m^2D}{\pi}\delta_{\alpha\beta}\mathrm{Re}\left[i(\omega_1-\omega_2)\mathcal{D}\left((\omega_1-\omega_2);\boldsymbol{r},\boldsymbol{r}'\right)\right]. \tag{3.10}$$

Here, m is the electron mass, $D = 1/3v_\mathrm{F}l_\mathrm{mfp}$ is the diffusion constant, and the frequency transformed, real-time diffusion propagator $D(\hbar\omega;\boldsymbol{r},\boldsymbol{r}')$ satisfies

$$\left(-i\omega-D\nabla^2_{\boldsymbol{r}}\right)\mathcal{D}(\omega;\boldsymbol{r},\boldsymbol{r}')=\delta(\boldsymbol{r}-\boldsymbol{r}'). \tag{3.11}$$

The diffusion propagator is given by

$$\mathcal{D}(\omega;\boldsymbol{r},\boldsymbol{r}')=\int\frac{d^3q}{(2\pi)^3}\frac{e^{i\boldsymbol{q}\cdot(\boldsymbol{r}-\boldsymbol{r}')}}{-i\omega+Dq^2}. \tag{3.12}$$

The imaginary-time propagator is obtained by replacing $-i\omega$ with $|\omega|$. Computationally, the averaging is equivalent to multiplying all energies/frequencies in the single particle Green's functions, for example, ω and Δ_0, by the factor $[1+1/(2\tau_\mathrm{tran}W)]$, where

$$W=\sqrt{\omega_\nu^2+\left(\frac{\Delta_0}{\hbar}\right)^2},\quad\text{and}\quad W'=\sqrt{(\omega_\nu+\omega_\mu)^2+\left(\frac{\Delta_0}{\hbar}\right)^2},$$

and τ_tran is the transport scattering time. The quantities $\omega_\nu=(2\nu+1)\pi k_\mathrm{B}T/\hbar$ and $\omega_\mu=2\mu k_\mathrm{B}T/\hbar$, with ν and μ integers, are the Matsubara frequencies appearing in the computation of the polarization bubbles f_i, g_i, h_i, and k_i introduced in Section 1.6 [3, 20, 22].

After the averaging, the system becomes translationally invariant in space, in addition to being so in time. The second order contributions to the effective action introduced in (1.104)–(1.108) can be written as

$$S^{(2)}_\mathrm{eff}=S_\Delta+S_\mathrm{J}+S_\mathrm{L}+S_\mathrm{D}. \tag{3.13}$$

The various terms in the frequency-momentum representation are

$$\frac{S_\Delta}{\hbar}=\frac{k_\mathrm{B}T}{2}\sum_{\omega_\mu}\int\frac{d^3\boldsymbol{q}}{(2\pi)^3}\chi_\Delta\tilde{\Delta}^2_\mathrm{L}, \tag{3.14}$$

$$\frac{S_J}{\hbar} = \frac{k_B T}{2} \sum_{\omega_\mu} \int \frac{d^3 q}{(2\pi)^3} \chi_J \tilde{\Phi}^2 , \tag{3.15}$$

$$\frac{S_L}{\hbar} = \frac{k_B T}{2} \sum_{\omega_\mu} \int \frac{d^3 q}{(2\pi)^3} \chi_L \tilde{v}_s^2 \tag{3.16}$$

and

$$\frac{S_D}{\hbar} = \frac{k_B T}{2} \sum_{\omega_\mu} \int \frac{d^3 q}{(2\pi)^3} \chi_D \tilde{E}^2 , \tag{3.17}$$

where the space-imaginary time Fourier transform $\tilde{\Delta}_L(\omega, q)$ is defined as

$$\tilde{\Delta}_L(\omega, q) = \int_0^{\hbar\beta} d\tau \int d^3 r \Delta_L(r, \tau) e^{-iq\cdot r} e^{i\omega\tau} , \tag{3.18}$$

and similarly for the other quantities. The respective susceptibilities for the Δ (χ_Δ), Josephson (χ_J), London (χ_L), and Drude (χ_D) terms are expressible in terms of the polarization bubbles f_i, g_i, h_i, and k_i, with coupling $g = \lambda$

$$\begin{aligned} \chi_\Delta &= 2\left[\frac{1}{\lambda} + h_0(q) + f_0(q)\right] , \\ \chi_J &= 4e^2 f_0(q) , \\ \chi_L &= 8m^2 \Delta_0^2 \frac{1/\lambda + h_0(q) - \left[1 + \hbar^2\omega^2/\left(2\Delta_0^2\right)\right] f_0(q)}{(\hbar q)^2} , \\ \chi_D &= -2e^2 \frac{f_0(q) + g_0(q)}{q^2} . \end{aligned} \tag{3.19}$$

These quantities can be determined in closed form only for limiting cases. Note that in the dirty limit of $l_{mfp} \ll (\xi, \lambda_L)$ with which we are dealing, the contribution of the normal metal magnetic susceptibility,

$$\chi_m = -\frac{2\hbar^2 e^2}{m^2 c^2} [g_2(q) + f_2(q) - g_3(q) - f_3(q)] , \tag{3.20}$$

is small [22] and will be neglected. Here, m is the electron mass, and q is the momentum.

As we are particularly interested in quantum tunneling behavior at temperature much below T_c, it is of interest to provide the limiting expressions for $T \ll T_c$. We are interested in the dynamics in the low energy and momentum limit. In particular, at $T = 0$, for $\hbar\omega \ll 2\Delta_0$ and $\hbar D q^2 \ll 2\Delta_0$, one finds (with $D(E_F) \equiv N(0)$)

$$\chi_\Delta = D(E_F)\left[1 + \frac{1}{3}\left(\frac{\hbar\omega}{2\Delta_0}\right)^2 - \frac{\pi}{4}\frac{\hbar D q^2}{2\Delta}\right],$$
$$\chi_J = e^2 D(E_F)\left[1 - \frac{2}{3}\left(\frac{\hbar\omega}{2\Delta_0}\right)^2 - \frac{\pi}{4}\frac{\hbar D q^2}{2\Delta}\right],$$
$$\chi_L = \frac{\pi}{\hbar} D(E_F) D m^2 \Delta_0\left[1 - \frac{1}{4}\left(\frac{\hbar\omega}{2\Delta_0}\right)^2 - \frac{2}{\pi}\frac{\hbar D q^2}{2\Delta}\right],$$
$$\chi_D = \frac{\pi\hbar\sigma}{8\Delta_0}\left[1 - \frac{3}{8}\left(\frac{\hbar\omega}{2\Delta_0}\right)^2 - \frac{8}{3\pi}\frac{\hbar D q^2}{2\Delta}\right]. \tag{3.21}$$

In the limit of high frequencies $|\hbar\omega| \gg 2\Delta_0$ and momentum $\hbar D q^2 \gg 2\Delta_0$, the expressions are

$$\chi_\Delta = D(E_F)\ln\frac{\hbar(|\omega| + D q^2)}{\Delta_0},$$
$$\chi_J = \frac{4e^2 D(E_F)\Delta_0^2}{\hbar^2(|\omega|^2 + D^2 q^4)}\left[\ln\frac{\hbar(|\omega| + D q^2)}{\Delta_0} - \frac{D q^2}{|\omega|}\ln\frac{\hbar|\omega|}{2\Delta_0}\right]$$
$$\chi_L = \frac{4m^2 D(E_F) D\Delta_0^2}{\hbar^2(|\omega|^2 + D^2 q^4)}\left[-D q^2\ln\frac{\hbar(|\omega| + D q^2)}{\Delta_0} + |\omega|\ln\frac{\hbar|\omega|}{2\Delta_0}\right],$$
$$\chi_D = \frac{\sigma}{|\omega| + D q^2} = \frac{e^2 D(E_F) D}{|\omega| + D q^2}. \tag{3.22}$$

The complete expressions are given in the Appendix in references 3 and 20.

3.2.1.2 Specialization to One Dimension

At this point, we are in a position to take the final step and specialize to the case of a 1D superconductor. To account for the behavior in a long, 1D nanowire, the free electromagnetic field energy can be rewritten in terms of the capacitive and inductive energies. Since the nanowire is extremely narrow in the transverse directions, with $\mathsf{d} \sim (\mathsf{w}, \mathsf{h}) \ll \lambda_L$, where $\lambda_L = \sqrt{\hbar c^2/[\pi e^2 D(E_F) D\Delta_0]} \sim 100\,\text{nm}$ is the London penetration depth characterizing the magnetic response of the superconductor, only the component along the direction of the wire length needs to be retained. We thus have

$$\frac{1}{\hbar}\int d\tau d^3 r \frac{\mathbf{E}^2 + \mathbf{B}^2}{8\pi} \rightarrow \frac{1}{2\hbar}\int d\tau dx\left(C V^2 + \frac{A^2}{L c^2}\right), \tag{3.23}$$

where C is the capacitance per unit length, and L is the inductance per unit length. For thin 1D nanowires, C takes the form $C = \epsilon/[2\ln(2\mathsf{d}_s/\mathsf{d})]$ and L the form

$L = (2/c^2)\ln(2d_s/d)$ where d_s is the distance to nearby metal electrodes. In the frequency-momentum representation, the effective action takes the form

$$\frac{S_{1D}}{\hbar} = \frac{A}{2}\int \frac{d\omega}{2\pi}\frac{dq}{2\pi}\left[\frac{(C+C')}{\hbar A}|\tilde{V}|^2 + \frac{1}{\hbar L c^2 A}|\tilde{A}|^2 + \chi_D|\tilde{E}|^2 \right.$$
$$\left. + \chi_J|\tilde{\Phi}|^2 + \frac{\chi_L}{4m^2}|\tilde{v}_s|^2 + \chi_\Delta|\tilde{\Delta}_1|^2\right], \tag{3.24}$$

with A being the cross section, and $\beta = 1/(k_B T)$. In this (ω, q) representation, we have

$$\tilde{E} = -iq\tilde{V} - i\left(\frac{\omega}{c}\right)\tilde{A},$$
$$\tilde{\Phi} = \tilde{V} - \left(\frac{i\omega}{2e}\right)\tilde{\varphi},$$
$$\tilde{v}_s = \left(\frac{1}{2m}\right)\left[i\hbar q\tilde{\varphi} + \left(\frac{2e}{c}\right)\tilde{A}\right]. \tag{3.25}$$

Here, $\tilde{Q} = \tilde{Q}(\omega, q)$ denotes the space-time Fourier transform of the quantity $Q(x, \tau)$:

$$\tilde{Q}(\omega, q) = \int_0^L dx \int_0^{\hbar\beta} d\tau\, Q(x,\tau) e^{-iqx} e^{i\omega\tau},$$
$$Q(x,\tau) = \frac{L}{2\pi}\int dq \frac{\hbar\beta}{2\pi}\int d\omega\, \tilde{Q}(\omega, q) e^{iqx} e^{-i\omega\tau}, \tag{3.26}$$

where Q stands for V, A, E, Φ, v_s, or φ, and L the nanowire length. At low temperatures, C' is negligible compared to C. C' is related to the polarization contribution of the normal-fluid component, with $C' = Ae^2 D(E_F) n_n/n$, where the normal density $n_n = n - n_s$ approaches zero, and n_s is the superfluid density.

To cast this expression into a form involving only the order-parameter phase and amplitude fluctuations, in the path integral, the terms involving $\tilde{V}$ and $\tilde{A}$ are in quadratic form and can be integrated out exactly to yield

$$\frac{S_{1D}}{\hbar} = \frac{A}{2\hbar}\int \frac{d\omega}{2\pi}\frac{dq}{2\pi}\left[\hbar^2 \tilde{F}(\omega, q)|\tilde{\varphi}|^2 + \chi_\Delta|\tilde{\Delta}_1|^2\right]. \tag{3.27}$$

In this low temperature limit, the susceptibilities are real. In addition, the kinetic inductance L_{kin} is typically much larger than the geometric inductance $L_{kin} \gg L$; L_{kin} appears in the London term with susceptibility χ_L, for which

$$\frac{e^2 A}{(mc)^2}\chi_L \approx \frac{1}{L_{kin} c^2} = e^2 n_s \frac{A}{mc^2} = \pi e^2 D(E_F) D\Delta_0 \frac{A}{\hbar c^2}.$$

The expression for $\tilde{F}$,

$$\tilde{F} = \frac{\left[\frac{\chi_\mathrm{J}}{(2e)^2}\omega^2 + \frac{\chi_\mathrm{L}}{(2m)^2}q^2\right]\left[\frac{C}{\mathsf{A}^2 Lc^2} + \chi_\mathrm{D}\left(\frac{C\omega^2}{\mathsf{A}c^2} + \frac{q^2}{\mathsf{A}Lc^2}\right)\right] + \frac{\chi_\mathrm{J}\chi_\mathrm{L}}{(2m)^2}\left(\frac{C\omega^2}{\mathsf{A}c^2} + \frac{q^2}{\mathsf{A}Lc^2}\right)}{D_{\tilde{F}}}$$

$$+ \frac{\frac{e^2\chi_\mathrm{L}^2}{(2m^2c)^2}q^2\chi_\mathrm{J}}{D_{\tilde{F}}} + \frac{\frac{e^2\chi_\mathrm{L}^2}{(2m^2c)^2}q^2\left[\left(\frac{1}{\mathsf{A}Lc^2} + \frac{C}{\mathsf{A}} + \chi_\mathrm{D}q^2\right)\frac{e^2}{(mc)^2}\chi_\mathrm{L} + \frac{1}{\mathsf{A}}\chi_\mathrm{D}q^2 + \frac{C}{\mathsf{A}}\chi_\mathrm{D}\frac{\omega^2}{c^2}\right]}{\left(\frac{1}{\mathsf{A}Lc^2} + \chi_\mathrm{D}\frac{\omega^2}{c^2} + \frac{e^2}{(mc)^2}\chi_\mathrm{L}\right)D_{\tilde{F}}},$$

$$D_{\tilde{F}} = \left(\frac{C}{\mathsf{A}} + \chi_\mathrm{J} + \chi_\mathrm{D}q^2\right) \times \left[\frac{1}{\mathsf{A}Lc^2} + \chi_\mathrm{D}\frac{\omega^2}{c^2} + \frac{e^2}{(mc)^2}\chi_\mathrm{L}\right] - \chi_\mathrm{D}^2\frac{\omega^2}{c^2}q^2, \tag{3.28}$$

simplifies to

$$\tilde{F} = \frac{\left(\frac{\chi_\mathrm{J}}{(2e)^2}\omega^2 + \frac{\chi_\mathrm{L}}{(2m)^2}q^2\right)\left(\frac{C}{\mathsf{A}} + \chi_\mathrm{D}q^2\right) + \frac{\chi_\mathrm{J}\chi_\mathrm{L}}{(2m)^2}q^2}{\left(\frac{C}{\mathsf{A}} + \chi_\mathrm{J} + \chi_\mathrm{D}q^2\right)}, \tag{3.29}$$

with the stationary conditions for the voltage V and vector potential A given by

$$\begin{aligned}
\tilde{V} &= \frac{\chi_\mathrm{J}\left(\frac{1}{\mathsf{A}Lc^2} + \chi_\mathrm{D}\frac{\omega^2}{c^2} + \frac{e^2}{m^2c^2}\chi_\mathrm{L}\right) + \frac{e^2}{m^2c^2}\chi_\mathrm{D}\chi_\mathrm{L}q^2}{D_{\tilde{F}}}\left(\frac{i\hbar\omega}{2e}\tilde{\varphi}\right) \\
&\approx \frac{\chi_\mathrm{J}}{\frac{C}{\mathsf{A}} + \chi_\mathrm{J} + \chi_\mathrm{D}q^2}\left(\frac{i\hbar\omega}{2e}\tilde{\varphi}\right), \\
\tilde{A} &= \frac{\frac{e^2}{m^2c^2}\chi_\mathrm{L}\left(\frac{C}{\mathsf{A}} + \chi_\mathrm{J} + \chi_\mathrm{D}q^2\right) + \chi_\mathrm{D}\chi_\mathrm{J}\frac{\omega^2}{c^2}}{D_{\tilde{F}}}\left(\frac{-i\hbar cq}{2e}\tilde{\varphi}\right) \\
&\approx 0\,.
\end{aligned} \tag{3.30}$$

The expression for $\tilde{V}$ indicates that the usual Josephson relation $\dot{\varphi} = -2e/\hbar V$ holds only when $\chi_\mathrm{J} \gg C/\mathsf{A} + \chi_\mathrm{D}q^2$! For typical mean free path l_mfp in the 3–10 nm range, this condition requires $\omega/\Delta_0 \ll 1$ and $Dq^2/\Delta_0 \ll 1$ with $D \approx v_\mathrm{F}l_\mathrm{mfp}/3$.

To arrive at a tractable model for the estimation of important quantities, such as the action of the quantum phase slip, yielding the macroscopic quantum tunneling rate, and to gain an understanding of the quantum phase transitions occurring at $T = 0$, a final expansion in powers of ω and q^2 is performed, leading to

$$\begin{aligned}
\frac{S_\mathrm{1D}}{\hbar} &= \frac{\mathsf{A}}{2}\int\frac{d\omega}{2\pi}\frac{dq}{2\pi}\left[\left(\frac{\hbar C}{\mathsf{A}}\omega^2 + \frac{\hbar e^2}{m^2}\chi_\mathrm{L}q^2\right.\right. \\
&\quad\left.\left. - \frac{\hbar e^2\chi_\mathrm{L}\chi_\mathrm{D}}{m^2\chi_\mathrm{J}}q^4 + \hbar\chi_\mathrm{D}q^2\omega^2\right)\frac{|\tilde{\varphi}|^2}{(2e)^2} + \chi_\Delta|\tilde{\Delta}_1|^2\right] \\
&= \frac{\mathsf{A}}{2}\int\frac{d\omega}{2\pi}\frac{dq}{2\pi}\left\{\left[\hbar\frac{C}{\mathsf{A}}\omega^2 + \pi e^2 D(E_\mathrm{F})\Delta_0 Dq^2 - \hbar\frac{\pi\sigma D}{8}q^4 + \hbar\chi_\mathrm{D}q^2\omega^2\right]\right. \\
&\quad\left.\times\frac{|\tilde{\varphi}|^2}{(2e)^2} + \frac{D(E_\mathrm{F})}{\hbar}\left(1 + \frac{\hbar^2\omega^2}{12\Delta_0^2} + \frac{\pi\hbar Dq^2}{8\Delta_0}\right)|\tilde{\Delta}_1|^2\right\},
\end{aligned} \tag{3.31}$$

where the last equality is obtained by inserting the low temperature values of the susceptibilities χs listed in (3.21), and $\sigma = e^2 D(E_F) D$.

Starting from this expression, the low temperature properties of the macroscopic-quantum tunneling rate of the order parameter phase, that is, QPS rate, quantum phase diagram, transport nonlinear characteristics, and so on can be computed.

The expression in (3.31) enables one to estimate the contributions to the quantum phase-slip action from a core portion, of size-scale $x_0 \sim \xi = \sqrt{\hbar D/\Delta_0}$, and timescale $\tau_0 \gtrsim \hbar/\Delta_0$, as will be shown below, as well as the hydrodynamic contribution arising from the Mooij–Schön mode outside of the core space-imaginary time region. This linearly dispersing Mooij–Schön plasmon mode coming from the phase fluctuations yields a logarithmic contribution in space-imaginary time distance. Outside of the core, the amplitude is assumed to be constant, and thus amounts to an assumption that there, only phase fluctuations are important in the low temperature limit. Further, the logarithmic contribution leads to a logarithmic interaction between PS–PS pairs and PS–anti-PS pairs, which is repulsive for the former and attractive for the latter. Such logarithmic terms are familiar in the context of the Kosterlitz–Thouless–Berezinskii transition associated with topological defects in 2D XY and related models. Therefore, one can anticipate this type of quantum phase transitions in superconducting nanowires based on a phase-only model. The core action turns out to determine the fugacity of the phase slips. The quantum phase-slip event is thus a topological defect in the domain of space-imaginary time, while in real time, it launches the Mooij–Schön plasmon mode to help carry away and dissipate the excess energy.

3.2.2 Core Contribution to the QPS Action

The estimation of the actions S_{core} and S_{hyd} is facilitated by going into the space-imaginary time representation of the effective action. In (3.31), the $(Dq^2)^2$ turns out to be unimportant and will be omitted [3]. The action becomes

$$\frac{S_{1D}}{\hbar} = \frac{\mathsf{A}}{2} \int dx\, d\tau \left\{ \left[\frac{\hbar C}{(2e)^2 \mathsf{A}} \left(\frac{\partial \varphi}{\partial \tau} \right)^2 + \frac{\pi e^2 D(E_F) D \Delta_0}{(2e)^2} \left(\frac{\partial \varphi}{\partial x} \right)^2 \right.\right.$$
$$\left. + \frac{\hbar^2 \pi \sigma}{8(2e)^2 \Delta_0} \left(\frac{\partial^2 \varphi}{\partial x \partial \tau} \right)^2 \right] + \frac{D(E_F)}{\hbar}$$
$$\left. \times \left[\Delta_1^2 + \frac{\hbar^2}{12 \Delta_0^2} \left(\frac{\partial \Delta_1}{\partial \tau} \right)^2 + \frac{\pi \hbar D}{8 \Delta_0} \left(\frac{\partial \Delta_1}{\partial x} \right)^2 \right] \right\} . \tag{3.32}$$

The QPS core involves spatial dimensions at the small length scale of ξ and high-frequency $\hbar\omega > \Delta_0$ (or short imaginary-time $\tau < \hbar/\Delta_0$), reflecting the fast dynamics of the phase slip. Furthermore, the amplitude of the gap, which is proportional to the Cooper pair-density, n_s, is suppressed inside the core. Thus, the full expression, including the higher order term (third term in (3.32)), as well as the amplitude fluctuation term involving Δ_1 must be retained.

Ideally, the QPS is a saddle point solution of the action, or equivalently an instanton. In practice, the QPS core action is obtained from this effective action S_{1D} by a variational method using a trial solution, the form of which is constrained by the basic qualitative features a phase slip must possess: (i) far from the phase-slip core, for $x \gg x_0$, the gap should return to the value Δ_0, (ii) the phase φ should change by 2π over the phase-slip core. For a phase slip centered in space-imaginary time (0,0), possible trial forms include

$$\begin{aligned}
\Delta_1 &= -\Delta_0 \exp\left(-\frac{x^2}{2x_0^2} - \frac{\tau^2}{2\tau_0^2}\right) , \\
\varphi &= -\frac{\pi}{2\cosh\left(\frac{\tau}{\tau_0}\right)} \tanh\left[\frac{x}{x_0 \tanh(\tau/\tau_0)}\right] , \\
\text{or} &= -\frac{\pi}{2} \tanh\left(\frac{x\tau_0}{x_0\tau}\right) ,
\end{aligned} \tag{3.33}$$

where at the center of the core, $\Delta = \Delta_0 + \Delta_1 = 0$, or the gap is suppressed to zero. Strictly speaking, the treatment of the magnitude of the gap as a uniform part Δ_0, plus a small perturbing fluctuating part Δ_1 is not fully valid at or near the core center, $x = 0$ and $\tau = 0$ and $|\Delta_1| \to \Delta$. While this variational method may be expected to yield the correct functional form and order of magnitude, nevertheless, the separation of the magnitude and phase degrees of freedom in the perturbative approach we have utilized leads to an artificial divergence near the core center, which must be eliminated by choosing a suitable cut-off in the integration limit.

With the cut-off in place, the ansatz in (3.33) yields the following action in terms of the parameters x_0 and τ_0:

$$\begin{aligned}
\frac{S_{1D}(x_0,\tau_0)}{\hbar} = \hbar \Bigg\{ &\left[a_1 \frac{C}{e^2} + \frac{1}{2\hbar} a_2 \mathsf{A} D(E_F)\right] \frac{x_0}{\tau_0} + \frac{1}{2\hbar} a_3 \mathsf{A} D(E_F) D \Delta_0 \frac{\tau_0}{x_0} \\
&+ a_4 \frac{\hbar \mathsf{A}\sigma}{2e^2\Delta_0} \frac{1}{x_0\tau_0} + \frac{1}{2\hbar^2} a_5 \mathsf{A} D(E_F) \Delta_0^2 x_0 \tau_0 + a_6 \frac{C\mathsf{L}}{e^2\tau_0} \Bigg\} .
\end{aligned} \tag{3.34}$$

The a_js are numerical coefficients of order unity, and the last term, associated with the total wire capacitance over its entire length L, may be neglected for sufficiently short wires. For example, using $\varphi = -\pi/2 \tanh(x\tau_0/(x_0\tau))$, the coefficient a_1 is

$$a_1 = \frac{\pi^4}{24} \int_0^{\mathsf{L}/2} \frac{dx}{x} e^{-x\tau_0/(x_0\hbar\beta)} . \tag{3.35}$$

At small x, the logarithmic divergence is artificial, and a short-distance cut-off of the order of a fraction of the core size x_0 would be sensible. At long distances, this functional form has the disadvantage of an unphysically-long cut-off scale $x_0\hbar\beta/\tau_0$ as $T \to 0$. Instead, this cut-off should be of the order x_0, as is the case for the first φ trial function above in (3.33). However, that functional form is intractable analytically.

In the term with coefficient a_4, the physical parameters are intentionally left in a form containing the conductivity, σ, to specify its origin. This term comes from the dissipative contribution of normal-electron electromagnetic response, via the susceptibility χ_D. In short wires, neglecting the a_6 term and minimizing the above action leads to the values

$$x_0 \approx \left[\frac{a_3 a_4}{a_2 a_5} \frac{\hbar^2 \sigma D}{e^2 D(E_F) \Delta_0^2}\right]^{1/4} ,$$
$$\tau_0 \approx \left[\frac{a_3 a_4}{a_2 a_5} \frac{\hbar^2 \sigma D}{e^2 D(E_F) \Delta_0^2}\right]^{1/4} . \tag{3.36}$$

The appearance of the conductivity σ in the solution for quantities clearly displays the importance of the dissipative process in the core action. From this expression, it is apparent that only accounting for the gap magnitude reduction, without including the dissipative contribution in the dynamics of the phase, cannot capture the entire story, as can be seen by setting $\sigma = 0$ to remove the a_4 term, leading to the nonsensical result where $x_0 = \tau_0 = 0$, that is, the phase slips do not have a finite size.

By the inclusion of the a_4 term, we find that $x_0 \sim \sqrt{\hbar D/\Delta_0} = \xi$, and $\tau_0 \sim \hbar/\Delta_0$. Thus, for the core, high-frequency components contribute at frequencies $\omega \gtrsim \Delta_0/\hbar$.

For this short-wire limit where $\mathsf{L} \ll \xi[e^2 D(E_F)\mathsf{A}/(2C)]$, the core contribution to the QPS action takes the form

$$\frac{S_{\text{core}}}{\hbar} = \frac{1}{2}\alpha\pi D(E_F)\mathsf{A}\sqrt{\hbar D \Delta_0} = 2\pi\alpha\frac{\Delta F}{\Delta_0} = \alpha\frac{R_Q}{R_\xi} ,$$
$$\alpha = \frac{2}{\pi}\left(\sqrt{a_2 a_3} + \sqrt{a_4 a_5}\right) , \quad \Delta F = \frac{1}{4} D(E_F)\Delta_0^2 \mathsf{A}\xi ,$$
$$R_Q = \frac{h}{(2e)^2} \approx 6.453\,\text{k}\Omega , \quad R_\xi = \frac{\xi}{\sigma \mathsf{A}} = R_N \frac{\xi}{\mathsf{L}} , \tag{3.37}$$

where the numerical constant α is of order one, ΔF is the free energy barrier, R_ξ is the normal state resistance on the length scale of the superconducting coherence length ξ, and R_N is the normal state resistance of the entire wire. This form looks appealingly similar to the phenomenological expression proposed by Giordano for the quantum tunneling rate of a phase slip, (2.127), albeit with the energy associated with the Ginzburg–Landau time, $E_{GL} = \hbar/\tau_{GL}$, replaced by the superconducting energy gap Δ_0.

For very long wires, the minimization yields

$$\frac{S_{\text{core}}}{\hbar} = \alpha' \frac{R_Q}{R_\xi}\sqrt{\frac{2\mathsf{L}C}{\xi e^2 D(E_F)\mathsf{A}}} , \tag{3.38}$$

with α' as a different numerical factor.

3.2.3
Hydrodynamic Contribution to the Phase-Slip Action

We next turn to the hydrodynamic contribution outside of the core, where the fluctuation in the amplitude Δ_1 is neglected. This is justified by the fact that at low temperatures, the amplitude fluctuations are expected to be small compared to Δ_0. This yields the often used phase-only model. Furthermore, since we are interested in long wavelength and low energy dynamics, only the terms involving $|\partial\varphi/\partial\tau|^2$ and $|\partial\varphi/\partial x|^2$ will be kept. The fact that we are focusing on the long wavelength, low energy behavior means that the wire cannot be too short. The resulting action corresponds to that of the linearly dispersing ($\omega = c_{\text{pl}} q$) Mooij–Schön plasmon mode [1]

$$\begin{aligned}\frac{S_{\text{hyd}}}{\hbar} &= \frac{\mathsf{A}}{2}\int dx\,d\tau\left[\frac{\hbar C}{(2e)^2\mathsf{A}}\left(\frac{\partial\varphi}{\partial\tau}\right)^2 + \frac{\pi e^2 D(E_{\text{F}}) D\Delta_0}{(2e)^2}\left(\frac{\partial\varphi}{\partial x}\right)^2\right]\\ &= \frac{\mathsf{A}}{2}\int dx\,d\tau\frac{\pi e^2 D(E_{\text{F}}) D\Delta_0}{(2e)^2}\left[\left(\frac{\partial\varphi}{c_{\text{pl}}\partial\tau}\right)^2 + \left(\frac{\partial\varphi}{\partial x}\right)^2\right]. \end{aligned} \tag{3.39}$$

The speed of the Mooij–Schön mode is given by[1)]

$$c_{\text{pl}} = \sqrt{\frac{\pi e^2 D(E_{\text{F}}) D\Delta_0\mathsf{A}}{\hbar C}}\,. \tag{3.40}$$

Estimating the hydrodynamic contribution to the QPS action involves finding the saddle point solution outside of the QPS core, φ_s, with the constraint that the phase shift should be 2π when winding around the phase-slip center in (x, τ) space. Such a constraint is given by

$$\frac{\partial^2\varphi_s}{\partial x\partial\tau} - \frac{\partial^2\varphi_s}{\partial\tau\partial x} = 2\pi\delta(\tau)\delta(x)\,. \tag{3.41}$$

The resulting solution for φ_s is

$$\varphi_s = -\arctan\left(\frac{x}{c_{\text{pl}}\tau}\right), \tag{3.42}$$

with an action

$$\frac{S_{\text{hyd},s}}{\hbar} = \mu\ln\frac{[\min(\mathsf{L}, \hbar\beta c_{\text{pl}})]}{[\max(c_{\text{pl}}\tau_0, x_0)]}\,. \tag{3.43}$$

1) This expression is valid for both cgs and SI units. The cgs unit of capacitance per unit length is dimensionless. The conversion for capacitance is $1\,\text{cm} = (8.99\times10^{11})^{-1}\,\text{F}$.

The parameter μ will turn out to be important in determining the phase boundary in the KTB quantum phase transitions. It is given by[2)]

$$\mu = \frac{\pi h}{(2e)^2 c_{\text{pl}}(L + L_{\text{kin}})} \approx \frac{\pi \hbar}{(2e)^2}\sqrt{\frac{\pi e^2 D(E_{\text{F}}) D \Delta_0 C \mathsf{A}}{\hbar}} = \frac{R_{\text{Q}}}{2Z} ,$$
$$Z = \sqrt{\frac{L_{\text{kin}}}{C}} , \quad L_{\text{kin}} = \frac{\hbar}{\pi e^2 D(E_{\text{F}}) D \Delta_0 \mathsf{A}} , \tag{3.44}$$

The impedance Z represents the plasmon propagation impedance. The doubling of Z in the denominator is due to the propagation in both directions away from a phase-slip core.

The total action for a QPS is now

$$S_{\text{QPS}} = S_{\text{core}} + S_{\text{hyd},s} . \tag{3.45}$$

Let us estimate the relative contributions of the two terms for typical experimentally realizable nanowires. The expression for the parameters capacitance per unit length is $C = \epsilon[2\ln(2\mathsf{d}_s/\mathsf{d})]^{-1}$, and for the kinetic inductance per unit length, it is $L_{\text{kin}} = [\pi e^2 D(E_{\text{F}}) D \Delta_0 \mathsf{A}/\hbar]^{-1}$.

Since the long wire core action is enhanced by the factor $\sqrt{2\mathsf{L}C/(\xi e^2 D(E_{\text{F}})\mathsf{A})}$ compared to the short wires, we will make the comparison in the short-wire. For $Mo_{0.79}Ge_{0.21}$ nanowires, $\mathsf{A} \sim 50\,\text{nm}^2$, $\mathsf{d} \sim (\mathsf{w}, \mathsf{h}) \sim 7\,\text{nm}$, $l_{\text{mfp}} \sim 2\,\text{nm}$, $v_{\text{F}} \sim 0.6 \times 10^6\,\text{m/s}$, $k_{\text{F}} \sim 5 \times 10^9\,\text{m}^{-1}$, and $\Delta_0 \sim 1\,\text{meV}$. These yield $S_{\text{core}} \gtrsim 5$ at the minimum, but more typically $S_{\text{core}} \sim 20\text{–}100$ and $S_{\text{hyd},s} \sim 2$. Thus, the hydrodynamic contribution can be neglected. Similarly, for aluminum nanowires, $S_{\text{core}} \gtrsim 40$, while $S_{\text{hyd},s} \sim 2\text{–}10$. For wider wires, the situation is even more in favor of the core contribution.

3.2.4
Quantum Phase-Slip Rate

To compute the prefactor to the QPS rate, which multiplies the exponential suppression factor associated with the Euclidean action of the QPS, it is necessary to solve the equation governing the time-dependence when displaced away from the equilibrium saddle point solutions. Near T_c, we have seen in Section 2.4.7 that by using a the time-dependent Ginzburg–Landau (TDGL) equation with a Langevin thermal noise term, it was possible to ascertain the prefactor with reasonable accuracy [29] by first transforming the TDGL with a noise term into a Fokker–Planck equation, and by studying the eigenvalues and modes of the quadratic terms accounting for the deviation of the order parameter about the saddle. In some sense, the eigenmodes provide a quantitative measure of the "attempt frequency" in multidimensional space about the saddle point.

2) The expressions are valid for both cgs and SI units. The cgs unit of resistance or impedance is s/cm. The conversion is $1\,\text{s/cm} = 8.99 \times 10^{11}\,\Omega$.

From the onset, McCumber and Halperin recognized that the TDGL is only valid under rather restrictive conditions. Golubev and Zaikin, and Meidan, Oreg and Refael [5] further stressed that, in fact, the condition under which the TDGL equations are applicable is so restricted that the LAMH full expression is hardly applicable to real experimental nanowire systems, except in a meaninglessly small temperature region, despite the fact that empirically, it has been extensively used to fit experiment data with considerable success.

Instead, Golubev and Zaikin utilized the conceptual framework provided by the decay rate calculation for metastable states based on the instanton technique [30]. When the QPS phase-slip action is large, $S_{\text{QPS}}/\hbar \gg 1$, the instanton technique [20, 21] enables the QPS rate Γ_{QPS} to be expressed as an attempt frequency, Ω_{QPS}, multiplied by an exponential suppression factor $\exp(-S_{\text{QPS}}/\hbar)$:

$$\Gamma_{\text{QPS}} = \Omega_{\text{QPS}} \exp(-S_{\text{QPS}}) \, . \tag{3.46}$$

The free energy

$$F = -k_B T \ln Z \, ; \quad Z = \int \mathcal{D}\Delta \int \mathcal{D}\varphi \exp\left(-S\frac{[\Delta, \varphi]}{\hbar}\right) \tag{3.47}$$

evaluated with contributions from all possible QPS saddle point solutions, and after integrating out the quadratic deviations about such saddle points, may be written as

$$F = F_0 - i\hbar\frac{\Gamma_{\text{QPS}}}{2} \, , \tag{3.48}$$

where the rate is given by

$$\Gamma_{\text{QPS}} = k_B T \frac{\int \mathcal{D}\delta\Delta \int \mathcal{D}\delta\varphi \exp(-S^2_{1\text{QPS}}[\delta\Delta, \delta\varphi]/\hbar)}{\int \mathcal{D}\delta\Delta \int \mathcal{D}\delta\varphi \exp(-S^2_{0\text{QPS}}[\delta\Delta, \delta\varphi]/\hbar)} \exp\left(-\frac{S_{\text{QPS}}}{\hbar}\right) \, . \tag{3.49}$$

Here, the superscript 2 denotes the second order term quadratic in the deviations $\delta\Delta, \delta\varphi$ from the saddle point solutions, and subscripts 0QPS and 1QPS denote the action with no QPS or with one QPS present.

When the phase slips are rare, that is, if the QPS rate Γ_{QPS} is small, the occurrence of phase-slip events are far separated in imaginary time, and phase-slip interaction in the time coordinate may be neglected. Under this condition, a quadratic expansion about configurations with a single phase slip will be adequate. We transform to the normal-mode coordinates of the one-phase slip configurations, taking care to account for the two zero-modes, that is, modes for which the energy does not change with a change in the direction of their respective normal coordinates, namely, those modes with an eigenvalue equal to zero in the quadratic action corresponding to a translation along the nanowire in space or in imaginary time (see Section 2.4.7). In terms of these 1QPS normal mode coordinates, the matrix elements representing the second derivatives appearing in the action for the 0-QPS case will be denoted by M_{ij}, while for the 1QPS case, the corresponding matrix is of course diagonal with eigenvalues λ_i. Such a transformation enables the quadratic action to be written in terms of the matrix elements and coordinates q_i, which represent the deviation of the ith coordinate from the 1QPS saddle point. For example, in this representation, the second order 1QPS action is given by $S^2_{1\text{QPS}} = \sum_i \lambda_i q_i^2$. Separating out the contributions of the zero-modes in the 1QPS action, the prefactor to the exponential in the rate now reads

$$\Omega_{\text{QPS}} = k_B T \frac{\int \mathcal{D}\delta\Delta \int \mathcal{D}\delta\varphi \exp\left(-S^2_{1\text{QPS}}[\delta\Delta, \delta\varphi]/\hbar\right)}{\int \mathcal{D}\delta\Delta \int \mathcal{D}\delta\varphi \exp\left(-S^2_{0\text{QPS}}[\delta\Delta, \delta\varphi]/\hbar\right)} = \int_0^{\mathsf{L}_{q1}} \int_0^{\mathsf{L}_{q2}} \sqrt{\frac{\det M_{ij}}{(2\pi)^2 \prod_{i=3} \lambda_i}} , \qquad (3.50)$$

where L_{q1} and L_{q2}, are, respectively, the sizes of the spaces for the two zero-normal-modes in (x, τ)-coordinates, and will be set equal to L, the nanowire length, and $\hbar\beta$.

Despite the fact that exact trajectories for the QPS solutions parameterized in the coordinates $[\delta\Delta(x, \tau), \delta\varphi(x, \tau)]$ are not available for the QPSs, Golubev and Zaikin argued that for fast modes with frequencies and momenta far exceeding the instanton size, those factors cancel out in the quotient, as these modes are insensitive to the presence of a phase slip, and thus are common to both the 0QPS and 1-QPS actions. This simplification enables all but the zero-modes to be canceled, leading to

$$\Omega_{\text{QPS}} = k_B T \frac{\int \mathcal{D}\delta\Delta \int \mathcal{D}\delta\varphi \exp\left(-S^2_{1\text{QPS}}[\delta\Delta, \delta\varphi]/\hbar\right)}{\int \mathcal{D}\delta\Delta \int \mathcal{D}\delta\varphi \exp\left(-S^2_{0\text{QPS}}[\delta\Delta, \delta\varphi]/\hbar\right)} = \int_0^{\mathsf{L}_{q1}} \int_0^{\mathsf{L}_{q2}} \sqrt{\frac{\det M_{(i,j)\le 2}}{(2\pi)^2 \prod\limits_{i=3} \lambda_i}} , \qquad (3.51)$$

where the limit of 2 in the indices corresponds to the number of zero-modes in this case.

A further argument is made that, similar to the case explored in Vainstein *et al.* for local Lagrangian systems with kinetic and potential energy terms, here, despite the appearance of nonlocal interaction terms, the determination for the 0-QPS action is approximated as

$$\det M_{(i,j)\le 2} \approx M_{11} M_{22} \approx \frac{(\sqrt{2 S_{\text{QPS}}/\hbar})^2}{(s_{0,1} s_{0,2})^2} , \qquad (3.52)$$

where $s_{0,i} = (x_{0,i}, \tau_{0,i})$ represents the shift in the ith zero-mode direction for the 0-QPS action equals the size of instanton in that same normal mode direction, that is ith mode direction, so that $|s_{0,1}| \approx x_0$, and $|s_{0,2}| \approx c_0 \tau_0$. Putting everything together, one arrives at the final value for the prefactor, that is,

$$\Omega_{\rm QPS} \approx b\frac{S_{\rm QPS}L}{\hbar x_0 \tau_0}\,, \tag{3.53}$$

with b a numerical factor. The final expression for the QPS rate is thus

$$\Gamma_{\rm QPS} \approx \frac{\Delta_0}{\hbar}\frac{R_{\rm Q}}{R_\xi}\left(\frac{L}{\xi}\right)^2 e^{-S_{\rm QPS}/\hbar}\,. \tag{3.54}$$

3.2.5
Quantum Phase-Slip Interaction and Quantum-Phase Transitions

The hydrodynamic contribution to the QPS action, with its logarithmic dependence on the system size as the temperature goes to zero ($\hbar\beta \to \infty$) is reminiscent of the size dependence of topological defects in 2D systems, which exhibit the Kosterlitz–Thouless–Berezinskii (KTB) type of phase transition [11, 12]. In the 2D XY model, or in 2D superconducting films, the low temperature phase has bound vortex-antivortex defects. While above the transition temperature, the free, unbound vortices or antivortices proliferate.

A similar scenario occurs here in the $(1 + 1)$ dimension of space-imaginary time (x, τ), with the quantity μ, introduced in (3.44), playing the role of $J/(k_{\rm B}T)$ in the case of the 2D XY model. Here, J is the interaction energy scale of the logarithmic between vortices.

Thus, it is natural to include solutions to the action $S_{\rm 1D}$, which account for pairs of PS and anti-PS, which represent topological defects in (x, τ) space.

The solution for a pair of PS-type defects, located at (x_1, τ_1) and (x_2, τ_2) with spacing further apart than the size of the core $(x_0, c_{\rm pl}\tau_0)$, is a superposition of the type introduced in (3.42). This solution, $\varphi_{s,2{\rm PS}} = \varphi_s(x-x_1, \tau-\tau_1)+\varphi_s(x-x_2, \tau-\tau_2)$, satisfies the equation

$$\frac{\partial^2\varphi_s}{\partial x\partial\tau} - \frac{\partial^2\varphi_s}{\partial\tau\partial x} = 2\pi\sum_{i=1}^{2} s_i\delta(\tau-\tau_i)\delta(x-x_i)\,. \tag{3.55}$$

The "charge" of a given phase slip s_i equals $+1$ or -1, depending on whether the defect is a QPS or an anti-QPS. The pair action for two phase slips $\varphi_{s,2{\rm PS}}$ is then

$$\frac{S_{\rm 2QPS}}{\hbar} = 2\frac{S_{\rm core}}{\hbar} - \mu s_1 s_2 \ln\left(\frac{r_{12}^2}{\xi^2}\right)\,, \quad r_{12} = \sqrt{(x_1-x_2)^2 + c_{\rm pl}^2(\tau_1-\tau_2)^2}\,. \tag{3.56}$$

Thus, opposites attract with a negative hydrodynamic contribution to the action, while like "charges" repel. Such charges are equivalent to what occurs in a 2D Coulomb gas model with logarithmically interaction rods of charges s_1 and s_2.

For n phase slips, this generalizes to

$$\varphi_{s,n{\rm PS}} = \sum_i^n \varphi_s(z-z_i)\,, \quad z \equiv (x,\tau)\,, \tag{3.57}$$

with an action

$$\frac{S_{\mathrm{nQPS}}}{\hbar} = n\frac{S_{\mathrm{core}}}{\hbar} - \mu \sum_{i \neq j} s_i s_j \ln \frac{r_{ij}}{\xi} \,. \tag{3.58}$$

The partition function accounting for multiple phase-slip pairs, in analogy to a 2D Coulomb gas, now reads

$$Z = \sum_{n=0}^{\infty} \frac{1}{(2n)!} \left(\frac{f_{\mathrm{QPS}}}{2}\right)^{2n} \prod_{i=1}^{2n} \int_{x_0}^{\mathrm{L}} \frac{dx_i}{x_0} \int_{\tau_0}^{\hbar\beta} \frac{d\tau_i}{\tau_0}$$
$$\times \sum_{s_i=\pm 1} \exp\left(- \frac{S_{2n\mathrm{QPS}} - 2n S_{\mathrm{core}}}{\hbar} + \frac{\phi_0 I}{c} \sum_i s_i \tau_i \right) \,, \tag{3.59}$$

in the presence of a bias current I, where the QPS fugacity f_{QPS} is related to the QPS rate, that is,

$$f_{\mathrm{QPS}} = (x_0 \tau_0/\mathrm{L}) \Gamma_{\mathrm{QPS}} \approx \frac{S_{\mathrm{core}}}{\hbar} e^{-S_{\mathrm{core}}/\hbar} \,, \tag{3.60}$$

subjected to the periodic condition that $\varphi_s(x,\tau) = \varphi_s(x, \tau + \hbar\beta)$ which implies that the total charge for a given configuration must be neutral: $\sum_i^n s_i = 0$.

This Coulomb gas type of partition function is amenable to the renormalization group analysis of the KTB transition introduced by Anderson, Yuval, and Hamann [31, 32]. The familiar scaling equations are

$$\frac{\partial \mu}{\partial l} = -\pi^2 \mu^2 f_{\mathrm{QPS}}^2 \,, \quad \frac{\partial f_{\mathrm{QPS}}}{\partial l} = (2-\mu) f_{\mathrm{QPS}} \,,$$
$$l = \ln\left(\frac{\rho}{\xi}\right) = \frac{\ln \sqrt{x^2 + c_{\mathrm{pl}}^2 \tau^2}}{\xi} \,. \tag{3.61}$$

Near $T = 0$ where $\hbar\beta \to \infty$, the KTB quantum transition in space-imaginary time thus occurs at $\mu = \mu^*$, for

$$\mu^* = 2 + 4\pi f_{\mathrm{QPS}} \approx 2 \,. \tag{3.62}$$

This conclusion holds when the scaling can proceed to sufficiently long length scales (see below for an estimate), that is, when the nanowire is sufficiently long such that finite-size corrections can be neglected. In analogy to the KTB transition for a 2D superconducting film or in a 2D *XY* model, for $\mu > \mu^*$, the QPS–anti-QPS pairs remain bound together, and free QPSs do not proliferate. This describes the superconducting state. In the opposite case where $\mu < \mu^*$, the QPS–anti-QPS become unbound, and free QPSs proliferate, leading to the destruction of superconductivity and a normal electron metallic state.

The length scale for comparison is the plasmon thermal wavelength, or roughly the distance the plasmon travels in the thermal time, $l_{\mathrm{pl}} \sim c_{\mathrm{pl}} \hbar\beta$. Thus, for long wires, the condition $L \gg c_{pl}\hbar\beta$ must be satisfied. It is interesting to note that dissipation does not enter into the criterion for the transition point $\mu = \mu^* \approx 2$. The parameter μ is determined solely by the propagation impedance $Z \sim \sqrt{L_{kin}/C}$ of the plasmon mode (3.44). It is possible that this behavior could be modified by inclusion of the normal electrons in the phase-slip core, which provides a second channel for removing the dissipated energy in the QPS process, acting in parallel with the plasmons as has been suggested by Refael *et al.*, and Khlebnikov and Pryadko in Sections 3.4 and 3.5. The fact that at the lowest energies, the normal resistance does not appear in μ may be a result of a screening effect of QPS–anti-QPS, which is believed to occur for typical experimental nanowire systems, as was pointed out by Refael *et al.* [4]. Please see the next section. Nevertheless, it is worthwhile bearing in mind that there may be situations, in which the resistance of the normal core resistance could play an important role and modify the parameter μ.

For short wires where $L \ll c_{pl}\hbar\beta$, the scaling is cut off by the finite-size at $\rho = L$ (or $\ln \rho/\xi = \ln L/\xi$). Carrying out the renormalization up to this scale in x results in a fugacity $\tilde{f}_{QPS} = f_{QPS}(l = \ln L/\xi)$ and $\tilde{\mu} = \mu(l = \ln L/\xi)$. Using these as the starting value, the scaling is now carried out in the imaginary-time coordinate τ only. Analogous renormalization occurs for a single Josephson junction [21]. The scaling equations for the now 1D scaling are

$$\frac{\partial \mu}{\partial l} = -\pi^2 \mu^2 f_{QPS}^2 \,, \qquad \frac{\partial f_{QPS}}{\partial l} = (1-\mu) f_{QPS} \,,$$
$$l = \ln\left(\frac{\rho}{\xi}\right) = \ln \frac{\sqrt{x^2 + c_{pl}^2 \tau^2}}{\xi} \,. \tag{3.63}$$

For small initial fugacity $\tilde{f}_{QPS} \ll 1$, f_{QPS} scales to zero for $\tilde{\mu} > 1$, and QPSs do not proliferate, yielding a superconducting state. On the other side, for $\tilde{\mu} < 1$, f_{QPS} increases as the renormalization proceeds to longer length scales and a metal phase results as QPSs proliferate.

3.2.6 Wire Resistance and Nonlinear Voltage–Current Relations

The logarithmic interaction at a long space-imaginary time distance between QPSs, $\ln \rho/\xi$, gives rise to power law correlations and transport relations at low energies. This is well-known from the context of the KTB transition [21], and from tunneling into a Luttinger liquid in a 1D interacting fermionic system [33–35]. In the context of superconductivity, the Schmid transition occurs in the Josephson junction, which exhibits a $(1+0)$D, imaginary-time direction only, KTB quantum phase transition associated with dissipative quantum dynamics. The transport through the junction exhibits a voltage current relationship, which is the power law at low currents and temperature. Depending on whether the applied biasing current exceeds the thermal current scale of $2e/(\hbar\beta)$ or otherwise, the power law occurs in the voltage versus bias current, or in the linear resistance versus temperature at small currents [21]. Following Schön and Zaikin, the average voltage drop across the nanowire can be found from the decay rate of the nanowire configuration at a bias current, with a given initial phase difference between the ends, that is, given $\varphi(\mathsf{L}) - \varphi(0)$, to the configuration with one extra difference of 2π, and the reverse process, end with one less phase difference of 2π:

$$V = -\frac{\phi_0}{c}[\Gamma^{+2\pi}(I) - \Gamma^{-2\pi}(I)]\,, \quad \phi_0 = \frac{hc}{2e}\,. \tag{3.64}$$

The rate Γ_{QPS} is obtained from the imaginary part of the QPS contribution to the free energy:

$$\Gamma_{\text{QPS}} = 2\text{Im}\{F\}. \tag{3.65}$$

The leading QPS contribution to the free energy is from QPS–anti-QPS pairs obtained from the partition function in (3.59), yielding (with $x_0 \sim \xi$)

$$F_{\text{QPS}} = \frac{\mathsf{L} f_{\text{QPS}}^2}{x_0 \tau_0} \int_{\tau_0}^{\hbar\beta} \frac{d\tau}{\tau_0} \int_{x_0}^{\mathsf{L}} \frac{dx}{x_0} e^{\phi_0 I \tau/\hbar c - 2\mu \ln \rho/x_0}\,, \quad \rho = \sqrt{x^2 + c_{\text{pl}}^2 \tau^2}\,. \tag{3.66}$$

The forward ($+2\pi$) and reverse (-2π) decay rates are related by

$$\frac{\Gamma^{2\pi}(I)}{\Gamma^{-2\pi}(I)} = \exp\left(2\pi \frac{\hbar I}{(2e)k_{\text{B}}T}\right)\,. \tag{3.67}$$

For long wires with $\mathsf{L} \gg c_{\text{pl}}\hbar\beta$, the expression for the voltage drop is then

$$V = -\frac{\phi_0}{c\tau_0}\frac{\mathsf{L}}{x_0}\frac{\pi\Gamma(\mu - \frac{1}{2})}{\Gamma(\mu)\Gamma(2\mu - 1)} f_{\text{QPS}}^2 \sinh\left(\frac{\phi_0 I}{2ck_{\text{B}}T}\right) \times \left|\Gamma\left(\mu - \frac{1}{2} + \frac{i}{2\pi}\frac{\phi_0 I}{ck_{\text{B}}T}\right)\right|^2 \left(\frac{2\pi\tau_0 k_{\text{B}} T}{\hbar}\right)^{2\mu-2}\,. \tag{3.68}$$

Γ is the gamma-function. For completeness, the QPS core size, x_0, the imaginary timescale, τ_0, the QPS fugacity, f_{QPS}, and QPS interaction parameter, μ, are given by

$$\begin{aligned} x_0 \sim \xi &= \sqrt{D/(\hbar\Delta_0(T))}\,, \\ \tau_0 &\gtrsim \frac{\hbar}{\Delta_0}\,, \\ f_{\text{QPS}} &\approx \frac{S_{\text{core}}}{\hbar} e^{-S_{\text{core}}/\hbar}\,, \\ \mu &= \frac{\pi\hbar c^2}{(2e)^2 c_{\text{pl}}(L + L_{\text{kin}})} \approx \frac{\pi\hbar c}{(2e)^2}\sqrt{\frac{\pi e^2 D(E_{\text{F}}) D\Delta_0(T) C\mathsf{A}}{\hbar c^2}}\,, \end{aligned} \tag{3.69}$$

where the approximation for μ holds when the superconductor kinetic inductance for a unit length L_{kin} far exceeds the geometrical inductance for a unit length L, as is usually the case in a nanowire.

This expression yields power law behavior for both the linear resistance R versus T, and for the nonlinear voltage versus current

$$R \propto \frac{h}{(2e)^2} f_{\mathrm{QPS}}^2 \left(\frac{2\pi\tau_0 k_{\mathrm{B}} T}{\hbar}\right)^{2\mu-3} , \quad k_{\mathrm{B}} T \gg \phi_0 I/c ,$$

$$V \propto \frac{\phi_0}{c\tau_0} f_{\mathrm{QPS}}^2 \left[\frac{I}{(2e)/\tau_0}\right]^{2\mu-2} , \quad k_{\mathrm{B}} T \ll \phi_0 I/c . \tag{3.70}$$

These predicted power laws can be tested in experiments and are the subject of intense efforts in the experimental studies of superconducting nanowires [36].

The above scaling equation is valid in the superconducting phase, for $\mu > \mu^* \approx 2$. The linear resistance for the superconducting phase thus approaches zero ($R \to 0$) as $T \to 0$. This linear resistance is ideally measured at a current which approaches zero ($I \to 0$). This is an important prediction of the theory! Thus, below T_c, a 1D superconducting nanowire will always exhibit a finite residual resistance. At high temperatures close to T_c, the resistance is dominated by the thermally activated process, which generates phase slips, TAP, with a resistance given approximately by the LAMH expression in (2.125). At lower temperatures, when thermal phase slips are exponentially suppressed, QPSs begin to dominate. At all finite temperatures, the residual resistance remains finite, but eventually goes to zero as absolute zero is approached. This predicted behavior differentiates superconductivity in 1D nanowires from its behavior in higher dimensions, where in the limit of zero current, the resistance vanishes for an infinite-size sample!

For $\mu = 2$, the resistance becomes constant in temperature, and the wire is in the normal phase. For small values of μ corresponding to the normal phase, the phase slips proliferate, and the approximations made in the effective action, which does not include many-body effects for a strongly interacting gas of QPSs, become inadequate, and the scaling relations presented are no longer applicable.

The above expressions are applicable for sufficiently long wires. Since the plasmons are only well-defined on timescales shorter than the thermal timescale, ideally, the scaling behaviors are strictly obeyed when $\mathsf{L} \gg c_{\mathrm{pl}}\hbar\beta$, where the quantity $c_{\mathrm{pl}}\hbar\beta = c_{\mathrm{pl}}\hbar/(k_{\mathrm{B}}T)$ may be thought of as a thermal wavelength of the plasmons $\lambda_{\mathrm{pl}}(T)$. For experimental systems, λ_{pl} can be quite long at temperatures below 500 mK. For example, in aluminum nanowires, it is $\sim$ 25 μm at 300 mK. Thus, long wires exceeding $\sim$ 50 μm in length will have some intermediate temperature range significantly below T_c, where the long wire expressions may apply. Careful studies in such long wires are important in testing the proposed theoretical models.

Regarding the other extreme, in short wires where $\mathsf{L} \ll \hbar\beta$, the scaling as a function of $l = \ln \rho/\xi = \ln\left[\sqrt{x_0^2 + (c_{\mathrm{pl}}\tau_0)^2}/\xi\right]$ must be cut off in the spatial direction once the sample size L is reached. Starting from the microscopic space-imaginary timescale to L, the parameters scale to values $\tilde{\mu} = \mu(l = \ln \mathsf{L}/\xi)$ and $\tilde{f}_{\mathrm{QPS}} = f_{\mathrm{QPS}}(l = \ln \mathsf{L}/\xi)$ (3.63). Further scaling takes place in the imaginary-time space only, and thus in effectively a $(1+0)$ dimensional space-imaginary time, and the voltage–current relation becomes (with further scaling of μ denoted by μ')

$$V = \frac{\phi_0}{c\tilde{\tau}_0} \frac{2}{\Gamma(2\mu)} \tilde{f}^2_{\rm QPS} \sinh\left(\frac{\phi_0 I}{2ck_{\rm B}T}\right) \left|\Gamma\left(\mu' + \frac{i}{2\pi}\frac{\phi_0 I}{ck_{\rm B}T}\right)\right|^2 \times \left(\frac{2\pi\tilde{\tau}_0 k_{\rm B} T}{\hbar}\right)^{2\mu'-1} , \quad (3.71)$$

where $\tilde{\tau}_0 \sim \mathsf{L}C/(e^2\mu')$. The transition between the superconducting phase and the normal phase now occurs at $\mu' = 1$.

Accordingly, at any finite temperature in either the superconducting phase or the normal phase, there is a finite resistance at zero bias current. On the superconducting side, this linear resistance goes to zero as a power law in temperature:

$$R \propto \frac{h}{(2e)^2} \tilde{f}^2_{\rm QPS} \left(\frac{2\pi\tilde{\tau}_0 k_{\rm B} T}{\hbar}\right)^{2\mu-2} , \quad k_{\rm B}T \gg \phi_0 I/c , \quad (3.72)$$

with the voltage behaving as

$$V \propto \frac{\phi_0}{c\tau_0} \tilde{f}^2_{\rm QPS} \left[\frac{I}{(2e)/\tilde{\tau}_0}\right]^{2\mu-1} , \quad k_{\rm B}T \ll \phi_0 I/c . \quad (3.73)$$

3.3 Short-Wire Superconductor–Insulator Transition: Büchler, Geshkenbein and Blatter Theory

The results obtained for the QPT and scaling behavior by Golubev and Zaikin [3, 10] for short wires neglected the influence of the leads connected to the ends of the nanowire. When the leads are superconducting, they impose boundary conditions on the phase field $\varphi(x, \tau)$ at the two ends. Büchler, Geshkenbein, and Blatter investigated the effect of the boundary on the QPT in short wires [10]. They sought to explain the experimental findings of Bezryadin *et al.* [37] Bollinger *et al.* [38], and Lau *et al.* [39], where, in the case of short wires in the $Mo_{0.79}Ge_{0.21}$ nanowire system with $\mathsf{L} \lesssim 200$ nm, a superconducting behavior is observed when the normal state resistance of the nanowire $R_{\rm N}$ is less than the quantum resistance $R_{\rm Q}$, that is, $R_{\rm Q}/R_{\rm N} > 1$, while for $R_{\rm Q}/R_{\rm N} < 1$, the nanowire exhibits insulating behavior at low temperatures. In their model, Büchler *et al.* assumed that dissipation in short wires only acted between the ends of the nanowires, in the form of a shunt resistor $R_{\rm s}$ across the entire length, in analogy to the case of a resistively-shunted Josephson junction. Therefore, the superconductor–insulator transition (SIT) in dissipative quantum systems found by Schmid [40] is relevant in this model.

The model analyzed by Büchler *et al.* is depicted in Figure 3.1. Outside of the QPS core, an effective, phase-only model was assumed, keeping the lowest order terms in frequency ω and momentum q in (3.31), Section 3.2.1.2 and integrating out the Gaussian action from the amplitude fluctuation in Δ_1. The phase-only model, including the dissipative boundary term and current drive term, is thus

$$\frac{S_{1D}}{\hbar} = \frac{\mathsf{A}}{2}\int dx\,d\tau\left[\frac{\hbar C}{(2e)^2\mathsf{A}}\left(\frac{\partial\varphi}{\partial\tau}\right)^2 + \frac{\pi e^2 D(E_F) D\Delta_0}{(2e)^2}\left(\frac{\partial\varphi}{\partial x}\right)^2\right] + \frac{S_{diss}}{\hbar} + \frac{S_{bias}}{\hbar}\,,$$

$$\frac{S_{diss}}{\hbar} = \frac{R_Q}{2\pi R_s}\int\frac{d\omega}{4\pi}|\omega||\tilde{\varphi}_-(\omega)|^2\,,$$

$$\frac{S_{bias}}{\hbar} = \frac{1}{-(2e)}\int_0^{\hbar\beta} d\tau\, I\varphi_-\,. \tag{3.74}$$

The dissipative term reflecting the shunt resistance R_s across the wire involves the value of φ at the boundary. This differs from the form in the Zaikin–Golubev theory as it only involves the boundary values rather than the value along the nanowire. This is an assumption which enables one to reproduce the experimental findings of an SIT, and which will find justification from a more microscopic view point in the theory of Meidan, Refael, *et al.*, discussed in the next section (Section 3.4). There, when QPS–anti-QPS pairs occuring simultaneously, but separated in space (so-called dipoles) proliferate, their mutual interaction "screens" the conversion process between the superfluid and normal channels within the wire, and in effect, decouple them, leaving the shunting resistance to only act between the ends of the nanowire.

The nanowire has its ends at $x_L = -\mathsf{L}/2$ and $x_R = \mathsf{L}/2$, where subscripts "L" and "R" denote left and right, respectively. Such a placement rather than at "0" and "L" facilitates a decomposition of the φ field into an antisymmetric combination under spatial inversion:

$$\varphi_-(x,\tau) \equiv \varphi(x,\tau) - \varphi(-x,\tau)\,, \tag{3.75}$$

and a symmetric combination

$$\varphi_+(x,\tau) \equiv \varphi(x,\tau) + \varphi(-x,\tau)\,. \tag{3.76}$$

The term $\tilde{\varphi}_-(\omega)$ in (3.74) above is then the Fourier transform of $\varphi_-(\tau)$, the antisymmetric combination at the boundaries

$$\varphi_-(\tau) = \varphi_-(x,\tau)|_{x=\mathsf{L}/2} = \varphi(\mathsf{L}/2,\tau) - \varphi(-\mathsf{L}/2,\tau)\,. \tag{3.77}$$

$\varphi_-(\tau)$ represents the difference in phase between the two ends and couples to the environment via the boundary condition imposed there. The symmetric combination $\varphi_+(\tau)$

$$\varphi_+(\tau) = \varphi_+(x,\tau)|_{x=\mathsf{L}/2} = \varphi\left(\frac{\mathsf{L}}{2},\tau\right) + \varphi\left(-\frac{\mathsf{L}}{2},\tau\right) \tag{3.78}$$

accounts for charge accumulation in the wire, and the position-dependent $\varphi_+(x,\tau)$ represents the fluctuation of local charge from neutrality.

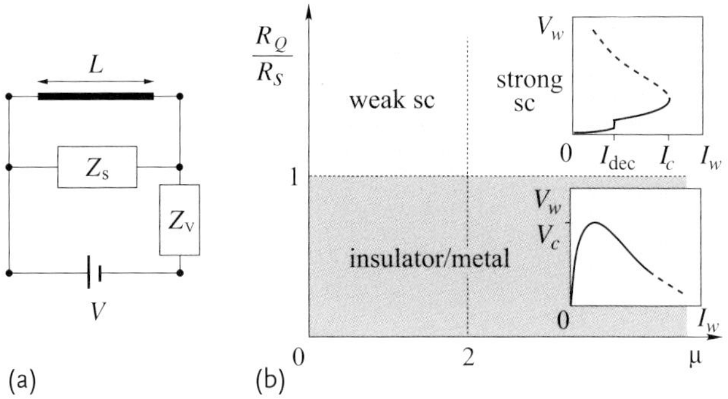

Figure 3.1 (a) Schematic diagram of a short, 1D superconducting nanowire driven by a constant currents source, consisting of a fixed voltage and a large impedance $Z_V(R_V)$, giving a bias current $I_{bias} = V/Z_V$. A shunt-impedance $Z_s(R_s)$ acts across the nanowire, and defines the environment. (b) Phase diagram with superconducting and insulating phases separated by a quantum phase transition at $R_Q/R_s = 1$. The superconducting phase splits into weak and strong regimes with a crossover at $\mu \approx 2$, the leftover of the superconductor–insulator transition in the infinitely long wire. The insets indicate the wire's I–V characteristic at $T = 0$. In the superconducting phase ($R_Q/R_s > 1$), the algebraic characteristic is dominated by the environment at small currents below the deconfinement current ($I < I_{dec}$). A highly conducting shunt (small R_s) with $R_Q/R_s \gg \mu$ enables one to probe the wire for currents exceeding I_{dec}. I_c denotes the critical current. The insulating phase exhibits a Coulomb gap behavior below the critical voltage V_c. From [10].

The above action for describing the dynamics outside of the QPS core with the inclusion of the terms S_{diss} and S_{bias} yield the boundary conditions

$$\frac{(-2e)}{2\pi}\frac{R_Q}{R_s}\int |\omega|\tilde{\varphi}_- e^{-i\omega t} d\omega + \frac{(-2e)}{2\pi}\mu c_{pl}\left(\frac{\partial\varphi}{\partial x}\right)_- = I\,, \quad \left(\frac{\partial\varphi}{\partial x}\right)_+ = 0\,;$$
$$\left(\frac{\partial\varphi}{\partial x}\right)_\pm \equiv \left.\frac{\partial\varphi_\pm(x,\tau)}{\partial x}\right|_{x=L/2}, \tag{3.79}$$

where $\mu = R_Q\sqrt{C/L_{kin}}/2$ (3.44). The first condition expresses current conservation – the sum of the normal current (first term on the left-hand-side) and the supercurrent (second term) equals the externally supplied current, while the second overall represents charge neutrality.

Similar to the case of long wires, accounting for QPS interaction enables the leading contributions of the partition function to be written in the following form (see (3.59)):

$$Z = Z_G \sum_{n=0}^{\infty}\left(\frac{(f_{QPS})^n}{n!2^n}\right)^2 \int \prod_{m=1}^{2n} \frac{d\tau_m}{\tau_0}\frac{dx_m}{c_{pl}\tau_0} \exp\left[\sum_{i\neq j} s_i s_j G(x_i,\tau_i;x_j,\tau_j)\right]. \tag{3.80}$$

Here, as before, $s_i = \pm 1$ represents the charge of the QPS, with ± 1 corresponding to a QPS and an anti-QPS, respectively. The contribution Z_G represents a Gaussian part. The kernel G must account for the influence of the boundary and is modified from the simple logarithm encountered for a long wire (3.58).

What remains is to examine the form of G in the limits of small ($R_Q/R_s \ll \mu$) and large ($R_Q/R_s \gg \mu$) dissipation. It will turn out that unlike the long wire case (3.61), where $\mu > 2$ causes the fugacity f_{QPS} to flow toward small values under renormalization, ensuring bound QPS–anti-QPSs and superconducting order, while $\mu < 2$ yields the opposite behavior of QPS proliferation and a metallic phase, here, the transition is determined by R_Q/R_s, with a phase boundary at the value $R_Q/R_s = 1$, in analogy to the Schmid transition for a dissipative quantum system [40]. The renormalization equations become

$$\frac{\partial f_{QPS}}{\partial l} = \left(1 - \frac{R_Q}{R_s}\right) f_{QPS}\,, \qquad \frac{\partial R_Q/R_s}{\partial l} = 0\,. \tag{3.81}$$

Thus, for large dissipation, that is, $R_Q/R_s > 1$, the shunt protects the superconducting nanowire, and superconducting order results as f_{QPS} scales toward small values. Insufficient dissipation, where $R_Q/R_s < 1$, enables free QPS to proliferate, and an insulator results.

To motivate how this comes about, it is instructive to first examine the behavior at low frequencies and subsequently, examine the consistency of the solutions in the different regimes to account for additional contributions from higher frequencies. At low frequencies, the boundary condition indicates that the normal current $(2e)/(2\pi)\ R_Q/(R_s)|\omega|\tilde{\varphi}_-$ in the Fourier transfer of (3.79) is negligible, and under vanishing drive $I \to 0$, no supercurrent enters or leaves the wire: $(-2e)/(2\pi)\mu c_{pl}\partial\varphi_-(x,\tau)/\partial x|_{L/2} = 0$. This indicates a Neumann boundary condition, for which the spatial derivative of φ_- at the ends vanishes. In Figure 3.2b, the solution corresponding to this Newmann boundary condition is depicted. At large separation between phase slips, where $|\tau_2 - \tau_1| > L/(\pi c_{pl})$, mutual screening of the QPS–anti-QPS pair is replaced by the screening by image charges, and a saturation of the action results, and the vortices become asymptotically free! Their contribution to the action is minimized by separating them toward to the wire ends, where their contribution to G vanishes.

What remains is the contribution from the environment through the values of φ_- at the ends. The imposed boundary condition modifies the QPS saddle point solutions. These may be obtained from the infinitely long wire saddle point configurations for the QPS and anti-QPS pair (see (3.55)) via a conformal mapping, giving

$$\begin{aligned}\varphi_{-,\mathrm{pair}}(\tau) = {} & 2\arctan\left\{\frac{\sinh\left[\frac{\pi c_{pl}}{L}(\tau-\tau_1)\right]}{\cos\frac{\pi x_1}{L}}\right\} \\ & - 2\arctan\left\{\frac{\sinh\left[\frac{\pi c_{pl}}{L}(\tau-\tau_2)\right]}{\cos\frac{\pi x_2}{L}}\right\}\,,\end{aligned} \tag{3.82}$$

where (x_1, τ_1) corresponds to the space-imaginary time position of the QPS, and (x_2, τ_2) to that of the anti-QPS. This leads to the following environmental contribution, G_e, toward G:

$$G_e(x_1, x_2; \tau_2 - \tau_1) \approx \frac{R_Q}{R_s} \ln \left[\frac{\cos^2 \frac{\pi}{2\mathsf{L}}(x_1 + x_2) + \left(\frac{\pi c_{pl}}{2\mathsf{L}}\right)^2 (\tau_2 - \tau_1)^2}{\cos \frac{\pi}{\mathsf{L}} x_1 \cos \frac{\pi}{\mathsf{L}} x_2} \right] . \quad (3.83)$$

The largest contribution comes from QPS–anti-QPS pairs that are located at the ends of the nanowire, with

$$G \approx G_e\left(x_1 = x_2 \approx \pm\frac{\mathsf{L}}{2}; \tau_2 - \tau_1\right) \approx \frac{R_Q}{R_s} \ln \frac{(\tau_2 - \tau_1)^2}{\tau_0^2} , \quad (3.84)$$

where we have inserted the short time cut-off τ_0. This dependence on the imaginary time only maps the problem onto the Schmid problem of a dissipative quantum mechanical system, with the resulting scaling behavior of (3.81) above.

The above solution corresponding to the Neumann boundary condition is valid for $R_Q/R_s \ll \mu$ when dissipation is small. In the opposite limit of large dissipation $R_Q/R_s \gg \mu$, the contribution exhibits different regimes of behaviors depending on the imaginary-time difference $\tau_2 - \tau_1$. For short times, high-frequency components must be considered, and the boundary condition for large ω and vanishing current drive I in (3.79) signifies a Dirichlet condition: $\varphi_- = 0$. The solutions are periodic in the wire length L, as shown in Figure 3.2a. The interaction between the QPS–anti-QPS pair now takes the form

$$G_{pair}(x_1, x_2; \tau_2 - \tau_1) = \mu \ln \left\{ \left(\frac{\mathsf{L}}{\pi c_{pl} \tau_0} \right)^2 \left(\sinh^2 \left[\frac{\pi c_{pl}}{\mathsf{L}} (\tau_2 - \tau_1) \right] + \sin^2 \left[\frac{\pi}{\mathsf{L}} (x_1 - x_2) \right] \right) \right\} . \quad (3.85)$$

Since the normal current is approximately zero with $\varphi_- = 0$ at the ends in accordance with (3.79), the dissipative term contribution vanishes in this high-frequency

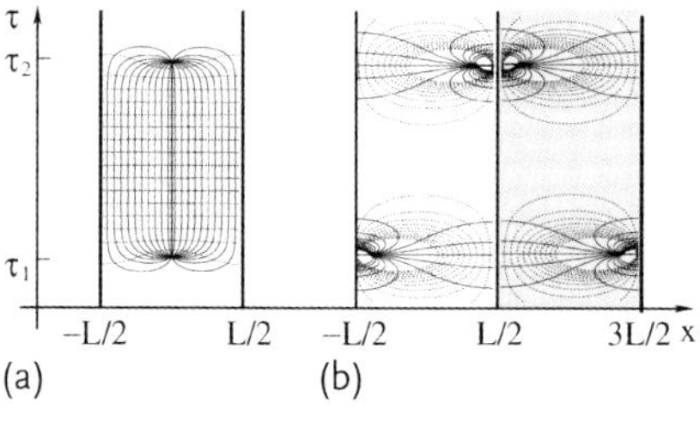

Figure 3.2 Phase-slip solutions: (a) with periodic boundary conditions (solid lines denote constant φ contours), and (b) with Neumann boundary conditions at intervortex distance $\mathsf{L}/c_{pl} < \bar{\tau}$. (a) For $R_Q/R_s \gg \mu$, defects are screened mutually; the 2π phase drop appears along the x-axis and drives a large current (dashed lines) through the highly conductive shunt; the resulting string confines defect pairs. (b) With $R_Q/R_s \ll \mu$, defects are screened individually by their mirror images (gray underlaid area); the 2π phase drop appears along the τ-axis and sets up displacement currents in the wire that cannot escape through the shunt; hence, charge accumulates at the boundary, resulting in a large voltage over the shunt. The defects are asymptotically free. From [10].

limit. Therefore, $G \approx G_{\text{pair}}$ starts out as a logarithm in space-time distance for short imaginary time

$$|\tau_2 - \tau_1| < \frac{R_Q}{R_s}\frac{\mathsf{L}}{\pi c_{\text{pl}}} : \quad G(|\tau_2 - \tau_1|) \sim \mu \ln \frac{c_{\text{pl}}^2(\tau_2 - \tau_1)^2 + (x_2 - x_1)^2)}{(c_{\text{pl}}\tau_0)^2}, \tag{3.86}$$

goes over to a linear dependence in intermediate times

$$\frac{\mathsf{L}}{\pi c_{\text{pl}}} < |\tau_2 - \tau_1| < \frac{R_Q}{R_s}\frac{\mathsf{L}}{\pi c_{\text{pl}}} : \quad G(|\tau_2 - \tau_1|) \sim \frac{2\pi\mu c_{\text{pl}}}{\mathsf{L}}(\tau_2 - \tau_1), \tag{3.87}$$

and ends with the long imaginary time-dependence for

$$|\tau_2 - \tau_1| > \frac{R_Q}{R_s}\frac{\mathsf{L}}{\pi c_{\text{pl}}} : \quad G \sim \frac{R_Q}{R_s} \ln \frac{(\tau_2 - \tau_1)^2}{\tau_0^2}. \tag{3.88}$$

Therefore, the final form at low energies and long imaginary time is solely determined by the environment through the shunt resistor R_s. The resultant phase diagram is shown in Figure 3.1b, with $R_Q/R_s = 1$ as the phase boundary between a superconducting and an insulating state in the SIT. The details within each phase are now more complex. Interested readers are referred to the original paper [10].

3.4 Refael, Demler, Oreg, Fisher Theory – 1D Josephson Junction Chains and Nanowires

Starting from a rather different model, Refael, Demler, Oreg, and Fisher (RDOF) [4] reached conclusions about the behavior of 1D superconducting nanowires similar to those of Zaikin, Golubev, *et al.* in the limit of very low temperatures. In addition, their model suggested more complex crossovers at intermediate temperatures. RDOF analyzed the behavior of a linear chain of Josephson junctions. The junctions are resistively shunted to model dissipation by the normal component of electrons, as shown in Figure 3.3, where the shunt resistor R runs in parallel to the Josephson junction in each cell. In addition, a conversion resistance, r, determines the rate of normal to Cooper pair conversion.

This model exhibits a rich phase diagram, as the Josephson coupling is varied between strong and weak; the dissipation parameters R and r affect different aspects of the phase diagram. Two different superconducting phases are possible, one with global phase coherence between junctions but phase and charge fluctuations on individual superconducting grains (SC-1), and the other with local phase order on individual grains (SC-2). In the end, the lattice parameter (a) of the discrete 1D chain is taken to zero to reach the continuum limit to model a continuous nanowire. For the nanowire, this model yielded an effective action which contains the phase degree-of-freedom only at the lowest temperature, similar to what was found by Zaikin, Golubev, *et al.*, leading to a KTB transition between an SC-1 phase and the normal metal (SMT) tuned by the superconducting stiffness via the wire diameter (stiffness proportional to parameter μ, (3.44)). At intermediate temperatures, a crossover to the more exotic SC-2 may occur for certain ranges of parameters.

The inclusion of the Cooper pair to normal electron conversion resistance r provided insight regarding the screening of dissipation by QPS–anti-QPS dipoles. This enabled the construction of a phenomenological model, in which for short wires, the shunt-resistance acts across the entire wire, in a manner analogous to a single, resistively-shunted Josephson junction [5, 10, 41, 42], as the screening from dipoles decouples the superconducting and normal channels within the nanowire. Consequently, an SIT takes place rather than an SMT. This treatment thus provides a justification for the assumptions of the Büchler, Geshkenbein, Blatter model, discussed in the previous section, Section 3.3, on short wires. Moreover, the resistance generated by the QPS–anti-QPS proliferation modifies the shunt resistance in a self-consistent way so that on the superconducting side, the linear residual resistance drops precipitously.

The resulting scenario provides a possible explanation of the experimental observations of Bezryadin *et al.* [37] and Bollinger *et al.*, [38], who found that in short $Mo_{0.79}Ge_{0.21}$, there is an apparent SIT at a critical normal state resistance R_N value equal to the quantum resistance $R_Q = h/(2e)^2 = h/(4e^2)$, where R_N acts as an effective shunt resistance R_s across the ends of the wire. This scenario is in agreement with the conclusion of Büchler, Geshkenbein, and Blatter (Section 3.3) which

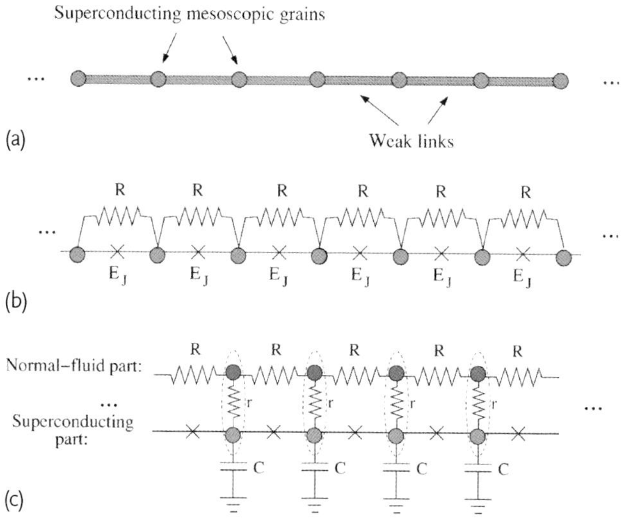

Figure 3.3 (a) A linear system of superconducting grains connected by weak links. The grains are coupled by the flow of both superconducting Cooper pairs and normal electrons. (b) The system in (a) is modeled by a 1D chain of superconducting grains connected by Josephson junctions and shunt resistors. (c) Two-fluid model: each mesoscopic grain is represented as a combination of a superconducting (SC) and a normal (N) part, depicted schematically as separate grains. Charge relaxation between the SC and N grains takes place via a conversion resistance r. From [4].

was obtained by assuming the presence of an effective shunting between the wire ends. Here, a more microscopic justification for that model emerges, in the form of screening of the dissipative QPS interaction by dipoles – pairs of QPS–anti-QPS.

This section is organized as follows: The details of the 1D Josephson junction chain model and the effective low-energy action are introduced in Section 3.4.1. The renormalization flow of the model is presented, as well as the connection of this discrete model to the case of a continuous nanowire. The scaling behavior of the 1D Josephson junction chain and the behavior in the continuum limit is discussed in Section 3.4.2. In the last subsection, Section 3.4.3, the behavior of short wires in relation to the SIT is presented.

3.4.1
Discrete Model of 1D Josephson Junction Chains

The model system in Figure 3.3 consists of a chain of identical mesoscopic superconducting grains which are linked by a coupling that allows the flow of both Cooper pairs and normal electrons. Such a system is typically modeled as an array of resistively shunted Josephson junctions (RSJJ). Each grain is envisioned to consist of two physically overlapping parts containing a superconducting grain (S) and a normal grain (N). Models of this type for 2D arrays have been extensively studied [43–45], also for 1D chains [46–48].

In the model considered by Refael *et al.*, the two types of grains are coupled capacitively, and in addition through a conversion process between Cooper pairs and the normal electrons, which is parameterized by the conversion resistance r [4, 41, 49]. In the previous studies of 1D chains, the analysis performed was for $r = 0$, equivalent to a conversion rate that is infinitely fast. This parameter r introduces interesting new behaviors in the phase diagram.

After an understanding is achieved for the model of a discrete Josephson junction chain, to make the connection with a continuous nanowire, the lattice constant (a) will be taken as zero.

The strategy for going over to a system of nanowires from the discretized two-fluid model is based on relations defined by the generalized Josephson equation for the superfluid current, Ohmic behavior for the normal current, and a relation for conversion:

$$\begin{aligned} \frac{\partial I_S}{\partial t} &= -Y \nabla V_S \,, \\ I_N &= -\sigma \nabla V_N \,, \\ i_{NS} &= \gamma (V_N - V_S) \,. \end{aligned} \tag{3.89}$$

The conversion current per unit length i_{NS} is related to the difference in normal and superfluid electrochemical potentials, V_N and V_S, respectively, via the conversion conductance per unit length $\gamma = 1/(ra)$.

Based on these ideas, which extend from the discrete model introduced below, Refael *et al.* arrived at similar conclusions as the Zaikin–Golubev theory at low temperature. In particular, they found a superconductor to normal metal transition (SMT), with a KTB transition between the SC-1 (SC^*) phase and a normal metallic phase. This results holds for long wires. Under unusual conditions of where phase-slip dipoles – QPS–anti-QPS pairs with small spatial separation – do not proliferate, the SC-2 (FSC) superconducting phase may occur. The two superconducting phases were uncovered in previous studies of discrete, 1D Josephson junction chains, in which the conversion resistance $r = 0$ [46–48].

In the SC-1 phase, dissipation is weak, the dissipation becomes screened by small-sized dipoles of phase slips and antiphase slips located on adjacent sites. At low energies, dissipation is nearly entirely screened out, and both the phases and the charges on individual grains exhibit sizable fluctuations. Nevertheless, the phase difference between grains does exhibit long-range order. The SC-2 phase occurs when dissipation is strong, and quantum fluctuations are suppressed on each individual grain to allow individual grains to be superconducting. The phases on neighboring grains are not strongly correlated, however, and superconducting correlations decay algebraically with distance. Because the SC-2 phase requires strong dissipation across each grain when the conversion resistance is large ($r > R_Q$), and thus the Cooper pair to normal electron conversion rate becomes slow, the normal grain resistance R across each individual junction becomes less effective in shunting the Josephson junction and can no longer stabilize superconductivity locally. Consequently, the SC-2 phase cannot occur for $r > R_Q$.

When $r \neq 0$, there exists a new temperature scale T^*. This scale is associated with the conversion process and may be parameterized as

$$T^* = \frac{\hbar}{C_{NS}} \left(\frac{1}{r} + \frac{1}{R} \right) . \tag{3.90}$$

Here, $C_{NS} \approx [D_S + D_N]^{-1}$ is the superconducting channel to normal channel capacitance on a grain, with D_S, D_N the inverse quantum compressibilities, and R the resistance in the normal channel. The significance of T^* is that below this scale, QPS proliferate. Moreover, in short wires connected to superconducting leads, due to the fast propagation time for the phase fluctuations (Mooij–Schön plasmon modes), QPS interaction may be modeled within a mean field treatment by an effective renormalized QPS impedance. This is significant as it offers one avenue for explaining experimental observations in short wires, in which Bezryadin, Bollinger *et al.* have uncovered an SIT, rather than an SMT.

The quantum Hamiltonian for the discrete 1D Josephson junction chain is written as a sum of three parts: the grain Hamiltonian, H_Q, the Josephson coupling between grains, H_J, and the dissipative part, H_{dis}, which includes the terms modeling the shunt resistance, R, and the Cooper pair to normal conversion resistance, r. To express the Hamiltonian in explicit form, Refael and coworkers introduced the following variables for the ith Josephson junction cell: the electrostatics potential ϕ_i, chemical potentials $\mu_{S,i}$ and $\mu_{N,i}$, electrochemical potentials $V_{S,i}$ and $V_{N,i}$, charges $Q_{S,i}$ and $Q_{N,i}$, respectively, for the superconducting and normal electron components, as well as the phases φ_i and χ_i for the superconducting and normal electrons. The charges and phases are conjugate variables, and thus satisfy the commutation relationships when quantized:

$$[Q_{S,i}, \varphi_i] = 2ie\delta_{ij}\,, \quad [Q_{N,i}, \chi_i] = ie\delta_{ij}\,,$$
$$[Q_{N,i}, \varphi_i] = [Q_{S,i}, \chi_i] = 0\,. \tag{3.91}$$

The charges, electrostatic potential, and electrochemical potentials are related by

$$\phi_i = \frac{Q_{S,i} + Q_{N,i}}{C}\,,$$
$$V_{S,i} = \phi_i + D_S Q_{S,i}\,, \quad V_{N,i} = \phi_i + D_N Q_{N,i}\,, \tag{3.92}$$

with C the self-capacitance of each Josephson junction grain, and the inverse compressibilities D_S and D_N related to the quantum contributions to the effective inverse capacitances. The inclusion of the inverse compressibilites is important in setting the timescale for charge relaxation between the superconducting and normal electrons, as well as the important energy scale $k_B T^*$.

The Hamiltonian is written as

$$H = H_Q + H_J + H_{\text{diss}}\,. \tag{3.93}$$

The grain Hamiltonian, H_Q, is given by

$$H_Q = \sum_i H_{Q_i}\,,$$
$$H_{Q_i} = \frac{1}{2}\left(\frac{1}{C} + D_S\right)\sum_i Q_{S,i}^2 + \left(\frac{1}{C} + D_N\right)\sum_i Q_{N,i}^2$$
$$+ \left(\frac{1}{C}\right)\sum_i Q_{S,i} Q_{N,i}\,. \tag{3.94}$$

The Josephson coupling between cells is

$$H_J = -E_J \sum_{ij} \cos(\varphi_i - \varphi_j)\,, \quad E_J = \frac{\hbar I_J}{2e}\,, \tag{3.95}$$

where the Josephson energy E_J is related to the Josephson current I_J. The dissipation has two components: ohmic dissipation within a grain characterized by the resistance R, and Cooper pair to normal electron conversation resistance r:

$$H_{\text{diss}} = \sum_{ij} H_{\text{diss}}(r, 2\chi_i - \varphi_j) + \sum_{ij} H_{\text{diss}}(R, 2\chi_i - 2\chi_j)\,. \tag{3.96}$$

The phases are periodic in the imaginary time τ, which has an interval $(0, \hbar\beta)$ as usual. The quantities R and r denote, respectively, the shunting resistor within a grain (Josephson cell), and the Cooper pair to normal electron conversion resistance. The origin of such ohmic dissipation is assumed, rather than derived from first principle.

The quadratic dependence on the charges $Q_{S,i}$ and $Q_{N,i}$ enable them to be integrated out in the partition function expressed in terms of the Euclidean action, yielding the starting point of the analysis of the quantum phase transitions, that is,

$$Z = \int \prod_i \mathcal{D}\varphi_i \mathcal{D}\chi_i \exp\left(-\frac{S_{1DJJ}}{\hbar}\right) ,$$
$$S_{1DJJ} = S_Q + S_{diss,r} + S_{diss,R} + S_J . \tag{3.97}$$

The various contributions to the actions are

$$\frac{S_Q}{\hbar} = \frac{1}{2}\int_0^{\hbar\beta} d\tau \sum_i \frac{1}{D_S D_N} \frac{1}{1/D_S + 1/D_N + C} \times \left\{\left[\frac{\partial\varphi_i(\tau)}{\partial\tau} - \frac{\partial\chi_i(\tau)}{\partial\tau}\right]^2 + C D_N \left[\frac{\partial\varphi_i(\tau)}{\partial\tau}\right]^2 + C D_S \left[2\frac{\partial\chi_i(\tau)}{\partial\tau}\right]^2\right\} , \tag{3.98}$$

$$\frac{S_{diss,r}}{\hbar} = \hbar\beta \sum_{\omega_n} \left[\frac{|\omega_n|}{r} |\tilde{\varphi}_i(\omega_n) - 2\tilde{\chi}_i(\omega_n)|^2\right] , \tag{3.99}$$

$$\frac{S_{diss,R}}{\hbar} = \hbar\beta \sum_{\omega_n} \sum_{ij} \left[\frac{|\omega_n|}{R} |2\tilde{\chi}_i(\omega_n) - 2\tilde{\chi}_j(\omega_n)|^2\right] , \tag{3.100}$$

and

$$\frac{S_J}{\hbar} = -E_J \int_0^{\hbar\beta} d\tau \sum_{ij} \cos(\varphi_i - \varphi_j) . \tag{3.101}$$

In the $T = 0$ limit, the sum over the Matsubara frequencies $\hbar\beta \sum_n \tilde{f}(\omega_n)$, where $\tilde{f}$ is any function of the frequencies $\omega_n = 2\pi n k_B T/\hbar$, may be replaced by the integral $\int_0^\infty d\omega \tilde{f}(\omega)/(2\pi)$.

With the inclusion of the Cooper pair to normal electron conversion, embodied in the term $S_{diss,r}$ in the action, the phase diagram is significantly more complex compared to the case $r \to 0$, for which the effective charges in the chain of the Josephson junction yielding the superconducting and normal currents are perfectly mixed. In Figure 3.4, we present the phase diagram for the system of 1D-chains of Josephson junctions, for different values of r/R_Q, as a function of the superconducting stiffness parameter $K \equiv 2\pi\sqrt{E_J/E_C}$ ($E_C = (2e)^2/C$ is the dominant Coulomb energy scale) and R/R_Q. Note that the stiffness K is essentially the parameter μ in the Zaikin–Golubev *et al.* theory (3.44), with the correspondence of $K/2 \leftrightarrow \mu$.

The SC-1 phase occurs when the Josephson coupling between grains is strong. In this limit, the action can be cast into a more convenient form. Refael *et al.* first derived a Coulomb gas representation of the n-QPS contribution to the action, n

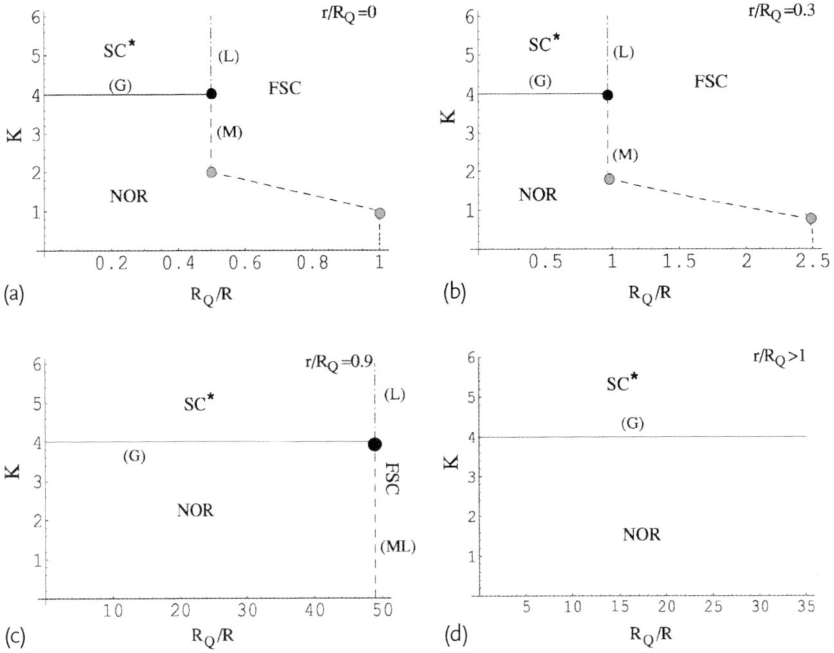

Figure 3.4 Phase diagram of the quantum two-fluid model of Figure 3.3c as a function of the shunt resistance R and the superconducting stiffness $K = 2\pi\sqrt{E_J/E_C}$, for various values of the Cooper pair to normal electron conversion resistance r. The phase boundaries between the SC-2 (FSC), SC-1 (SC*), and normal phases are of differing nature: (G) indicates global (solid line), (L) local (dashed solid line), and (M) mixed (dashed lines). These come together at a multicritical point (black dot). The FSC-NOR phase boundary has sections with three different characters, separated by bicritical points (gray dots). From [4].

being the number of QPSs:

$$\frac{S_{\text{nQPS}}(\{s_i\})}{\hbar} = \frac{1}{2}\sum_{i \neq j}^{n} s_i s_j\, G(x_i - x_j, \tau_i - \tau_j)\,,$$

$$G(x,\tau) \approx -2\pi\sqrt{\frac{E_J}{E_C}}\ln\frac{\sqrt{x^2 + (c_{\text{pl}}\tau)^2}}{c_{\text{pl}}\tau_0} - \frac{R_Q}{\sqrt{R^2 + 4rR}}\,e^{-|x|/\lambda_Q}\ln\frac{|\tau|}{\tau_0}\,. \tag{3.102}$$

Here, the indices i or j label the ith or jth phase slip, x_i is the position of the grain at which the ith phase slip occurs along the Josephson junction chain, τ_i the time variable for that phase-slip event, s_i is the charge of the QPS, with $+1$ for a PS and -1 for an anti-PS. The approximation for G holds in the limit $T \ll T^*$, and the second, spatially localized term containing the factor $e^{-|x|/\lambda_Q}$, with

$$\lambda_Q = a\left[\ln\frac{2r}{R + 2r - \sqrt{R^2 + 4rR}}\right]^{-1}, \tag{3.103}$$

reflects the contribution introduced by the finite Cooper pair to normal conversion resistance r. The quantity $\sqrt{R^2 + 4rR}$ represents an effective resistance seen by one-phase slip on a junction due to its shunt resistance R and the two conversion resistances r, one on each side, the latter in parallel with the left and right chain resistive networks. In previous studies, this spatially localized term is completely local, as $\lambda_Q \to 0$ for $r \to 0$, and the exponential factor becomes equivalent to a delta function $e^{-|x|/\lambda_Q} \to \delta(x)$. The above form follows from a standard Villain transformation [50] of the Josephson contribution $\exp \int d\tau\, E_J\{\cos[\varphi_{i+1}(\tau) - \varphi_i(\tau)]\}$. Using the Villain transformation, the QPS partition function is written as [4]

$$Z_{QPS} = \sum_{\{s_i\}} \left(\frac{f_{QPS}}{2}\right)^n \exp\left[-\frac{S_{nQPS}(\{s_i\})}{\hbar}\right] . \tag{3.104}$$

A dual representation of the action in a sine-Gordon form was derived next, which decouples the contributions of the two terms in $G(x, \tau)$, by the introduction of Hubbard–Stratonovich fields, θ_i and ψ_i located on the junctions between grains $(i, i+1)$ [49, 51]. The variables ψ_i enable the decoupling of the normal degrees of freedom, while $\theta_i + \psi_i$ decouple the superconducting degrees of freedom. The sine-Gordon dual representation is given by

$$\begin{aligned}\frac{S_{SG}}{\hbar} = &\int_0^{\hbar\beta} d\tau \left[\frac{E_C}{8\pi^2\hbar} \sum_i (\theta_{i+1} - \theta_i)^2 + \frac{\hbar}{2E_J} \sum_i \left(\frac{\partial\theta_i}{\partial\tau}\right)^2\right] \\ &+ \frac{\hbar\beta}{2\pi} \sum_n \frac{|\omega_n|}{2} \left[\frac{R}{R_Q} \sum_i |\psi_i(\omega_n)|^2\right. \\ &+ \left.\frac{r}{R_Q} \sum_i |\psi_{i+1}(\omega_n) - \psi_i(\omega_n)|^2\right] \\ &- f_{QPS} \int_0^{\hbar\beta} \frac{d\tau}{\tau_0} \sum_i \cos(\theta_i + \psi_i) .\end{aligned} \tag{3.105}$$

In addition, the importance of phase-slip dipoles is emphasized, the simplest of which occurs on adjacent sites and at nearly the same imaginary-time τ. The effect of the dipoles is captured in the dipole action:

$$\frac{S_{QPS,dpl}}{\hbar} = \sum_{p=1}^{\infty} f_{dpl,p} \sum_i \int \frac{d\tau}{\tau_0} \cos\{[\theta_i(\tau) + \psi_i(\tau)] - [\theta_{i+p}(\tau) + \psi_{i+p}(\tau)]\} . \tag{3.106}$$

This term is added explicitly, as it is generated in the renormalization iterations, starting from zero value of the fugacities $f_{dpl,p}$. The partition function, in this strong Josephson coupling limit, now reads

$$Z = \int \mathcal{D}[\theta]\mathcal{D}[\psi] \exp\left\{-\frac{S_{SG}[\theta, \psi]}{\hbar} - \frac{S_{QPS,dpl}[\theta, \psi]}{\hbar}\right\} . \tag{3.107}$$

Renormalization group equations can be derived, which yield the type of phase diagram shown in Figure 3.4. The varieties of transitions can be quite complex, involving the normal, SC-1, and SC-2 phases. The SMT between the SC-1 and normal phases is of greatest relevance to nanowires at the lowest temperatures, while a crossover to the SC-2 may be relevant at intermediate temperature. Regarding the normal phase in the SC-1 to normal transition, it suffices to consider the relevant parameters f_{QPS}, the single QPS fugacity, and the $f_{\text{dpl},p}$ dipole fugacity for the small-size and least costly dipole, namely, $f_{\text{dpl},p=1}$. In the normal phase, the renormalization flow drives these to larger values, and phase slips proliferate; as the smallest dipoles proliferate the larger ones will follow. On the other side of the transition, in the SC-1 phase, f_{QPS} is small and irrelevant, and individual, unbound QPSs do not proliferate. However, the dipole fugacities $f_{\text{dpl},p}$ are relevant; they proliferate, but not to the extent that they bind into quadrupoles. For most values of starting parameters, the fluctuation in ψ is small, and the QPS dipoles screen out the dissipation at the low energies of long wavelengths, and dissipation becomes irrelevant. The effective action can then be cast into the form for standard 1D chains of Josephson-coupled grains with self-capacitance in terms of θ only:

$$\frac{S_{\text{SC-1}}}{\hbar} \approx \int \frac{d\omega}{2\pi} \int \frac{dq}{2\pi} \left(\frac{\tilde{E}_C q^2}{8\pi^2 \hbar} + \frac{\hbar}{2\tilde{E}_J} \right) \theta^2 - f_{\text{QPS}} \int_0^{\hbar\beta} d\tau \sum_i \cos\theta_i(\tau) \,, \tag{3.108}$$

where a continuum limit is permitted as we are only interested in long wavelength behavior, at scales much longer than the grain separation a. Here, the tildes over the parameters indicate that their values have been renormalized. The RG scaling equations for the stiffness K and fugacity f_{QPS} are

$$\begin{aligned} \frac{dK}{dl} &= -\frac{\pi^2}{2} K^2 f_{\text{QPS}}^2 , \\ \frac{d f_{\text{QPS}}}{dl} &= f_{\text{QPS}} \left(2 - \frac{K}{2} \right) , \\ K &= \sqrt{\frac{E_J}{(E_C + 4\pi\hbar/\tau_0 \, f_{\text{dpl}})}} \,, \quad dl = -\frac{d\Omega}{\Omega} \,, \end{aligned} \tag{3.109}$$

with $d\Omega$ the high-energy frequency shell to be integrated out in the renormalization process. These scaling equations are identical to those of Zaikin–Golubev *et al.* for long wires (3.61), with the correspondence $K/2 \leftrightarrow \mu$ (3.44). The fixed line delineating the boundary between the SC-1 and normal phases occurs at $K_c = 4 + O(f_{\text{QPS}})$, as shown in Figure 3.4.

3.4.2
Resistance of the Josephson Junctions and the Nanowire

In a single Josephson junction shunted by a resistance R_s, the scaling equation for fugacity has the familiar form [21, 40]

$$\frac{d f_{QPS}}{dl} = \left(1 - \frac{R_Q}{R_s}\right) f_{QPS}\,, \quad l = \ln \frac{\Omega_0}{\Omega}\,, \tag{3.110}$$

where Ω is the highest remaining energy scale, and the initial starting scale $\Omega_0 \sim \sqrt{E_J E_C}/\hbar$ is on the order of the plasma frequency of the Josephson junction.

For $R_Q/R_s < 1$, the fugacity f_{QPS} grows, and QPSs proliferate, and superconductivity is destroyed by excessive phase fluctuations. On the other side of the QPT, for $R_Q/R_s > 1$, f_{QPS} decreases and scales toward zero. The QPS remain bound as QPS–anti-QPS pairs, and the junction is superconducting. The junction resistance can be found in a manner similar to the nanowire case (Section 3.2.6) [21], and is given, in the limit of the zero bias current, by

$$R_{JJ} \sim R_Q f^2_{QPS,0} \left(\frac{k_B T}{\hbar \Omega_0}\right)^{2R_Q/R_s - 2}\,, \tag{3.111}$$

where $f_{QPS,0}$ is the initial bare fugacity at the scale of the lattice, a. On the superconducting side, the resistance scales to zero as a power law of T for T approaching absolute zero. At the transition, $R_Q/R_s = 1$, the resistance is constant, behaving as a normal metallic conductor. For $R_Q/R_s < 1$, the junction becomes insulating, and the current is shunted through the shunt resistor.

In the case of the 1D JJ chain, the resistivity should account for an extra factor of the renormalized size of the QPS, $a \to a(\hbar\Omega_0/(k_B T))$, and for the vicinity of the SC1 $-$ N transition relevant to nanowires, scales as

$$\rho_{1DJJ} \sim \left[\frac{R_Q}{a}\left(\frac{\hbar\Omega_0}{k_B T}\right)\right] f^2_{QPS,0} \left(\frac{\hbar\Omega_0}{k_B T}\right)^{2-K} = \left(\frac{R_Q}{a}\right) f^2_{QPS,0} \left(\frac{k_B T}{\hbar\Omega_0}\right)^{K-3}\,. \tag{3.112}$$

On the superconducting side, where $K \geq 4$, the resistivity scales to zero as $T \to 0$. On the insulating side $K < 4$, there is a nonmonotonic regime where $4 > K > 3$, where the resistivity first decreases with decreasing T, before crossing over at a lower temperature, and reversing the direction and thus increasing, and approach the value R/a for the R-shunted 1D chain, as $T \to 0$.

In the crossover intermediate temperature regime delineated below, the SC-2 phase may be relevant. There, the scaling associated with the resistance across individual junctions behaves as

$$\rho_{SC\text{-}2} = \frac{R_{JJ}}{a} \sim \left(\frac{R_Q}{a}\right) f^2_{QPS,0} \left(\frac{k_B T}{\hbar\Omega}\right)^{K + 2R_Q/\sqrt{R^2+4rR} - 3}\,, \tag{3.113}$$

where the quantity $R_Q/\sqrt{R^2 + 4rR}$ is the parameter for the spatially localized logarithmic QPS interaction in imaginary time in the QPS action (3.102).

In a nanowire, one should take the limit where the lattice size $a \to 0$. Therefore, while taking care to maintain the finite superconducting coherence length ξ, multiple QPSs can now occur arbitrarily close to each other. The finite ξ means that the QPS is now spread over many lattices (with $a \to 0$). The dissipation interaction represented by the localized logarithmic interaction is the only mechanism which suppresses the proliferation of QPS dipoles at low energies. In taking a to zero, with a small size cut-off on the order of the Fermi wavelength $\lambda_F \ll \xi$, the minimum-size dipoles can have QPS–anti-QPS pairs at very small spatial separations, drastically improving screening to the dissipative interaction. Therefore, this continuum limit favors dipole proliferation and the SC1 phase is the only superconducting phase, which is relevant in a smooth and uniform nanowire at sufficiently low energies and temperatures.

Perhaps most dramatic and interesting, it is possible to envision a scenario in which the SC-2 phase may become relevant for nanowires. In this phase, the dissipation is relevant, and the scaling power law exponent reflects its importance in (3.113) above. This crossover to an SC-2 phase occurs at intermediate temperatures. To go from an SC-1 phase at low temperatures to an SC-2 phase at an intermediate temperature T_D (3.116) below, one must examine how the smallest-size, $p = 1$ dipole fugacity, given by $f_{dpl,1}$, scales, since the single QPS fugacity f_{QPS} is irrelevant in either superconducting phases. In the SC-1 phase, dipoles proliferate and $f_{dpl,1}$ is relevant, while in the SC-2, it is irrelevant.

The dissipation mediates a contribution to the $p = 1$ action of

$$2\beta \ln \frac{|\tau_2 - \tau_1|}{\tau_0}, \quad \beta = 2\frac{R_Q}{\sqrt{R^2 + 4rR}} \times \frac{\sqrt{R^2 + 4rR} - R}{2r}. \tag{3.114}$$

The corresponding scaling equation for $f_{dpl,1}$ is

$$\frac{d f_{dpl,1}}{dl} \approx f_{dpl,1}(1-\beta), \quad f_{dpl,1} = f_{dpl,lo}\, e^{(1-\beta)l}, \tag{3.115}$$

where $f_{dlp,lo} \sim f_{QPS}^2$ is the value following an initial rapid regime. We are interested in the region where $\beta < 1$ below the critical value where the SC-2 state becomes unstable. The crossover occurs at the intermediate temperature scale T_D, where $f_{dpl,1}$ becomes comparable to Ω. Inverting the relationship

$$\frac{f_{dpl,1}(T)}{f_{QPS}^2} \sim \frac{\Omega(T)}{\Omega_0} \sim \frac{k_B T_D}{\hbar \Omega_0}, \tag{3.116}$$

yields

$$k_B T_D \sim \hbar \Omega_0 (f_{QPS})^{-2/(1-\beta)}. \tag{3.117}$$

This assumes that $f_{dpl,1}$ initially increases rapidly and becomes of the order of f_{QPS}^2 before crossing over to the scaling in (3.115) above.

In the next section, the issue of the short distance cut-off for dipoles will be further discussed, in relation to the SIT transition.

The KTB SC-1 to normal metal SMT described by the renormalization flow (3.110) was predicated on the screening of the dissipation at low energies and long wavelengths. For finite length nanowires, there is a cut-off imposed by the wire length. Such issues, as well as the behavior of the propagation of the Mooij–Schön plasma mode, subject to boundary conditions if the 1D JJ chain or nanowire is connected to large superconducting leads, will modify the behavior of the QPT, as described below and in Section 3.5.3. The length scale of comparison is the thermal wavelength of the plasmon mode, $c_{\mathrm{pl}}\hbar/(k_{\mathrm{B}}T)$. The issue is whether it is appropriate to account for the effect of dissipation from the individual shunt resistors, by shunting the nanowire between its ends with the total resistance, as was suggested by Büchler *et al.* [10], to account for the SIT observed in short $\mathrm{Mo}_{0.79}\mathrm{Ge}_{0.21}$ nanowires [37, 38].

3.4.3
Mean Field Theory of the Short-Wire SIT

Building on their analysis of the 1D Josephson junction chain with a nonzero Cooper pair to normal electron conversion resistance ($r \neq 0$), Meidan, Refael and coworkers proposed a mean field theory to account for the behaviors of short wires [5, 41, 42]. Making use of the insight gained from an understanding of the screening effect of QPS dipoles discussed in the previous section, they conclude that local coupling to the dissipative interaction between the superconducting and normal electrons can be screened out at low energies, leaving an effective shunt resistor across the entire short nanowire. This means that an additional effective capacitor in series with r now blocks any current from escaping the superconducting channel into the normal channel at low frequencies with the wire. This conclusion provides justification of the assumptions of the work of Büchler *et al.* [10], where they assumed such a scenario, leading to a QPT analogous to a single resistively-shunted Josephson junction for short nanowires, for which $\mathsf{L} \ll c_{\mathrm{pl}}\hbar/(k_{\mathrm{B}}T)$. Moreover, the effective contribution to the overall shunt resistance by the QPSs sharpens the drop-off of the nanowire resistance below T_{c}, in a manner consistent with experimental findings.

Experimentally, Bezryadin and co-workers found a significant difference in the behavior of $\mathrm{Mo}_{0.79}\mathrm{Ge}_{0.21}$ nanowires that are short, with lengths between 50–200 nm, from the longer nanowires with lengths from 200 nm, up to 1 µm in length [37–39]. The coherence length for such nanowires is typically between 10–20 nm. Focusing on the short nanowires, it was found that those with normal state resistance R_{N} exceeding R_{Q} ($R_{\mathrm{N}} > R_{\mathrm{Q}}$) will become insulating as the temperature goes toward absolute zero, while, nanowires with $R_{\mathrm{N}} < R_{\mathrm{Q}}$ exhibit superconducting behavior down to the lowest measured temperatures $T \sim 50\,\mathrm{mK}$. Therefore, experimentally, there appears to be a superconductor–insulator transition (SIT) at $R_{\mathrm{N}} \approx R_{\mathrm{Q}}$, rather than a superconductor-metal transition (SMT). Moreover, the linear resistance below T_{C} for the superconducting nanowires drops off toward zero resistance much more sharply than the power law-type behavior predicted in the standard KTB-type QPT.

In their mean field analysis, Meidan *et al.* [5, 42] and Refael *et al.* [41] constructed a phenomenological theory, which showed that the behavior of short nanowires in the regime of relevance, specifically, $\xi < \mathsf{L} < \xi\sqrt{N_\perp}$, parallels that of a single Josephson junction exhibiting the Schmid-type SIT [40, 52]. Here, $N_\perp$ is the number of transverse electronic channels (or modes) in the width and height directions.

A single Josephson junction, shunted by a resistor R_s, is well-known to exhibit an SIT transition at $R_s \approx R_Q$ [53]. When $R_s > R_Q$, phase fluctuation is large, as Coulomb blockade sets in, and the Cooper pair number becomes a good quantum number on the grain. In the opposing situation of $R_s < R_Q$, a supercurrent can be sustained through the junction, and the junction is superconducting.

The physical significance of $N_\perp$ in this context lies in the fact that it is the ratio of the velocity of the propagation of the low energy collective modes associated with the phase fluctuations, that is, the Mooij–Schön plasmon mode, to the velocity of propagation for the disturbance of the order parameter magnitude. From (3.31) and also (3.32) and (3.40) in the small q and ω limit, the respective velocities are

$$c_{\mathrm{pl}} = \sqrt{\frac{\pi e^2 D(E_{\mathrm{F}}) D\Delta_0 \mathsf{A}}{\hbar C}}\,, \quad c_\Delta = \sqrt{\frac{3\pi D\Delta_0}{2\hbar}}\,,$$

$$\frac{c_{\mathrm{pl}}}{c_\Delta} = \sqrt{\frac{2e^2 D(E_{\mathrm{F}}) 3\mathsf{A}}{C}} \sim \sqrt{\frac{k_{\mathrm{F}}^2 \mathsf{A}}{\pi^2}} = \sqrt{N_\perp}\,. \tag{3.118}$$

Because the nanowire is short and the phase fluctuations propagate at such a high speed that the plasmon mode can traverse the entire length of the nanowire in a time much shorter than that given by the inverse of the energy gap, that is, $\mathsf{L} \ll c_{\mathrm{pl}}\hbar/\Delta = \xi\sqrt{N_\perp}$, one expects that phase slips occurring within the nanowires are largely overlapping and indistinguishable from each other.

Whereas in the situation of very short nanowires, $\mathsf{L} \ll \xi$, the system can be mapped into a Josephson junction [5, 10, 41, 42], the present case of $\xi < \mathsf{L} < \xi\sqrt{N_\perp}$ involves a more sophisticated model. Building on their analysis with a finite r discussed in Section 3.4.1, Refael and Meidan *et al.* were able to establish that the fast propagation of the phase disturbance leads to an interaction between them, so that in the presence of a dense population of phase-slip dipoles, the interaction acts to modify the resistance shunting other phase slips. More significantly, it is argued that at sufficiently low temperatures, the superconducting and normal electrons decouple due to such phase-slip interactions, rendering the local dissipation unimportant. Consequently, the shunt resistance only acts between the ends of the nanowire! The situation thus becomes analogous to a resistively shunted Josephson junction.

A two-cell Josephson junction is instructive [41, 49]. There, for an infinitely fast Cooper pair to normal conversion, corresponding to $r = 0$, each junction acts independently in the $\omega \to 0$ limit, and each undergoes a Schmid transition when its respective shunt resistance equals R_Q. The two junctions become coupled when $r \neq 0$, but QPS dipoles – a QPS on one junction and an anti-QPS on the other at nearly the same time – cannot block the supercurrent through the junctions, as the voltage pulses they generate along the superconducting channel are of opposite polarities, and thus cancel. However, the frequent occurrence of phase slips generates a large effective resistance between the super fluid and normal channels within each junction cell; this resistance blocks the conversion current which runs in opposite directions on the two sites. This blockade is equivalent to a decoupling of the superconducting and normal electrons. Thus, the two junctions act as if they are shunted in the normal channel across the entire length of the two-junction system.

In a long 1D Josephson junction chain of lattice constant a, shunt resistance R on each junction and conversion resistance r, the QPS interaction leads to a pair contribution to the effective action (3.102)

$$\frac{S_{2\text{QPS}}}{\hbar} = -s_1 s_2 \left\{ 2\pi\sqrt{\frac{E_\text{J}}{E_\text{C}}} \ln\left[\frac{\sqrt{x^2+(c_\text{pl}\tau)^2}}{a}\right] + \left(\frac{R_\text{Q}}{\sqrt{R^2+4rR}}\right) e^{-|x|/\lambda_\text{Q}} \ln\frac{|c_\text{pl}\tau|}{a} \right\}, \qquad (3.119)$$

where $s_i = +1\ (-1)$ corresponds to the effective charge of a QPS (anti-QPS), and the charge relaxation length $\lambda_\text{Q} = \max\{a\sqrt{r/R} = \sqrt{1/(\gamma\rho)}, a\}$. Such dipole terms must be explicitly included and are generated in the renormalization procedure from the single QPS terms. The action is transformed into a sine-Gordon representation by a dual transformation, into ψ and θ fields which, in the combination $e^{i(\psi_i+\theta_i)}$, creates a QPS on junction i. The sine-Gordon representation is given by (3.105)–(3.107)

$$\begin{aligned}\frac{S}{\hbar} &= \int_0^{\hbar\beta} d\tau \left[\frac{E_\text{C}}{8\pi^2\hbar}\sum_i(\theta_{i+1}-\theta_i)^2 + \frac{\hbar}{2E_\text{J}}\sum_i\left(\frac{\partial\theta_i}{\partial\tau}\right)^2\right] \\ &+ \frac{\hbar\beta}{2\pi}\sum_n\frac{|\omega_n|}{2}\left[\frac{R}{R_\text{Q}}\sum_i|\psi_i(\omega_n)|^2\right. \\ &\left.+\frac{r}{R_\text{Q}}\sum_i|\psi_{i+1}(\omega_n)-\psi_i(\omega_n)|^2\right] - f_\text{QPS}\int_0^{\hbar\beta}\frac{d\tau}{\tau_0}\sum_i\cos(\theta_i+\psi_i) \\ &+ \sum_{p=1}^{\infty} f_{\text{dpl},p}\sum_i\int\frac{d\tau}{\tau_0}\cos\{[\theta_i(\tau)+\psi_i(\tau)]-[\theta_{i+p}(\tau)+\psi_{i+p}(\tau)]\},\end{aligned} \qquad (3.120)$$

where for large frequencies ω, $f_{\text{dpl},2} \sim f_\text{QPS}^2$.

The above effective action has a KTB transition between the SC1 and normal states at $K = 2\pi\sqrt{E_\text{J}/E_\text{C}} \approx 4$. The QPS dipoles proliferate when

$$2R_\text{Q}\left[1-\exp\left(-\frac{a}{\lambda_\text{Q}}\right)\right] < R\,. \qquad (3.121)$$

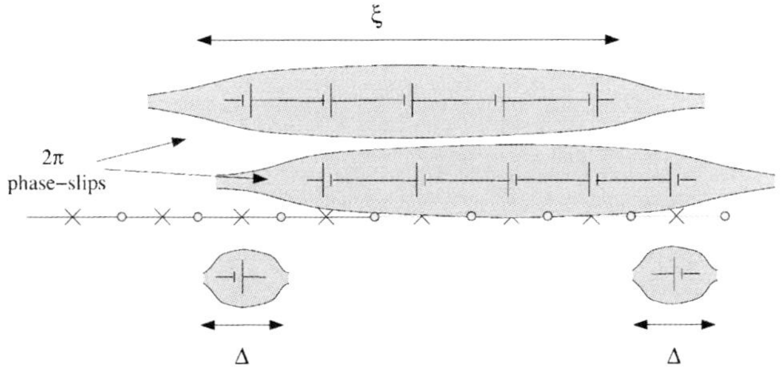

Figure 3.5 To describe a continuous wire, starting with the discrete 1D chain, the lattice constant a is taken to zero ($a \to 0$), while keeping the coherence length and size of the phase slip, ξ, fixed. This leads to a scenario where the phase slips are spread out over a large number of ξ/a junctions. The phase slips are thus able to form dipoles with spatial separation $x < \xi$. The voltage signs symbolize the voltage drop caused by a phase slip; thus phase slips have one polarity, while antiphase slips have the opposite polarity. From [41].

In the continuum limit of $a \to 0$ to model a continuous nanowire, the factor in brackets indicates that the dipoles always proliferate. As this limit is taken, the QPS size in the 1D Josephson junction chain also shrinks to zero. To account for the finite size of a QPS core in a nanowire, which is of order ξ, the superconducting coherence length, an envelope function of extent ξ, is employed. This signifies that the QPS is spread over a large number of $\sim \xi/a$ junctions. Please see Figure 3.5. This spatial smearing can be incorporated by using a smearing function

$$\cos(\psi(x,\tau) + \theta(x,\tau)) \to \cos\left(\frac{\xi}{a}\right) \sum_s w(s)[\psi(x+s,\tau) + \theta(x+s,\tau)] \,, \quad (3.122)$$

where $w(x)$ is an envelope function of extent ξ.

In the end, the requirement of the envelope does not alter the form of the second term in the QPS action relevant for dipoles, which is logarithmic in imaginary-time separation between the QPS–anti-QPS pair ($-\ln c_{\mathrm{pl}}|\tau_2 - \tau_1|/\xi$)

$$- s_1 s_2 \int dx_1 \int dx_2 w(x_2 - x) w(x_1 - x') \left[K \ln \frac{\sqrt{(x_2 - x_1)^2 + c_{\mathrm{pl}}^2 (\tau_2 - \tau_1)^2}}{\xi} \right.$$
$$\left. + \, \alpha \exp\left(-\frac{|x_2 - x_1|}{\lambda_Q}\right) \ln c_{\mathrm{pl}} \frac{|\tau_2 - \tau_1|}{\xi} \right] . \quad (3.123)$$

Refael *et al.* argued that in effect, the cut-off in the spatial separation between the QPS–anti-QPS pair is the Fermi wavelength, λ_F, which is much smaller than ξ ($\lambda_F \ll \xi$). The dissipative interaction should thus have an ultraviolet cut-off at high frequencies given by $\hbar c_{\mathrm{pl}}/\lambda_F$. In the continuum limit, the condition for dipole proliferation in (3.119) above becomes

$$\frac{2R_Q}{R}(1 - e^{-a/\lambda_Q}) \to \frac{R_Q}{R_\xi} \frac{s_{\min}^2}{[\max\{\lambda_Q, \xi\}]^2} < 1 \,. \quad (3.124)$$

Here, $s_{\min}$ is the minimum spatial separation between a QPS and the anti-QPS, and $R_\xi = \rho\xi$ is the nanowire resistance on the length scale of ξ. There is thus a critical value for a given nanowire with its characteristics R_ξ and Cooper pair to normal conversion resistance r. Because $s_{\min}$ is of order λ_F, Refael *et al.* concluded that for experimental nanowire systems at temperatures well below T_c, and more specifically, for the $Mo_{0.79}Ge_{0.21}$ nanowires studied by Bezryadin, Bollinger *al.* [37, 38], at low temperatures, dipoles always proliferate, and the Cooper pair to normal conversion is always suppressed within the nanowire interior.

Armed with this important insight, which justifies modeling a short wire as a superconducting wire with a parallel normal electron channel, which is decoupled from it, but with a shunt resistor R_s connected between the wire ends, Refael *et al.* further incorporated the effective resistance caused by phase-slip proliferation into the effective shunt resistance. In addition, the resistance of the quasiparticles thermally excited across the superconducting gap, R_{qp},

$$R_{qp} = \frac{1}{\sigma_{qp}}\frac{\mathsf{L}}{a} = \frac{m}{e^2\tau_{qp}n_{qp}}\frac{\mathsf{L}}{a} \approx R_N\sqrt{\frac{k_B T}{2\pi\Delta_0}}\,e^{\Delta_0/(k_B T)}\,, \tag{3.125}$$

which runs in parallel with the QPS resistance R_{QPS}, as well as the propagation impedance of the leads, modeled as transmission lines of impedance Z_{leads}, yielded a shunt resistance of

$$R_s[f_{QPS}(l)] = Z_{leads} + \frac{R_{QPS}[f_{QPS}(l)]R_{qp}(T)}{R_{QPS}[f_{QPS}(l)] + R_{qp}(T)}\,, \qquad l = -\ln\frac{\Omega}{\Omega_0}\,. \tag{3.126}$$

The scaling equation for a nanowire effectively shunted by $R_s[f_{QPS}(l)]$ is, as before,

$$\frac{d f_{QPS}}{dl} = \left\{1 - \frac{R_Q}{R_s[f_{QPS}(l)]}\right\} f_{QPS}(l)\,, \tag{3.127}$$

or equivalently

$$\frac{d f_{QPS}^2}{dl} = 2\left\{1 - \frac{R_Q}{R_s[f_{QPS}(l)]}\right\} f_{QPS}^2\,. \tag{3.128}$$

Making use of the relation $R_{QPS} \propto (\mathsf{L}/\xi) f_{QPS}^2$, the renormalization of this resistance is now

$$\begin{aligned}\frac{d R_{QPS}}{dl} &= 2\left\{1 - \frac{R_Q}{R_s[f_{QPS}(l)]}\right\} R_{QPS} \\ &= 2\left\{1 - \frac{R_Q}{Z_{leads} + \frac{R_{QPS}[f_{QPS}(l)]R_{qp}(T)}{R_{QPS}[f_{QPS}(l)] + R_{qp}(T)}}\right\} R_{QPS}\,.\end{aligned} \tag{3.129}$$

To obtain $R_{QPS}(T)$ for comparison with experiment, this flow equation is integrated from the ultraviolet cut-off, defined as

$$\Omega^* = \frac{\Delta(T^*)}{\hbar} = \frac{k_B T^*}{\hbar}\,, \tag{3.130}$$

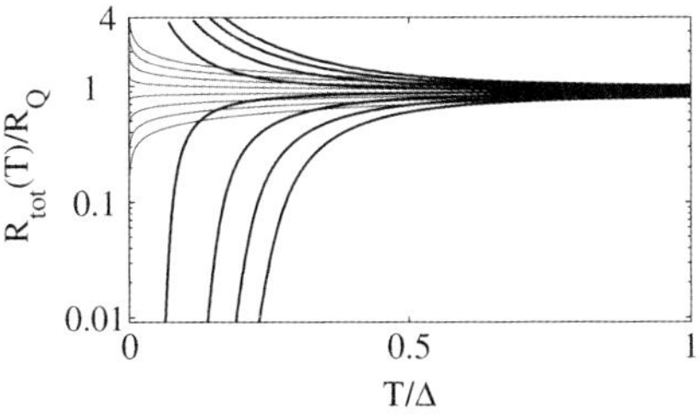

Figure 3.6 Reduced resistance R_{tot}/R_Q as a function of renormalized temperature T/Δ for different R^*/R_Q, ranging from 0.8 (lowest dark curve) to 1.01 (top dark curve) in increasing steps of 0.03. R^* denotes the value of R_{tot} at the ultraviolet cut-off temperature scale $T^* = \hbar\Omega^*/k_B$. The environmental resistance R_{leads} is set at 0.1 R_Q. Gray traces depict R_{tot} (see (3.131)) with $R_{QPS}(T) = R^*(T/\Delta)^{2(R_Q/R_s-1)}$, and a fixed $R_s = R^* + R_{leads}$. The insulating plots are cut off at a temperature, T_0, for which the phase-slip fugacity $f_{QPS}(T_0) = 1$. From [5].

to the infrared cut-off, $\Omega(T) = k_B T/\hbar$.

In a DC measurement, the lead electrodes behave as a capacitor in parallel to the nanowire, and the experimentally measured resistance is the total resistance R_{tot} of the nanowire from the dissipation mechanisms within it:

$$R_{tot}(T) = \frac{R_{QPS}(T)\, R_{QP}(T)}{R_{QPS}(T) + R_{QP}(T)} \,. \tag{3.131}$$

In Figure 3.6, we reproduce the behavior of $R_{tot}(T)$ scaled by R_Q as a function of temperature. R_{leads} is a fitting parameter taken to be $R_{leads} = 0.1\,R_Q$ in the plot, and $R^* \equiv R_{tot}(T^*)$ is the total normal state resistance at the ultraviolet cut-off scale. What is striking is the dramatic increased sharpness in the drop-off of the nanowire resistance on the superconducting side, below T_c. This is compared to the power law type of behavior in an ordinary, resistively-shunted Josephson junction. Thus, the inclusion of the interaction effects of QPS dipoles, which proliferate in nanowires, can explain the overall behaviors found experimentally in $Mo_{0.79}Ge_{0.21}$ nanowires by Bollinger *et al.*, including the superconducting–insulator transition at $R_N/R_Q \approx 1$, as well as the unusual, sharp drop-off of the linear resistance below T_c on the superconducting side. Thus, with the effective R_s determined in a self-consistent manner to include the screening properties of QPS dipoles, the experimentally observed behaviors in short wires can be reproduced. Below, in Section 3.5.3 an alternative formulation of the short-wire behavior, starting from a microscopic viewpoint, will be discussed.

3.5 Khlebnikov–Pryadko Theory – Momentum Conservation

The main issue addressed by the theory of Khlebnikov and Pryadko [6] is that of the need to account for the momentum unwound during a QPS process. A phase slip involves a change in the momentum of the Cooper pair wavefunction, as the nanowire system moves from a local free energy minimum to an adjacent free energy minimum. In the thermal activated case, as can be seen in the LA theory of phase slips within the Ginzburg–Landau picture near T_c (see Sections 2.4.3–2.4.7), the momentum index k changes in this process. A similar change in momentum must also take place in a macroscopic quantum-mechanical tunneling process, between adjacent saddle point solutions of the effective action.

This momentum change is sizable, on the order of πn_s, where n_s is the superfluid density. It is not *a priori* a foregone conclusion that the normal electrons at the phase-slip core, which are, in principle, capable of dissipating the energy released in a phase-slip event, are capable of unwinding such a large amount of momentum by transferring the momentum to the disorder potential in a completely efficient manner, as is assumed in the Zaikin–Golubev *et al.* theory. Alternatively, if the scattering is large so that the mean free path is short, a large momentum can be unwound, but the energy dissipation becomes inefficient. Khlebnikov and Pryadko argue that instead, the Mooij–Schön plasmon mode associated with phase-fluctuations outside of the phase-slip core is the channel which is ideally suited for carrying away the momentum and the energy dissipated in a phase-slip process. The efficient unwinding of the momentum and the dissipated energy becomes possible in the presence of disorder. In the absence of disorder, however, the Galilean invariance ensures the QPS rate to be zero at absolute zero in temperature.

In the ZG theory, the computation of the QPS rate was performed using instantons of a *disorder-averaged* action. Instead, Khlebnikov and Pryadko proposed to start from the opposite situation, where momentum conservation is explicitly accounted for, and perform a disorder average for a distribution of potentials with a finite range correlation afterward. This procedure has the effect of binding together QPS–anti-QPS pairs on the scale of the potential fluctuation correlation length, and helps drive the QPT into the dissipative quantum mechanics (DQM) universality class, which is relevant for the Schmid transition in single resistively-shunted Josephson junctions [40, 52, 53].

The above scenario of a KTB transition in an effective imaginary time, $(1 + 0)$-dimensional, theory is valid in the limit of long wires, satisfying the condition $L \gg c_{pl}\hbar\beta$, with c_{pl} the plasmon velocity. In the opposite limit of short wires connected to large superconducting leads, Khlebnikov finds that the interaction between QPSs takes a linear form, being proportional to the imaginary-time separation of each pair of QPSs [7]. This linear form, rather than a logarithmic form, arises due to the boundary condition on the phase field φ, at the two ends of the nanowire, and drastically reduces the tunneling rate. This reduction is essentially the result of impedance mismatch, as the much larger value of the superconductivity stiffness in the leads yields a higher plasmon velocity than in the nanowire, and a strong reflection, encoded in the boundary condition, occurs. This strong suppression of the QPS rate offers an alternative explanation for the sharp drop of the residual resistance below the superconducting temperature T_c, found in the experiments of Bezryadin *et al.* [37] and Bollinger *et al.* [38]. Derived from a microscopic approach, it offers a different perspective for understanding the sharpness of the drop, which is much more dramatic than predicted by the Schmid-type theory [40], advanced by Büchler *et al.* (Section 3.3). Furthermore, the predicted functional form: $\Gamma_{QPS} \sim \exp\{-\pi^2 K_s w/(2Lk_B T)\}$, differs from the prediction of the Meidan *et al.* phenomenological theory (Section 3.4.3).

We will begin by describing the essential aspects of the Khlebnikov–Pryadko theory addressing the central issue of momentum conservation, followed by a description of the short-wire case subjected to the boundary conditions imposed by hard, superconducting leads.

In order to make the computation tractable, a phase-only theory was developed, without inclusion of the core normal electrons as a parallel mechanism for dissipation. It is argued, however, that inclusion of core normal electron dissipation should not change the universality class of the quantum phase transition which emerges.

The 1D plasmon mode, in the presence of nearby ground planes which screen the long range Coulomb interaction, has a linear dispersion and a velocity $c_{pl} = 1/\sqrt{C L_{kin}}$, where $C \sim \epsilon/[2\ln(2d_s/d)]$ is the geometric capacitance per unit length of the SC nanowire, $L_{kin} = m/(e^2 n_s A) = \hbar/[\pi e^2 D(E_F) D\Delta_0 A]$ (3.44) is the kinetic inductance per unit length, and d_s is the distance to nearby ground planes, d the nanowire diameter, and A the nanowire cross section. The 1D superconducting nanowire thus acts as a transmission line, with a total characteristic impedance of $2Z = 2\sqrt{L_{kin}/C}$ (doubled for one transmission line on each side). The ratio $R_Q/2Z$ will control the QPT, which will be shown to be a transition between a superconducting phase and an insulation phase (SIT) in the dissipation quantum mechanics (DQM) universality class, analyzed by Schmid [40, 52]. The theory does not explicitly contain core dissipation by normal electrons. To account for normal electron core dissipation, Khlebnikov and Pryadko suggest that it is sufficient to replace the impedance 2Z, by the effective parallel impedance:

$$Z_{eff} = \frac{1}{1/2Z + 1/R_\xi}\,, \qquad (3.132)$$

where $R_\xi = R_N(\xi/L)$ is the normal state nanowire resistance on the scale of the coherence length ξ. They argue that the inclusion of the effect of the normal core should not modify the universality class. In comparison, within the Refael *et al.* theory, this possibility that core dissipation could modify the effective impedance, and hence the parameter $K = R_Q/Z$ (see paragraph following (3.101), and see (3.109); equivalent to 2μ (3.44) in the Zaikin–Golubev *et al.* notation), occurs in the intermediate temperature regime only within the SC-2 phase (3.113).

3.5.1
Gross–Pitaevskii Model and Quantum Phase Slips

The starting point of Khlebnikov and Pryadko's analysis is the Gross–Pitaevskii bosonic Euclidean Lagrangian density for the Cooper pair field ψ in 1D, in the presence of a disorder potential $V_{dis}(x)$:

$$\mathcal{L}_{1D} = \psi^\dagger \frac{\partial\psi}{\partial\tau} + \frac{1}{2M}\left|\frac{\partial\psi}{\partial x}\right|^2 + \frac{(2e)^2}{2C}|\psi|^4 - [\mu_{SC} + V_{dis}(x)]|\psi|^2\,. \qquad (3.133)$$

Here, μ_{SC} is the chemical potential. It is assumed that this form holds on length scales exceeding the coherence length ξ, that is, all quantities vary smoothly on the scale ξ. In the absence of the dissipation mechanism through the normal electrons in the phase-slip core, the QPS core size cannot be determined from the Euclidean action of this Lagrangian density; this core size is taken to be $\sim \xi$. The mass of the Cooper pair $M = 2m$, where m is the electron effective mass, and the superconducting electron density is given by

$$n_s = 2\langle \psi^\dagger \psi \rangle \,. \tag{3.134}$$

From the theory of Zaikin–Golubev (Section 3.2), it is clear that interaction between phase slips, particularly between a QPS–anti-QPS pair, plays an essential role in determining the QPT of this 1D nanowire system. The above Gross–Pitaevskii Lagrangian density is expanded about a uniform value to derive a phase-only effective Lagrangian density to highlight the importance of QPSs. Writing ψ in terms of a Cooper pair-density magnitude term $\rho = \rho_0 + \rho_1$, and a phase $\varphi = \varphi_0 + \varphi_1$, and with small fluctuations ρ_1 about the stationary point value ρ_0 and φ_0 due to weak disorder V_{dis} we have

$$\psi = \sqrt{\rho_0 + \rho_1}\, e^{i(\varphi_0 + \varphi_1)} \,. \tag{3.135}$$

After expanding in ρ_1 and integrating out the ρ_1 fluctuations, the effective phase-only Lagrangian becomes

$$\begin{aligned} \mathcal{L}_{1D} = {} & i\rho_0 \frac{\partial \varphi_1}{\partial \tau} + \frac{2C}{(2e)^2}\left(\frac{\partial \varphi_1}{\partial \tau} - iu\frac{\partial \theta_1}{\partial x} - iV\right)\left(\frac{\partial \varphi_1}{\partial \tau} - iu\frac{\partial \varphi_1}{\partial x} - iV\right) \\ & + \frac{\rho_0}{2M}\left(\frac{\partial \varphi_1}{\partial x} + \varphi_0'\right)^2 . \end{aligned} \tag{3.136}$$

This first term, which is a topological term [54, 55], is crucial and modifies the theory in a profound way. It gives a contribution to the Euclidean action, seen in (3.143) below, which may be viewed as an interference term between QPSs at different locations.

The stationary value $\varphi_0(x)$ accounts for a bias current, I, and thus has a uniform gradient

$$\varphi_0(x) = \varphi_0' x + \text{constant}\,, \quad I = (-2e)\varphi_0' \frac{\rho_0}{M} \,. \tag{3.137}$$

This bias current I also gives the superfluid velocity u:

$$u = \frac{I}{-2e\rho_0} \,. \tag{3.138}$$

The stationary point ρ_0 occurs at the minimum of the effective potential:

$$U(\rho) \equiv \frac{(2e)^2}{2C}\rho^2 - \mu_{SC}\rho - \frac{MI^2}{8e^2\rho} \,. \tag{3.139}$$

The critical current I_c is

$$I_c = \frac{1}{2e}\sqrt{\frac{C}{M}}\left(\frac{2}{3}\mu_{SC}\right)^{3/2}, \tag{3.140}$$

beyond which the effective potential $U(\rho)$ no longer has a minimum. The phase fluctuation plasmon mode propagates at the velocity

$$c_{pl} = \sqrt{\frac{\rho_0(2e)^2 A}{CM}} = \frac{1}{\sqrt{L_{kin}C}}, \quad L_{kin} = \left[(2e)^2\rho_0\frac{A}{M}\right]^{-1}. \tag{3.141}$$

To account for the importance of QPS–anti-QPS pair configurations to the effective action, the vortex-antivortex pair configuration in the 2D space and imaginary-time (x, τ) space is highlighted. As in the 2D *XY* model, the instanton solution for a QPS–anti-QPS pair is

$$e^{2i\theta_1} = \frac{x - x_2 + iv_+(\tau - \tau_2)}{x - x_2 - iv_-(\tau - \tau_2)} \times \frac{x - x_1 - iv_-(\tau - \tau_1)}{x - x_1 + iv_+(\tau - \tau_1)}, \tag{3.142}$$

where (x_i, τ_i) for $i = 1, 2$ are the respective space-imaginary time coordinates of the QPS and anti-QPS, and $v_\pm = c_{pl} \pm u \approx c_{pl}$, since for realizable superconducting nanowires, $c_{pl} \sim v_F \gg u$.

Integration of the Lagrangian density about the QPS–anti-QPS pair solution gives the following effective action:

$$\frac{S_{1D}}{\hbar} = \frac{1}{\hbar}[iP(x_1 - x_2) - E(\tau_1 - \tau_2)] + \left\{R_Q\sqrt{\frac{C}{L_{kin}}} \times \ln\frac{\sqrt{(x_1 - x_2)^2 + c_{pl}^2(\tau_1 - \tau_2)^2}}{\xi} + \left[-\frac{2\pi i C}{(2e)^2}\int_{x1}^{x_2} dx\, V(x)\right]\right\}, \tag{3.143}$$

where (3.44)

$$\frac{2\pi\hbar c_{pl} C}{(2e)^2} = R_Q\sqrt{\frac{C}{L_{kin}}} = 2\mu\ .$$

Since the theory does not contain normal electrons, the short-distance cut-off is set on the scale ξ, the expected QPS core size. The momentum to be unwound is $P = 2\pi\rho_0$, and comes from the topological term. It appears explicitly in the action, as does the energy released $E = \phi_0 I/c = hI/(2e) = uP$. The term in the second set of brackets, entering with a positive sign, reflects the logarithmic interaction of the QPSs with opposite charges, that is, a QPS and an anti-QPS.

3.5.2
Disorder Averaging, Quantum Phase Transition and Scaling for the Resistance and Current–Voltage Relations

The condition of weak disorder is parameterized by α_{dis}

$$\alpha_{\text{dis}} = \frac{V_{\text{dis}}}{M c_{\text{pl}}^2} = \frac{V_{\text{dis}} C}{(2e)^2} \frac{1}{\rho_0 \mathsf{A}} \ll 1 \,, \tag{3.144}$$

where V_{dis} gives the mean-squared amplitude of the disorder potential. This condition ensures that $\rho_1 \ll \rho_0$, enabling the expansion above, from which the effective action is derived.

Now that the requirement of momentum conservation and energy dissipation is explicitly contained in the action, the disorder averaging can be performed on the partition function. A finite range potential with correlation length l_{dis} is assumed. The disorder is taken to be Gaussian distributed with zero mean, $\langle V \rangle = 0$, and spatial correlation $\langle V(x) V(x') \rangle = V_{\text{dis}}^2 f(x - x')$. The correlation f is normalized, implying $f(0) = 1$, and for $|x - x'| \gtrsim l_{\text{dis}}$, $f(x - x') \to 0$.

The averaging involves the disorder potential term contained in the third set of brackets in (3.143):

$$\left\langle \exp\left\{ - \left[- \frac{2\pi i C}{(2e)^2} \int_{x1}^{x_2} dx\, V(x) \right] \right\} \right\rangle = \left\langle \sum_{n=0}^{\infty} \frac{[2\pi i C/(2e)^2]^n}{n!} \right.$$
$$\left. \times \prod_{i=1}^{n} \int_{x1}^{x_2} d\tilde{x}_1\, V(\tilde{x}_1) \int_{x1}^{x_2} d\tilde{x}_2\, V(\tilde{x}_2) \ldots \int_{x1}^{x_2} d\tilde{x}_n\, V(\tilde{x}_n) \right\rangle . \tag{3.145}$$

All terms containing an odd number of Vs, that is, for n odd, average to zero. For $n = 2n'$, n' a nonnegative integer, correlations of higher order beyond pair correlation ($n = 2$) are neglected. This implies that all distinct pairings of $\tilde{x}_i$s must be accounted for, yielding

$$\left\langle \exp\left\{ - \left[- \frac{2\pi i C}{(2e)^2} \int_{x1}^{x_2} dx\, V(x) \right] \right\} \right\rangle = \sum_{n'=0}^{\infty} \frac{[2\pi i C/(2e)^2]^{2n'}}{n'!} [\Phi(x_2 - x_1)]^{n'} \,,$$
$$= \exp\left[-\frac{1}{2\hbar^2} P^2 \alpha_{\text{dis}}^2 \Phi(x_2 - x_1) \right] ,$$
$$\Phi(x) = 2 \int_0^x dx' (x - x') f(x') \tag{3.146}$$

after disorder-averaging.

The quantity $\Phi(x)$ is proportional to the average square of the disorder-induced phase between a QPS and anti-QPS. For the finite-range disorder considered here,

Φ has a diffusive form at large distances $|x| \gg l_{\text{dis}}$:

$$\Phi(x) = 2l_{\text{ave}}|x| \,, \quad l_{\text{ave}} = \int_0^\infty dx\, f(x) \,. \tag{3.147}$$

For single-scale disorder, $l_{\text{ave}} \approx l_{\text{dis}}$.

This disorder in the action has the effect of binding together a QPS–anti-QPS pair with a characteristic scale in the space coordinate x of l_{pair} determined by the disorder length scale l_{ave}, and the disorder amplitude V_{dis}. In the limit of an extremely short-ranged or strong amplitude fluctuation, the scale l_{pair} approaches zero. With a short binding distance, the dynamics more and more closely approach a system with a logarithmic interaction in imaginary time only, $\ln c_{\text{pl}}|\tau_2 - \tau_1|$, in the long-wavelength limit beyond l_{pair}. This is then analogous to what gives rise to the Schmid-type KTB QPT in the resistively-shunted single Josephson junction. The universality class to which the QPT belongs is the dissipative quantum mechanics (DQM) class, associated with the dissipative quantum system. This result is analogous to what Giamarchi found for a 1D Luttinger liquid in the presence of superconducting correlations [9].

Therefore, the scaling flow is analogous to those presented in (3.63) for the Zaikin–Golubev *et al.* theory in the short-wire limit and is closely related to the Schmid transition of Büchler *et al.* and Refael *et al.* ((3.81) and (3.110)). The QPS rate Γ_{QPS} can be calculated in the standard way (see Section 3.2.6)

$$\begin{aligned}\Gamma_{\text{QPS}} &= \frac{C_1\hbar^2 c_{\text{pl}}^4 C^2}{(2e)^4\xi^3}\,\text{Im}\int\frac{dx_1 dx_2}{\mathsf{L}}\int d\tilde{\tau}\exp\left(-\frac{S_{1\text{D}}}{\hbar}\right)\,, \quad (\tilde{\tau}\equiv\tau_2-\tau_1)\\ &= \frac{C_1\hbar^2\pi^{3/2}c_{\text{pl}}^3 C^2}{(2e)^4\xi^2}\int dx\exp\left[\frac{iPx}{\hbar}-\left(\frac{1}{2\hbar^2}\right)\alpha_{\text{dis}}^2P^2\Phi(x)\right]\\ &\quad\times\left(\frac{E\xi^2}{2\hbar c_{\text{pl}}|x|}\right)^\nu\frac{J_\nu(E|x|/\hbar c_{\text{pl}})}{\Gamma(\nu+1/2)}\,.\end{aligned} \tag{3.148}$$

The exponent $\nu = R_{\text{Q}}\sqrt{C/L_{\text{kin}}}/2 - 1/2 = R_{\text{Q}}/2Z - 1/2 = \mu - 1/2$.

In the absence of disorder, $\alpha_{\text{dis}} = 0$, and the integral in the rate equals zero for any $E < c_{\text{pl}}P$. This holds for small current drives, for which the superfluid velocity $u < c_{\text{pl}}$. It occurs since there is not sufficient energy released to create a plasmon of momentum P. However, in the presence of disorder, $\alpha_{\text{dis}} \neq 0$, the disorder can take up the momentum change. In this limit, the Bessel function in the integral may be approximate by its small argument limit

$$J_\nu\left(\frac{E|x|}{\hbar c_{\text{pl}}}\right) \to \frac{(E|x|/2\hbar c_{\text{pl}})^\nu}{\Gamma(\nu+1)}\,, \tag{3.149}$$

where Γ is the gamma-function. This approximation corresponds to the disorder taking up all the momentum, while the plasmon does not carry any of it. The QPS rate becomes

$$\Gamma_{\text{QPS}} = \frac{C_1 \hbar^2}{\Gamma(\nu+1)\Gamma(\nu+\frac{1}{2})} \frac{C^2}{(2e)^4} \frac{(\sqrt{\pi} c_{\text{pl}})^3}{\xi^2} \left(\frac{E\xi}{2\hbar c_{\text{pl}}}\right)^{2\nu} l_{\text{dis}} \Lambda_{\text{dis}} ,$$
$$A_{\text{dis}} \equiv \int \frac{dx}{l_{\text{dis}}} e^{iPx/\hbar - (1/2\hbar^2) P^2 a_{\text{dis}}^2 \Phi(x)} . \tag{3.150}$$

Depending on the strength of the disorder V_{dis} its range l_{dis}, and the momentum P, the constant A_{dis} has different functional dependences on these parameters, spanning the opposing limits from where a single scattering event takes up the entire momentum, to when multiple scatterings off the disorder potential are required [6].

At temperatures $T \gg ZE/2R_Q = E\sqrt{L_{\text{kin}}/C}/2R_Q$, the characteristic size of the QPS–anti-QPS pair $\tau_2 - \tau_1 \approx \hbar/(2k_B T)$ [54, 55]. The above expression yields a simple power law form for the linear resistance,

$$R_{\text{QPS}} \propto R_Q \frac{l_{\text{dis}} A_{\text{dis}}}{\xi} \left(\frac{k_B T \xi C}{(2e)^2}\right)^{2\nu-1} , \tag{3.151}$$

and the nonlinear current–voltage relation behaves as

$$V \propto R_Q \frac{l_{\text{dis}} A_{\text{dis}}}{\xi} \left(\frac{\phi_0 \xi C}{c(2e)^2}\right)^{2\nu-1} I^{2\nu} \tag{3.152}$$

for $I \gg (2e)/(\hbar\beta)$. The transition to a nonsuperconducting state occurs at $2\nu - 1 = 0$, or $\nu = 1/2$. For $\nu < 1/2$, QPSs become unbound and proliferate. This is an SIT transition in the sense of Schmid [40]. It is closely related to the transition found by Giamarchi in an analysis of the interplay between superconducting correlations and weak disorder in the 1D Luttinger liquid system [9].

The detailed evaluation of the integral (3.148) at finite temperatures is rather involved, and requires sophisticated instanton techniques developed in the quantum field theoretic context [56]. Extending those techniques to addressing the problem of momentum conservation of phase-slip processes in superfluids, Khlebnikov makes use of the periodic instanton to compute the saddle point configuration, which gives the maximum QPS tunneling rate [54, 55]. Configurations in the $(1+1)$ space-imaginary time space, in which QPS and anti-QPS solutions at the lowest energy are arranged in alternating but periodic fashion in $\hbar\beta$, provide an approximation to the true period instanton solutions. The actual computation involves a microcanonical ensemble, in which the bosons, plasmons in the present case, are accounted for by use of a projection operation in energy, cast in a coherent state formulation of the plasmons. In the end, the energy of the initial field configuration before tunneling, E_i, and that of the final state configuration E_f, can be written in terms of the coherent state raising and lowering operators, and the plasmon propagation impedance, Z. Denoting the raising operators for the initial and final states by $a_k^\dagger$ and $b_k^\dagger$, respectively, and similarly for the lowering operators, the occupation numbers in the initial and final plasmon states are

$$\langle a_k^\dagger a_k \rangle = \frac{R_Q}{2Z|k|} \frac{\sinh^2(\omega_k \eta/2)}{\sinh^2(\hbar\omega_k \beta/2)} ,$$
$$\langle b_k^\dagger b_k \rangle = \frac{R_Q}{2Z|k|} \frac{\sinh^2(\hbar\omega_k(\beta - \hbar\eta/2))}{\sinh^2(\hbar\omega_k \beta/2)} , \tag{3.153}$$

where the plasmon disperses as $\omega_k = c_{\mathrm{pl}}|k|$. The initial and final state energies are, respectively,

$$E_\mathrm{i} = \int dk \omega_k \langle a_k^\dagger a_k \rangle \frac{R_Q k_B T}{Z} \left(1 - \frac{\pi\hbar\eta}{\beta} \cot \frac{\pi\hbar\eta}{\beta}\right) ,$$
$$E_\mathrm{f} = \int dk \omega_k \langle b_k^\dagger b_k \rangle \frac{R_Q k_B T}{Z} \left(1 - \frac{\pi(\beta - \hbar\eta)}{\beta} \cot \frac{\pi(\beta - \hbar\eta)}{\beta}\right) . \tag{3.154}$$

Here, $\eta = \tau_\mathrm{f} - \tau_\mathrm{i}$ is the difference in imaginary-time coordinates for the saddle point solution in the initial and final state configurations. This parameter η encodes information regarding the momentum change, disorder potential, coherence length, and so on, and may be deduced from experiment.

To make direct contact with experimentally measured linear resistance and nonlinear current–voltage (I–V) behavior in the low temperature limit, well below the superconducting temperature ($T \ll T_c$), one must account for the parallel channel for removal of energy by the QPS core normal electrons. In (3.132), it was argued that one way would be to use an effective impedance: $1/Z_{\mathrm{eff}} = 1/2Z + 1/R_\xi$, where R_ξ is the resistance of the core, which equals the normal state resistance on the scale of ξ. This scenario applies to wires that are either long, exceeding the thermal plasmon wavelength $c_{\mathrm{pl}}\hbar\beta$, or else are connected to normal metal leads at both ends, rather than to superconducting leads.

The strategy is then to find the fraction of the energy dissipated $\phi_0 I/c$ that is carried away by the plasmons. To do so, we thus write

$$(E_\mathrm{f} - E_\mathrm{i})2Z = \left(\frac{\phi_0 I}{c}\right) Z_{\mathrm{eff}} , \tag{3.155}$$

yielding

$$\frac{\phi_0 I}{c} = 2\left(\frac{R_Q}{Z_{\mathrm{eff}}}\right) \pi k_B T \cot \frac{\pi\hbar\eta}{\beta} , \quad \hbar\eta = \frac{\beta}{\pi} \arctan \frac{2(R_Q/Z_{\mathrm{eff}})\pi k_B T}{\phi_0 I/c} . \tag{3.156}$$

In the small and large currents I, we have, respectively, the following scaling behaviors

$$R_{\mathrm{QPS}} \propto T^{(R_Q/Z_{\mathrm{eff}})-2} ,$$
$$V \propto I^{(R_Q/Z_{\mathrm{eff}})-1} ,$$
$$\frac{R_Q}{Z_{\mathrm{eff}}} - 1 = 2\nu + \frac{R_Q}{R_\xi} . \tag{3.157}$$

3.5.3
Short Wires – Linear QPS Interaction and Exponential QPS Rate

To understand the sharp drop in R vs. T in short $Mo_{0.79}Ge_{0.21}$ nanowires connected to superconducting leads from a microscopic point of view, Khlebnikov analyzed a phase-only model [7] in which the boundary conditions imposed by the robust superconducting leads on the Mooij–Schön plasmon mode existing in the superconducting state, significally affected QPS tunneling rate. The theory applies to short and narrow wires, with width $\mathsf{w} \ll \mathsf{L}$, but not just in the 1D regime which requires $\mathsf{d} \sim (\mathsf{w}, \mathsf{h}) < \xi$. In the Euclidean action considered, the instantons representing the most likely path for tunneling, describes a vortex moving across the width of the nanowire. This approach provides an alternative to understanding the short-wire behavior, offering a different perspective from the approach taken by Meidan *et al.* and presented in Section 3.4.3.

Khlebnikov considered the tunneling of vortices across a narrow wire and the effect on the tunneling rate due to the plasmons subjected to the hard boundary conditions at the ends imposed by the robust leads; these leads are modeled as superconducting wires with a much larger stiffness K_s for the phase fluctuation mode (plasmon-mode) than in the wire itself: this large stiffness signifies a larger propagation velocity, and introduces a large impedance mismatch, which hinders the plasmons generated in a phase-slip process to exit the nanowire. Thus, the boundaries effectively cut off the plasmons at wavenumber $\sim 1/\mathsf{L}$ and reduce the tunneling probability by blocking the emission of longer wavelength plasmons due to the impedance mismatch (the plasmon velocities on the two sides, being proportional to the $\sqrt{K_s}$, are vastly different). At low levels of bias current, the rate was significantly reduced from the long-wire case, and the power law in temperature for the linear resistance is modified into

$$\Gamma_{QPS}(T) \propto \exp\left(-\frac{\pi^2 K_s \mathsf{w}}{2\mathsf{L} k_B T}\right) = \exp\left[-\frac{\pi^2 R_Q \Delta}{(4R_N) k_B T} \tanh\frac{\Delta}{2k_B T}\right], \tag{3.158}$$

where the equality follows after insertion of the dirty limit value of K_s. This exponential form reflects the modification of the QPS interaction from a logarithmic dependence on separation, to a linear dependence. Although $\hbar$ does not appear, and the functional dependence on the temperature T appears to be identical to a thermally activated process, this rate is due to a quantum-mechanical tunneling process in which the energy of the initial and final tunneling states determines the instanton action, instead of the free energy barrier height. For typical parameters, the rate is expected to be significantly enhanced relative to the thermally activated rate. Thus, it should, in principle, be possible to experimentally differentiate the two behaviors, despite the similarity in the temperature functional form. Because thermal dephasing can nullify the effect of the boundary conditions, the criterion delineating a short wire is $L \ll c_{pl}\hbar\beta$, or the length compared to the plasmon propagation distance in the thermal time.

The formal computation is based on a mapping of the phase-only action to a dual formulation in terms of vortices as charges, and the plasmon as a "photon." The mapping is conveniently carried out in the real-time formulation, after which one passes to the Euclidean action with imaginary time by substituting time $t \to -i\tau$ in order to facilitate the computation of the tunneling rate using the instanton solution. By mapping into charges and electromagnetic fields, the usual methods for solving electrodynamic problems can be employed to find the instanton solutions.

The phase-only Lagrangian density in real time for a wire of width w and length L is

$$\begin{aligned} \mathcal{L}_{\text{wire}} &= \frac{(2\pi\hbar)^2 C_{\text{w}}}{2(2e)^2}\left(\frac{\partial\varphi}{\partial t}\right)^2 - \frac{K_{\text{s}}}{2}\sum_{i=1}^{2}\left(\frac{\partial\varphi}{\partial x_i}\right)^2 , \\ &= \frac{K_{\text{s}}}{2}\sum_{i=0}^{2}\left[\frac{1}{c_{\text{pl}}^2}\left(\frac{\partial\varphi}{\partial t}\right)^2 + \left(\frac{\partial\varphi}{\partial x_i}\right)^2\right] , \\ K_{\text{s}} &= \frac{\phi_0^2}{c^2 L_{\text{kin,w}}} , \quad \phi_0 = \frac{hc}{2e} . \end{aligned} \tag{3.159}$$

The $(1+2)$ dimensional time-space coordinates are

$$(c_{\text{pl}}t; x, y) \equiv (x_0; x_1, x_2) , \quad c_{\text{pl}} = \sqrt{\frac{(2e)^2 K_{\text{s}}}{h^2 C_{\text{w}}}} = \frac{1}{\sqrt{C_{\text{w}} L_{\text{kin,w}}}} . \tag{3.160}$$

The parameter C_{w} is the capacitance per unit area, and $L_{\text{kin,w}}$ is the kinetic inductance for a unit area. To relate to the 1D parameters in the limit of very narrow wires, we have $C_{\text{w}} \to C/\text{w}$, and $L_{\text{kin,w}} \to L_{\text{kin}}\text{w}$.

The phase variable φ can be multivalued, or equivalently have branch cuts to accommodate the presence of vortices. Such a phase-only model emerges after the electromagnetic field degrees of freedom are integrated out. (Please see Section 3.2.1.2 and (3.23) for the analogous case of a 1D superconducting nanowire). The duality maps a vortex into a vortex charge q_λ, where $\lambda = \{0, 1, 2\}$ indexes the time and spatial components ($x_0 = c_{\text{pl}}t$; $x_1 = x$, $x_2 = y$). Associated with φ-field is a vortex current density, $J^\mu_{\text{vor}} : (J^0_{\text{vor}}; J^1_{\text{vor}}, J^2_{\text{vor}}) = (J^0_{\text{vor}}; J^x_{\text{vor}}, J^y_{\text{vor}})$, with J^0_{vor} being the vortex density. In relativity notation, we have

$$\begin{aligned} J^\mu_{\text{vor}} &= \frac{(-2e)}{2\pi}\epsilon^{\mu\nu\rho}\partial_\nu\partial_\rho\varphi \; : \\ \left(J^0_{\text{vor}}; J^x_{\text{vor}}, J^y_{\text{vor}}\right) &= \left([\partial_1\partial_2 - \partial_2\partial_1]\varphi; [\partial_2\partial_0 - \partial_0\partial_2]\varphi, [\partial_0\partial_1 - \partial_1\partial_0]\varphi\right) , \\ \partial_\nu : \quad (\partial_0; \partial_1, \partial_2) &\equiv \left(\frac{\partial}{\partial x_0}; \frac{\partial}{\partial x_1}, \frac{\partial}{\partial x_1}\right) , \end{aligned} \tag{3.161}$$

where $\epsilon^{\mu\nu\rho}$ is the totally antisymmetric tensor, and the sum of repeated indices is implied. The dual-representation vortex charge q_ρ has components

$$q_\rho : (q_0; q_1, q_2) \equiv (\partial_0\varphi; \partial_1\varphi, \partial_2\varphi) . \tag{3.162}$$

In terms of q_ρ, the vortex current J^μ_{vor} is

$$\begin{aligned} J^\mu_{\text{vor}} &= \frac{(-2e)}{2\pi}\epsilon^{\mu\nu\rho}\partial_\nu q_\rho \; : \\ \left(J^0_{\text{vor}}; J^x_{\text{vor}}, J^y_{\text{vor}}\right) &= (\partial_1 q_2 - \partial_2 q_1; \partial_2 q_0 - \partial_0 q_2, \partial_0 q_1 - \partial_1 q_0) . \end{aligned} \tag{3.163}$$

The charges q_ρ satisfy the equation of motion obtained from variation of the action for the Lagrangian density in (3.159)

$$\partial_0 q_0 - \partial_1 q_1 - \partial_2 q_1 - \left(\partial_0^2 - \partial_1^2 - \partial_2^2\right)\varphi = 0 \,. \tag{3.164}$$

This equation of motion is obeyed even in the presence of a nonzero topological term for which q_0 is modified by the addition of a constant term.

The action in the real-time formulation becomes

$$\begin{aligned}\frac{S_{\text{wire}}}{\hbar} &= \frac{1}{2} K_s \int d x_0 \int_0^W d x_1 \int_0^L d x_2 \left[(\partial_0\varphi)^2 - (\partial_1\varphi)^2 - (\partial_2\varphi)^2\right] \\ &= \frac{1}{2} K_s \int d x_0 \int_0^W d x_1 \int_0^L d x_2 \left(q_0^2 - q_1^2 - q_2^2\right) \,. \end{aligned} \tag{3.165}$$

Based on this action, we will focus on those configurations satisfying this equation of motion which describe the motion of vortices in real space in the presence of supercurrents and plasmon waves, and in imaginary time, in order to describe the instantons giving the most likely tunneling paths.

Using the duality map [7],

$$\begin{aligned}\exp\left(\frac{i S_{\text{wire}}}{\hbar}\right) &= \int \prod_{\mu\nu} \mathcal{D} f_{\mu\nu} \mathcal{D}\lambda \exp\left[i \frac{K_s}{4\pi} \int d x_0 d x_1 d x_2 \right. \\ &\quad \times \left.\left(-\frac{1}{4} f^{\mu\nu} f_{\mu\nu} - \frac{1}{2}\epsilon^{\mu\nu\rho} f_{\mu\nu} q_\rho + \frac{1}{2}\lambda\epsilon^{\mu\nu\rho}\partial_\mu f_{\nu\rho}\right)\right] , \\ &= \int \prod_{\mu\nu} \mathcal{D} f_{\mu\nu} \mathcal{D}\lambda \exp\left\{i \frac{K_s}{4\pi} \int d x_0 d x_1 d x_2 \right. \\ &\quad \times \exp\left[\frac{1}{2}\left(f_{01}^2 + f_{02}^2 - f_{12}^2\right) \right. \\ &\quad - \frac{1}{2}\left(f_{01} q_2 - f_{02} q_1 + f_{12} q_0 + f_{01} q_2 - f_{02} q_1 + f_{12} q_0\right) \\ &\quad \times \left.\left. - \lambda(\partial_0 f_{12} - \partial_1 f_{02} + \partial_2 f_{01})\right]\right\} , \end{aligned} \tag{3.166}$$

where $f^{\mu\nu}$ and $f_{\mu\nu}$, with their elements in matrix form given, respectively, by

$$\begin{aligned} f^{\mu\nu} &: \begin{pmatrix} 0 & f^{01} & f^{02} \\ -f^{01} & 0 & f^{12} \\ -f^{02} & -f^{12} & 0 \end{pmatrix} \\ f_{\mu\nu} &: \begin{pmatrix} 0 & f_{01} & f_{02} \\ -f_{01} & 0 & f_{12} \\ -f_{02} & -f_{12} & 0 \end{pmatrix} = \begin{pmatrix} 0 & -f^{01} & -f^{02} \\ f^{01} & 0 & f^{12} \\ f^{02} & -f^{12} & 0 \end{pmatrix} , \end{aligned} \tag{3.167}$$

$f_{\mu\nu}$ will have the appearance of the field tensor of a (1 + 2) photon field after the integration about the saddle point solution delineated below. Thus, $f_{\mu\nu}$ is an antisymmetric tensor in indices μ and ν and is required to satisfy the following boundary condition:

$$\begin{aligned}&\epsilon^{\mu\nu\rho} f_{\mu\nu} n_\rho|_b = -2n^\rho q_\rho , \quad \text{or}\\&(f_{01}n_2|_b - f_{02}n_1|_b + f_{12}n_0|_b - f_{10}n_2|_b + f_{21}n_0|_b - f_{20}n_1|_b)\\&\quad = 2(n_0q_0 + n_1q_1 + n_2q_2) ,\end{aligned} \tag{3.168}$$

with n_ρ as the normal to the boundary of the space-time volume.

Integration over λ gives

$$\epsilon^{\mu\nu\rho}\partial_\rho f_{\mu\nu} = 2(\partial_0 f_{12} - \partial_1 f_{02} + \partial_2 f_{01}) = 0 . \tag{3.169}$$

It can be shown that at the saddle point, λ is a constant, and $f_{\mu\nu}$ satisfies

$$f_\text{s}^{\mu\nu} = -\epsilon^{\mu\nu\rho} q_\rho = -\epsilon^{\mu\nu\rho}\partial_\rho\varphi , \tag{3.170}$$

so that the nonzero elements are

$$\begin{aligned}&f_\text{s}^{01} = -\partial_2\varphi = -\frac{\partial\varphi}{\partial y} , \quad f_\text{s}^{02} = +\partial_1\varphi = +\frac{\partial\varphi}{\partial x} ,\\&f_\text{s}^{12} = -\partial_0\varphi = -\frac{\partial\varphi}{c_\text{pl}\partial t} .\end{aligned} \tag{3.171}$$

The real-time action in terms of the saddle point values $f_{\text{s},\mu\nu}$ is

$$\begin{aligned}\frac{S_\text{dual}}{\hbar} &= \frac{1}{16\pi} K_\text{s} \int dx_0 dx_1 dx_2\, f_\text{s}^{\mu\nu} f_{\text{s},\mu\nu}\\&= \frac{1}{8\pi} K_\text{s} \int dx_0 dx_1 dx_2 \left\{\left[f_\text{s}^{12}\right]^2 - \left[f_\text{s}^{01}\right]^2 - \left[f_\text{s}^{02}\right]^2\right\} .\end{aligned} \tag{3.172}$$

Variation of this action gives the Maxwell equation

$$\partial_\mu f_\text{s}^{\mu\nu} = -2\pi J_\text{vor}^\mu . \tag{3.173}$$

At this point, we pass into imaginary time (dropping the subscript s denoting the saddle point solution in $f^{\mu\nu}$, but adding subscript E to denote Euclidean), with

$$\begin{aligned}&t \to -i\tau , \quad x_0 \to -i x_{\text{E},0} ,\\&\partial t \to i\partial\tau , \quad \partial 0 \to i\partial_{\text{E},0} ,\\&J_\text{vor}^0 = -i J_\text{E}^0 ,\\&f_\text{E}^{12} = -f_{\text{E},12} = -\partial_0\varphi = -i\partial_{\text{E},0}\varphi ,\\&f_\text{E}^{10} = -f_{\text{E},10} = i\partial_2\varphi = i\partial_y\varphi ,\\&f_\text{E}^{20} = -f_{\text{E},20} = -i\partial_1\varphi = -i\partial_x\varphi .\end{aligned} \tag{3.174}$$

The Euclidean action, $S_\text{E,wire}$, for which the instanton solutions will yield the QPS rate, is

$$\frac{S_\text{E,wire}}{\hbar} = -\frac{1}{8\pi} K_\text{s} \int dx\,dy\,d\tau \left\{\left[f_\text{E}^{12}\right]^2 + \left[f_\text{E}^{10}\right]^2 + \left[f_\text{E}^{20}\right]^2\right\} . \tag{3.175}$$

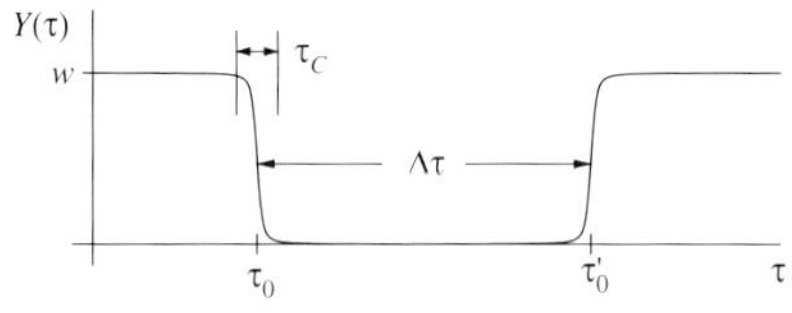

Figure 3.7 The transverse position in the y-direction (width-direction) as a function of the imaginary time. $\Delta\tau$ denotes the instanton–anti-instanton (QPS–anti-QPS) separation in τ, and τ_C the duration of an instanton. From [7].

Since the $f_E^{\mu\nu}$'s are all purely imaginary, the action is positive semidefinite.

The solution for the Maxwell fields f_E^{12}, f_E^{10}, and f_E^{20} will give the instanton saddle point solutions. Here, we will only relate the physical considerations and sketch the computational aspect, and refer the reader to the original article for details. The calculation was performed in the case where tunneling events are rare, and is intended to describe the superconducting state; it does not apply to the insulating state, where the tunneling rates are large, and QPSs proliferate.

Maxwell's equations are solved for a narrow rectangular strip of width w (y-direction) much smaller than its length L (x-direction). Since the strip is narrow, vortices are likely to pass across the entire width at nearly the same x position, and without reversing course in the y-direction. For an instanton–anti-instanton pair, the situation is depicted in Figure 3.7.

Standard Fourier transform in the y-coordinate and imaginary-time space, along with an approximately Dirichlet boundary condition on f_E^{12} and f_E^{10} are imposed in order to minimize the contribution to the action from the superconducting leads connected to the two ends of the nanowire, which are characterized by the much stiffer constant $K_{s,\text{lead}} \gg K_s$. The vortex current takes the form

$$
\begin{aligned}
J_{\text{vor}}^1 &= J_{\text{vor}}^x = 0\,,\\
J_{\text{vor}}^2 &= J_{\text{vor}}^y = \frac{i}{c_{\text{pl}}}\partial_\tau y_v(\tau)\delta(x-x_v)\delta[y-y_v(\tau)]\,,\\
J_{\text{vor,E}}^0 &= i J_{\text{vor}}^0 = i\delta(x-x_v)\delta[y-y_v(\tau)]\,,
\end{aligned}
\tag{3.176}
$$

with $[x_v, y_v(\tau)]$ as the trajectory of the vortex center and with x_v as a constant. The solution is obtained following a Fourier expansion in y and in imaginary time, indexed respectively by integers l and n, with Matsubara frequencies $\Omega_n = 2\pi n k_B T/\hbar$ for the τ transform.

Because of the large stiffness $K_{s,\text{lead}}$ in the long leads, the phase fluctuation amplitude must be small to minimize the total action, which includes the contributions from the leads to the left ($-\infty < x < 0$) and to the right ($L < x < \infty$) of the nanowire. Continuity of the phase fluctuation and its spatial and temporal derivatives at the boundaries $x = 0$ and $x = L$ then lead to Dirichlet boundary conditions for the spatial and imaginary-time partial derivatives of the phase field $\varphi(x)$. This is a crucial qualitative feature, not captured in the treatment of short wires in the Zaikin–Golubev theory. The Dirichlet boundary then applies to hyperbolic sine and cosine dependences along the x-direction.

The most singular terms of interest end up being the $l = 0$ mode in the y-direction for the quantities f_E^{12} and f_E^{20}. This contribution corresponds to that which is uniform along y. Thus, the behavior is relevant to 1D nanowires, for which the spatial variation along y can be neglected. From the $l = 0$ and $n = 0$ mode, it is possible to compute the topological winding number of the instanton saddle point solution, in analogy with what was done in the LAMH theory discussed in Chapter 1. This mode contributes an action dominated by the contributions from the leads

$$\frac{S_{00}}{\hbar} = \frac{\phi_0}{8\pi} K_{\mathrm{s,lead}} \mathsf{w} k_B T \int dx (|F_{00}|^2 - |F_{gs,00}|) = -2\pi \frac{I\phi_0 \Delta\tau}{\hbar c} . \qquad (3.177)$$

$\Delta\tau$ is the temporal separation between the instanton and anti-instanton (or QPS–anti-QPS) pair. F_{00} is the $l = 0$ and $n = 0$ Fourier component for f_E^{20} in the saddle point solution, and $F_{gs,00}$ is the corresponding Fourier component in the absence of the instanton–anti-instanton pair, in the ground state of the system of constant current density along x, but with the same winding number.

More interestingly, the contributions of the $l = 0$ but $n \neq 0$ modes to the action are logarithmic in the QPS–anti-QPS separation for long wires, but becomes linear for short wires. The contribution is approximate by the expression

$$\frac{S_{0n}}{\hbar} = \frac{1}{8\pi\hbar c} K_s \mathsf{w} \int_0^{1/(c_{\mathrm{pl}}\tau_C)} \frac{dk}{k} [1 - \cos(k\Delta\tau)] \coth(k\mathsf{L}/2) . \qquad (3.178)$$

Here, τ_C is the short-time cut-off related to the core time of the QPS, or equivalent to the duration of the instanton. The $[1 - \cos(k\Delta\tau)]$ factor indicates that the main contributions come from k values exceeding $1/(c_{\mathrm{pl}}\Delta\tau)$. If $\Delta\tau \sim \hbar\beta$, then for long wires satisfying the condition $\mathsf{L} \gg c_{\mathrm{pl}}\Delta\tau \sim c_{\mathrm{pl}}\hbar\beta$, the main contribution comes from a region where the coth factor approaches one, as its argument far exceeds unity, while the $[1 - \cos(k\Delta\tau)]$ factor remains ~ 1. The integral yields a logarithmic dependence on $\Delta\tau$, with $S_{0n} \sim \ln(\Delta\tau/\tau_C)$. This is the same result obtained originally in the long wire limit.

For short wires such that $\mathsf{L} \ll c_{\mathrm{pl}}\Delta\tau \sim c_{\mathrm{pl}}\hbar\beta$, on the other hand, the coth factor is as yet unable to saturate toward 1, in the region of k where the $[1 - \cos(k\Delta\tau)]$ factor is near unity. The dependence crosses over to a *linear dependence* on $\Delta\tau$ [7]. A linear dependence was also encountered in the short-wire theory of Büchler, Geshkenbein, and Blatter in Section 3.3, in the limit of large dissipation $R_Q/R_s \gg \mu$. Minimizing the action for the total contributions from the $l = 0$ and $n = 0$ and $n > 0$ modes, the saddle point value of $\Delta\tau$ yields a relationship between this imaginary timescale and the applied current I:

$$\Delta\tau = \frac{K_s \mathsf{w}}{\phi_0 I} , \quad \phi_0 = \frac{hc}{2e} . \qquad (3.179)$$

Accordingly, the linear dependence for short wires occurs for small currents $I \ll K_s \mathsf{w} c/\phi_0 \mathsf{L}$. Minimizing the full expression (rather than the approximate expression) for the action in this small current limit yields the final result

$$\frac{S_{0n}}{\hbar} = \hbar \pi^2 K_s \frac{w}{2Lk_B T} + O(\ln L) . \tag{3.180}$$

This action yields a linear resistance due to QPS, R_{short},

$$R_{short} \sim e^{-S_{0n}/\hbar} \approx \exp\left(-\frac{\pi^2 K_s w}{2Lk_B T}\right) . \tag{3.181}$$

This exponential dependence with an activated form in $1/(k_B T)$ should be distinguished from the TAP-type solution (Section 2.4.7 (2.125)). It arises from the modification of the interaction between QPS–anti-QPS imposed by the hard boundaries of the stiff superconducting leads, which suppresses the tunneling rate by blocking the emission of plasmons necessary for carrying away the energy dissipated during the phase-slip process. Although the temperature has a functional form similar to TAP, its origin is quantum-mechanical tunneling, and the coefficient to the $1/T$ term in the exponent depends on the energy of the system in the initial and tunneling configurations, rather than the free energy barrier between adjacent minima. The suppression of tunneling is proposed to explain the sharp drop-off of the resistance for short wires on the superconducting side of the transition, below T_c observed by Bezryadin, Bollinger *et al.* [37, 38]. This scenario offers an alternative to the proposal of Refael, Demler, Oreg, [41] and Meidan *et al.* [5, 42]. Note that here, the interaction between QPS–anti-QPS (or instanton–anti-instanton) pairs has not been accounted for, in contrast to the Refael, Meidan picture. The interaction is important to drive the SIT, where pairs proliferate. On the superconducting side, the number of pairs can be rare or dense, depending on the QPS fugacity.

The dual formulation of QPS is an important outgrowth of the theoretical analysis of the tunneling process, enabling the uncovering of new physics associated with QPSs. These will be described in detail in the following Chapter 4. There, we will discuss the apparent SIT in the Khlebnikov picture, as well as the intriguing consequences of duality, predicted independently by Mooij and Nazarov [57] and by Khlebnikov [8].

3.6
Quantum Criticality and Pair-Breaking – Universal Conductance and Thermal Transport in Short Wires

Sachdev, Werner, and Troyer proposed an entirely different physical picture for superconducting nanowires based on quantum criticality [13]. They argue that in the vicinity of a quantum superconductor-metal transition (SMT) at $T = 0$, the conductance of short wires (compared to correlation lengths in the vicinity of the quantum critical point) has a singular contribution, which is a universal function of its length L, the thermal energy $k_B T$, and the energy scale of the measurement frequency $\hbar\omega$. The form of the universal scaling depends on the boundary condition imposed on the nanowire. Thus, the cases where the nanowire ends are connected to large superconducting leads, (SS), one end to a superconducting lead with the other to a normal lead (SN), and both ends to normal leads (NN), respectively, represent different universality classes.

This SMT is related to the superconductor–insulator transition (SIT) in $d = 2$ for disordered superconducting thin films proposed by Feigel'man and Larkin and others [58–64]. A key distinguishing feature from the phase-only theories described in previous sections in this chapter is that within the region where this critical theory is expected to apply as discussed below, both amplitude and phase fluctuations of the order parameter are equally important and must be fully taken into account. The applicability of this type of quantum critical theory is envisioned to arise, when there is a quantum critical point in the phase diagram, arising from added terms in the Hamiltonian that destabilize superconductivity.

For their quantum critical theory to be applicable, it is essential to have pair-breaking processes to drive the fluctuations. In a 3D bulk system, the first analysis of the pair-breaking effect on the stability of superconducting order was carried out in the spirit of BCS mean field approach by Abrikosov and Gorkov [65], when there are paramagnetic impurities present. Del Maestro, Sachdev *et al.* [66] motivated the physical scenario giving rise to their quantum critical theory by building on the analysis of fluctuation contributions in reduced dimensions in the presence of pair-breaking put forth by Shah, Lopatin, and Vinokur beyond what occurs in 3D bulk systems [67, 68]. Pair-breaking arises in various ways, including the presence of magnetic impurities on the surface or within the nanowire by the application of a magnetic field along the nanowire, or any mechanism that breaks time-reversal invariance [69]. An excessively large rate of pair-breaking destabilizes superconducting order, driving the system into a normal metallic state. In Figure 3.8, we

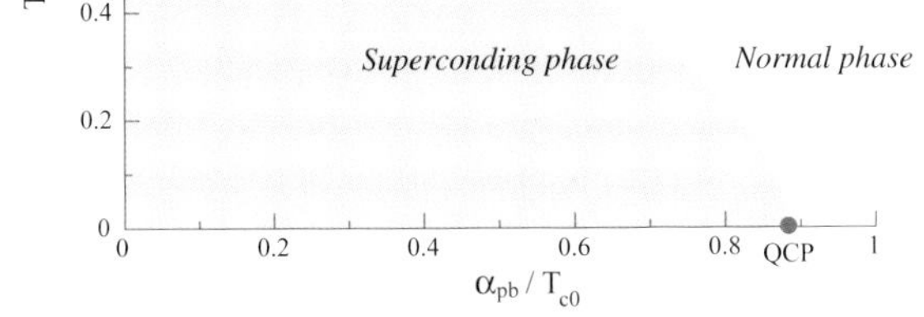

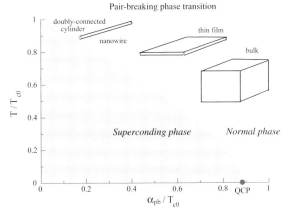

Figure 3.8 Phase diagram showing the pair-breaking transition from the superconducting to the normal state. A superconducting quantum critical point occurs at a critical pair-breaking strength $\alpha_{pb,c} = k_B T_{c,0}/\hbar = 0.889$. Here, $T_{c,0}$ denotes the superconducting transition temperature in the absence of pair-breaking. Different pair-breaking systems are shown as insets. From [67].

present Shah and Lopatin's calculated phase diagram in the context of the nanowire system.

The relevance of pair-breaking to experimental systems of superconducting nanowire wires comes from two directions. As discussed by Shah and Lopatin, pair-breaking interactions can be introduced in two ways in a nanowire: by introducing paramagnetic impurities on the surface of the nanowire, such as was done by Rogachev *et al.* [70], or by applying a magnetic field parallel to the nanowire. Moreover, even without such artificial means, it is possible that dangling-bonds at the surface of the nanowire may behave as paramagnetic impurities. There are many examples of this type of behavior in very thin devices such as what has been found in graphene [71], or Sc atoms in Al/Al_xO_y tunnel junctions [72].

The physics of the Sachdev, Troyer, and Werner model [13] is that of a Cooper pair superconducting electron fluid, strongly overdamped by some pair-breaking mechanism, which breaks up the Cooper pairs into normal quasiparticles. The Euclidean action is given by

$$\begin{aligned}
S_{qc} &= S_{1D} + S_{bdry} , \\
\frac{S_{1D}}{\hbar} &= \frac{A}{\hbar}\int_0^L dx \int_0^{\hbar\beta} d\tau \left(\tilde{D}\left|\frac{\partial \Psi}{\partial x}\right|^2 + \alpha|\Psi|^2 + \frac{u}{2}|\Psi|^4 \right) \\
&\quad + \frac{A}{\hbar}\int_0^L dx \frac{\hbar\gamma}{\beta}\sum_n |\omega_n||\tilde{\Psi}(x,\omega_n)|^2 , \\
\frac{S_{bdry}}{\hbar} &= \int_0^{\hbar\beta} d\tau \{ C_L|\Psi(0,\tau)|^2 + C_R|\Psi(L,\tau)|^2 \\
&\quad - \mathrm{Re}[H_L\Psi(0,\tau) + H_R\Psi(L,\tau)\} , \\
\tilde{\Psi}(x,\omega_n) &= \int d\tau \Psi(x,\tau) e^{i\omega_n\tau} ,
\end{aligned} \tag{3.182}$$

where $\Psi(x,\tau)$ is the Cooper pair order parameter, and as usual, $\beta = \hbar/(k_B T)$. The bosonic Matsubara frequencies $\omega_n = 2\pi n k_B T/\hbar$, as usual, and right- and left-lead indices are labeled "R" and "L", respectively. The constants C prescribe the boundary conditions on the superconducting order. For a normal lead (N), $C > 0$ with $H = 0$, while for a superconducting lead (S), $H \neq 0$ as the superconductivity in the lead imposes an ordering field on the nanowire at the connection point.

For dirty superconductors such as those encountered in experiments, the parameters in (3.182) are as follows. The diffusion constant $\tilde{D} \sim D = (1/3)v_F l_{mfp}$. The pair-breaking parameter is denoted by α_{pb} and appears in the Cooper pair propagator in the metallic state $1/(\tilde{D}k^2 + |\omega_n| + \alpha_{pb})$, at wavevector $\boldsymbol{k}$ and Matsubara frequency (Fourier transform of imaginary-time representation). In a nanowire, this important parameter α_{pb} can be tuned by changing the nanowire diameter. The parameter u in the quartic term ensures stability and accounts for Cooper pair repulsion. Its inclusion [73] also differentiates from an earlier analysis based on the assumption that pair-breaking effect and reduced dimensionality give rise to Gaussian fluctuations only [67, 68]. Such an assumption amounts to setting $u = 0$. The Gaussian approximation is valid in the high temperature limit [73]. In the dirty limit relevant for nanowires, $u \approx 2.9v_F/(\hbar N_\perp)$ with $N_\perp = 2k_F^2 A/(3\pi)$ as the number of transverse channels. The dissipation parameter, γ, can be related to the relaxation parameter in a time-dependent Ginzburg–Landau equation, with $\gamma \approx 3/(2k_F l_{mpf})$; the dissipative term is thus motivated by the TDGL type of relaxation, and involves a time-derivative of the Cooper pair field $\Psi(x, \tau)$. In the Matsubara frequency representation above, the dissipation term contains the factor $|\omega_n||\Psi(x, \omega_n)|^2$. In imaginary-time representation, this form yields a long-range interaction in imaginary time behaving as $1/\tau^2$. This interaction is to be contrasted with the situation in dissipative Josephson junctions. There, the dissipative term involves the difference in superconducting phase φ on adjacent sites, in the form of a factor $|\omega_n||\varphi_i - \varphi_j|^2$, where (i, j) are site indices. In the continuum limit, this translates to a spatial derivative of the φ field. This distinction is essential, and places the quantum-critical theory discussed here in a different universality class of nonrandom, pair-breaking class, instead of the Kosterlitz–Thouless–Berezinskii class [13].

This action automatically incorporates the quantum fluctuations that lead to the Aslamazov–Larkin (AL) corrections to the conductivity; the AL correction are the dominant fluctuation corrections to transport in the quantum critical regime at low temperature. In analogy to the analysis carried for 2D Josephson island arrays and dirty thin films ($\xi_{BCS} \gg l_{mfp}$) it is important to address issues of the enhanced fluctuations in reduced dimensions ($D < 3$), which can lead to the destruction of superconductivity correlations which would otherwise exist in a BCS mean field analysis [58–64]. This analysis of enhanced fluctuations also goes beyond those based on the duality between a Cooper pair and the vortex formulation in 2D, where the 2D SIT occurs at the symmetry point delineated by $R_Q/R_{sqr} = 1$, where R_{sqr} is the resistance of the film per square in the normal state [44, 74].

This theory, absent the boundary terms, is identical in form to the Hertz–Millis–Moriya theory describing the Fermi liquid to spin-density wave transition [75–77]. There, an O(3) field describes the diffusive paramagnons order parameter instead of the Cooper pair field Ψ.

The quantum critical point occurs at $T = 0$, at a critical value of $\alpha_{pb,c}$ for fixed D, u, and γ. Away from the $k_B T = 0$, $\alpha_{pb,c}$ point in the $k_B T$ vs. α_{pb} plane, large quantum fluctuations occur in its vicinity, on a length scale corresponding to a characteristic correlation length and a characteristic timescale in the imaginary-time direction. These scales diverge as the critical point is approached. Two critical exponents characterize how this divergences occurs: the dynamical exponent z, which measures the fluctuation timescale in terms of the correlation length, and the correlation length exponent ν, which measures the rate of divergence of the spatial correlation. In the Gaussian approximation, $z = 2$ and $\nu = 1/2$. In the critical region, $k_B T \gg |\alpha_{pb} - \alpha_{pb,c}|^{z\nu}$. In Figure 3.9, we show the phase boundary in the $k_B T$–α_{pb} plane.

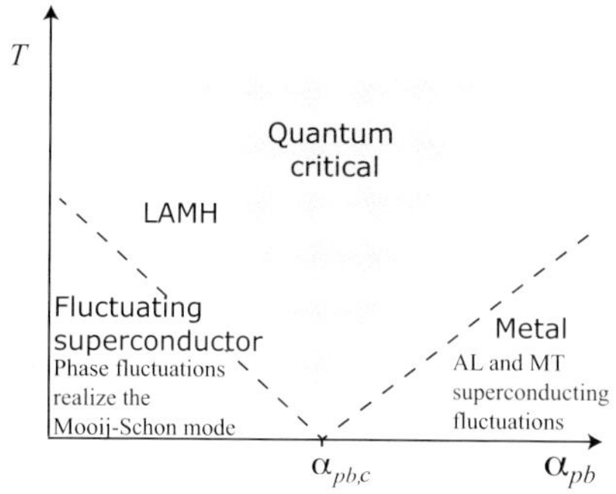

Figure 3.9 Crossover phase diagram of the superconductor-metal transition in a quasi-1D superconductor. The "Metal" is described by the perturbative theory of Shah and Lopatin [66]. The "Quantum-critical" region is described by (3.182). The Mooij–Schön plasmon mode is present everywhere, but strongly couples to superconducting fluctuations only in the "Fluctuating superconductor" regime, a regime (3.182) does not describe. The dashed lines are crossover boundaries occurring at $k_B T \sim \hbar|\alpha_{\rm pb} - \alpha_{\rm pb,c}|^{z\nu}$, where z is the dynamical critical exponent, and ν is the correlation length exponent. In the Gaussian approximation, $z = 2$ and $\nu = 1/2$. From [66].

In the figure, the values of D, u, and γ are held fixed [66]. The parameters in the figure have been fixed as

$$\gamma \to 1\,; \quad \tilde{D} \to \gamma\,\tilde{D}\,; \quad \alpha_{\rm pb} \to \gamma\,\alpha_{\rm pb}\,; \quad u \to \gamma^2 u\,; \quad \Psi \to \frac{\Psi}{\sqrt{\gamma}}\,. \tag{3.183}$$

The quantum critical region has a crossover boundary in the $k_B T - \alpha_{\rm pb}$ plane given by the $k_B T \propto |\alpha_{\rm pb} - \alpha_{\rm pb,c}|^{\nu z} \sim |\alpha_{\rm pb} - \alpha_{\rm pb,c}|$ in the Gaussian approximation ($z\nu = 1$). On the low $\alpha_{\rm pb}$ side, emergence of the strong coupling of the superconducting order parameter to the Mooij–Schön phase fluctuation (plasmon) mode yields an effective action similar to the phase-only theories discussed previously. However, the normal metal phase lies on the other side of the quantum critical point along the $\alpha_{\rm pb}$ axis, and a transition to that phase must pass through the quantum critical region and does not occur directly. When all other parameters are fixed, the quantum critical point is encoded in the pair-breaking parameter $\alpha_{\rm pb}$, that is, the coefficient of the $|\Psi|^2$ term in the action $S_{\rm 1D}$. For a larger dissipation, γ, the corresponding value of $\alpha_{\rm pb}$ at the critical point, $\alpha_{\rm pb,c}$, becomes smaller.

An important property of the above action is that in the vicinity of the quantum critical point, the theory obeys hyperscaling [78]. Transport, including electrical and thermal transport, has a singular contribution, which is a universal scaling function in the distance from the critical point, in the temperature–$\alpha_{\rm pb}$ plane.

The conductance takes the universal form

$$g(\omega) = \frac{(2e)^2}{h} F_{\rm LR}\left[c_1 \hbar\omega {\rm L}^z, c_1 k_B T {\rm L}^z, c_2 {\rm L}^{1/\nu}(\alpha_{\rm pb,c} - \alpha_{\rm pb})\right]\,, \tag{3.184}$$

where "L" and "R" denote the type of leads connected to the left and right ends of the nanowire, respectively, with "S" being superconducting and "N" normal metal.

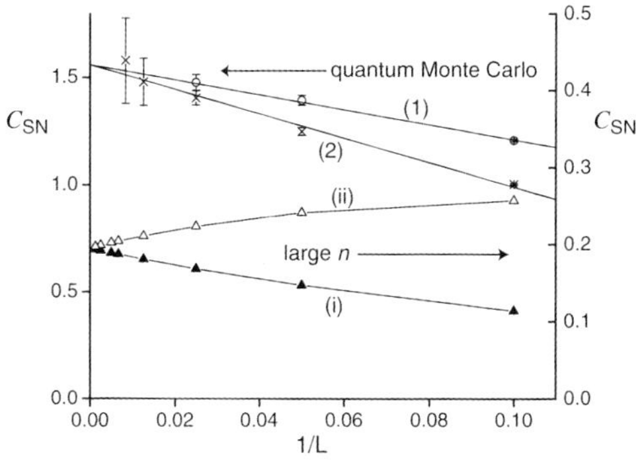

Figure 3.10 Extrapolation of the DC conductance C_{SN}, corresponding to one superconducting lead and one normal lead at the end of the 1D narrow wire. The extrapolation is to the universal scaling limit of $L \to \infty$. The curves labeled (1) and (2) are results based on an *n*-component order-parameter field for large *n*, and correspond to different parameters, showing the universality as $1/L \to 0$. For the quantum Monte Carlo calculations, the model used is a discretized lattice Ising-spin model. The different curves (i) and (ii) are results for different parameters values. From [13].

In the quantum critical region, in the short-wire limit defined by

$$L < L_T \sim (c_1 k_B T)^{-1/z} , \quad L < (c_2 |\alpha_{pb,c} - \alpha_{pb}|)^{-\nu} , \tag{3.185}$$

this universal form reduces to

$$g(\omega) = \frac{(2e)^2}{h} F_{LR}[c_1 \hbar\omega L^z, 0, 0] . \tag{3.186}$$

In Figure 3.10, we present the extrapolation of the DC conductance C_{SN} for the case of mixed boundaries, as the size of the system is increased. This numerical computation is based on a discretized lattice version of the theory [13].

Similar scaling forms for the thermal conductivity are discussed by Del Maestro *et al.* [66].

Up to this point, disorder has been accounted for only in terms of the dirty limit parameters in the action, in γ, $\tilde{D}$, u, and so on. This amounts to an assumption of homogeneity on the scale of the coherence length. Disorder can also lead to variations on a longer scale, and thus positional dependence in the value of the parameters. Moreover, quenched randomness can have profound effects on the nature of the quantum criticality as disorder correlations are of infinite range in the imaginary-time direction! It turns out that disorder induced variations represent a relevant perturbation in the renormalization group sense, and an exotic behavior in the quantum criticality emerges.

The quantum critical behavior in this dissipative system rather miraculously turns out to be identical to that of the random-transverse-field Ising model (RTFIM), which does not contain any dissipation and has a discrete symmetry [14, 16]. Hoyos *et al.* [16] showed that randomness causes the quantum critical behavior to exhibit unusual dynamical behavior. The dynamic scaling is no longer of a power law variety, but is instead, in the form of an activated behavior where the logarithm of the temporal fluctuation timescale behaves as a power law of the spatial correlation length. This behavior is associated with the infinite randomness fixed point (IRFP). In the RTFIM model, Fisher obtained many exact results [17, 18] by means of the strong-disorder renormalization group method [79]. Many of these exact results appear to carry over to the SMT in the 1D superconducting nanowire wire system, in the presence of quenched randomness.

Del Maestro *et al.* hypothesized, based on Hoyos *et al.*'s analysis, that the SMT (superconductor-metal transition) in their quantum critical model containing strong dissipation when random quenched disorder is present, is characterized by this IRFP fixed point. In Figure 3.11, we reproduce their numerical results for the frequency scale of critical fluctuations in the vicinity of the critical point. The exotic activated scaling is evident, indicating the type of behavior associated with the IRFP. One of the unusual consequences of the IRFP predictions is that the linear resistance does not decrease monotonically with decreasing temperature, but rather it first increases as $1/\sqrt{T}$, and then crosses over to a T^2 dependence. More recently, Del Maestro and colleagues [15] uncovered novel dynamical conductivity in the dirty superconductor limit, associated with an IRFP quantum critical point. At criticality, the real part of the conductivity diverges at $\mathrm{Re}[\sigma(\omega)] \sim [\ln(\omega_0/\omega)]^{1/\psi}$, where ψ is the tunneling exponent, and ω_0 is a reference frequency. In the vicinity of the critical point, $\mathrm{Re}[\sigma(\omega)]$ satisfies an unconventional activated scaling form:

$$\mathrm{Re}[\sigma(\delta, \omega)] = \frac{4e^2}{h}\left[\ln\left(\frac{\omega_0}{\omega}\right)\right]^{1/\psi} \Phi_\sigma\left(\delta^{\nu\psi}\ln\frac{\omega_0}{\omega}\right) . \qquad (3.187)$$

This type of highly unusual behavior should be amenable to experimental tests and verification in the next generation of nanowire experiments.

In their theory, Del Maestro, Sachdev *et al.* motivated the quantum critical behavior in superconducting nanowires based on the pair-breaking mechanism. This emphasis followed the analysis, performed by Shah and Lopatin [67] and Lopatin, Shah, and Vinokur [68], who investigated the quantum fluctuation correction to the conductivity due to pair-breaking on the metallic side. Interestingly, from an entirely different perspective, Khlebnikov recently developed a theory of a 1D superconducting nanowire with strong electron interaction [80] based on the duality between theories of gravity/gauge interactions [81] in order to model a superconducting nanowire under a large current-bias just below the depairing critical current. This theory was put forth to account for experimental studies on the statistical distribution of the switching current, that is, the current at which a superconducting nanowire goes normal during a ramp-up in the current, which has been performed at different laboratories in the past several years. In these studies, it was found that at sufficiently low temperatures, a single-phase slip event dissipates enough energy to initiate a thermal runaway, driving the nanowire into a normal state (Section 7.4.3). At intermediate low temperatures, the width of the statistical distribution appeared to be thermally-dominated, following a $\sim T^{2/3}$ power law behavior. This crosses over to a saturation in the width at even lower temperatures.

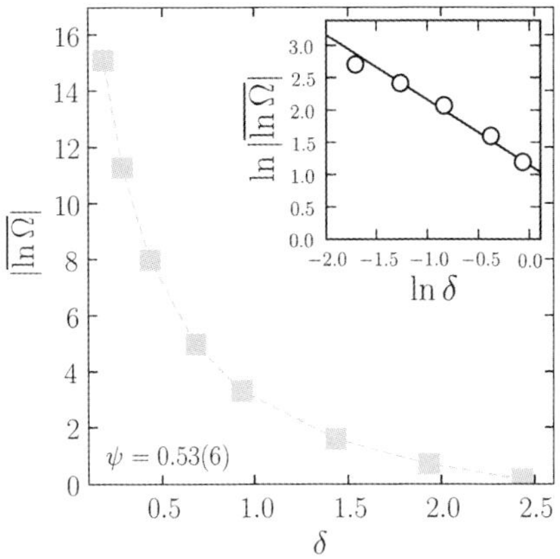

Figure 3.11 The finite-size scaled value of the disorder-averaged logarithm of the minimum excitation energy plotted against the distance from the critical point $\delta \equiv \alpha_{\text{pb}} - \alpha_{\text{pb,c}}$. The observed divergence is consistent with the scaling form $|\ln\overline{\Omega}| \sim \delta^{-\nu\psi}$, where $\nu = 1.9(2)$ is the correlation length critical exponent, and $\psi = 0.53(6)$ is the tunneling exponent. From [14].

To explain the experimental findings, Khlebnikov noted that the behavior appears to be controlled by an unstable Gaussian fixed point within an effective theory derived from the gravity/gauge duality. The gauge theory lives in $(3+1)$D space-time, with a color gauge group SU(N). The conformal invariance of the $(3+1)$D gauge theory leads to an attractive $1/r$ interaction between left and right moving electrons in $(1+1)$D, the relevant dimension for 1D superconducting nanowires. The duality enables one to compute the properties of the gauge theory modeling 1D nanowires by performing the calculations for gravity in higher dimensions for N large. Starting with 10-dimensional space-time, an effective action of the Ginzburg–Landau-type emerges for $(1+1)$D superconductivity. Intriguingly, the form of this effective action is precisely that put forth by Sachdev, Werner, and Troyer [13]. Thus, it may turn out that this type of critical theory is relevant to nanowires, even in the absence of an explicit pair-breaking mechanism, and beyond that caused by the presence of a large biasing current!

References

1 Mooij, J. and Schön, G. (1985) Propagating plasma mode in thin superconducting filaments. *Physical Review Letters*, **55** (1), 114–117.
2 Zaikin, A.D., Golubev, D.S., van Otterlo, A., and Zimányi, G.T. (1997) Quantum phase slips and transport in ultrathin superconducting wires. *Physical Review Letters*, **78**, 1552–1555.
3 Golubev, D.S. and Zaikin, A.D. (2001) Quantum tunneling of the order parameter in superconducting nanowires. *Physical Review B*, **64** (1), 014504.
4 Refael, G., Demler, E., Oreg, Y., and Fisher, D.S. (2007) Superconductor-to-normal transitions in dissipative chains of mesoscopic grains and nanowires. *Physical Review B*, **75** (1), 014522.

5 Meidan, D., Oreg, Y., and Refael, G. (2007) Sharp superconductor–insulator transition in short wires. *Physical Review Letters*, **98** (18), 187001.
6 Khlebnikov, S. and Pryadko, L.P. (2005) Quantum phase slips in the presence of finite-range disorder. *Physical Review Letters*, **95** (10), 107007.
7 Khlebnikov, S. (2008) Quantum phase slips in a confined geometry. *Physical Review B*, **77** (1), 014505, doi:10.1103/PhysRevB.77.014505.
8 Khlebnikov, S. (2008) Quantum mechanics of superconducting nanowires. *Physical Review B*, **78** (1), 014512.
9 Giamarchi, T. (1992) Resistivity of a one-dimensional interacting quantum fluid. *Physical Review B*, **46**, 342–349.
10 Büchler, H.P., Geshkenbein, V.B., and Blatter, G. (2004) Quantum fluctuations in thin superconducting wires of finite length. *Physical Review Letters*, **92** (6), 067007.
11 Berezinski, V.L. (1971) Destruction of long-range order in one-dimensional and two-dimensional systems having a continuous symmetry group I. Classical systems. *Soviet Journal of Experimental and Theoretical Physics*, **32**, 493.
12 Kosterlitz, J.M. and Thouless, D.J. (1973) Ordering, metastability and phase transitions in two-dimensional systems. *Journal of Physics C Solid State Physics*, **6**, 1181–1203.
13 Sachdev, S., Werner, P., and Troyer, M. (2004) Universal conductance of nanowires near the superconductor–metal quantum transition. *Physical Review Letters*, **92** (23), 237003.
14 Del Maestro, A., Rosenow, B., Müller, M., and Sachdev, S. (2008) Infinite randomness fixed point of the superconductor–metal quantum phase transition. *Physical Review Letters*, **101** (3), 035701.
15 Del Maestro, A., Rosenow, B., Hoyos, J.A., and Vojta, T. (2010) Dynamical conductivity at the dirty superconductor–metal quantum phase transition. *Physical Review Letters*, **105** (14), 145702.
16 Hoyos, J.A., Kotabage, C., and Vojta, T. (2007) Effects of dissipation on a quantum critical point with disorder. *Physical Review Letters*, **99** (23), 230601.
17 Fisher, D.S. (1992) Random transverse field Ising spin chains. *Physical Review Letters*, **69**, 534–537.
18 Fisher, D.S. (1995) Critical behavior of random transverse-field Ising spin chains. *Physical Review B*, **51**, 6411–6461.
19 Duan, J.M. (1995) Quantum decay of one-dimensional supercurrent: Role of electromagnetic field. *Physical Review Letters*, **74**, 5128–5131.
20 Arutyunov, K., Golubev, D., and Zaikin, A. (2008) Superconductivity in one dimension. *Physics Reports*, **464** (1/2), 1–70.
21 Schön, G. and Zaikin, A.D. (1990) Quantum coherent effects, phase transitions, and the dissipative dynamics of ultra small tunnel junctions. *Physics Reports*, **198**, 237–412.
22 van Otterlo, A., Golubev, D., Zaikin, A., and Blatter, G. (1999) Dynamics and effective actions of BCS superconductors. *The European Physical Journal B*, **10** (1), 131–143.
23 Keldysh, L.V. (1965) Concerning the theory of impact ionization in semiconductors. *Soviet Journal of Experimental and Theoretical Physics*, **21**, 1135.
24 Tomonaga, S.I. (1950) Remarks on Bloch's method of sound waves applied to many-fermion problems. *Progress of Theoretical Physics*, **5** (4), 544–569.
25 Luttinger, J.M. (1963) An exactly soluble model of a many-fermion system. *Journal of Mathematical Physics*, **4**, 1154–1162.
26 Kamenev, A. (2009) Introduction to the Keldysh formalism – TFP-Homepage: ba-tfp1.physik.uni-freiburg.de/Capri09/lectures/Kamenev...
27 Kamenev, A. and Levchenko, A. (2009) Keldysh technique and non-linear σ-model: basic principles and applications. *Advances in Physics*, **58**, 197–319.
28 Kamenev, A. (2007) Many-body theory of non-equilibrium systems. *arXiv:cond-mat/0412296*.
29 McCumber, D.E. and Halperin, B.I. (1970) Time scale of intrinsic resistive fluctuations in thin superconducting wires. *Physical Review B*, **1** (3), 1054–1070.

30 Vaĭnshteĭn, A.I., Zakharov, V.I., Novikov, V.A., and Shifman, M.A. (1982) Reviews of topical problems: ABC of instantons. *Soviet Physics Uspekhi*, **25**, 195–215.

31 Anderson, P.W., Yuval, G., and Hamann, D.R. (1970) Exact results in the Kondo problem. II. Scaling theory, qualitatively correct solution, and some new results on one-dimensional classical statistical models. *Physical Review B*, **1**, 4464–4473.

32 Anderson, P. (1959) Theory of dirty superconductors. *Journal of Physics and Chemistry of Solids*, **11**, 26–30.

33 Wen, X.G. (1992) Theory of the edge states in fractional quantum hall effects. *International Journal of Modern Physics B*, **6**, 1711–1762.

34 Kane, C.L. and Fisher, M.P.A. (1992) Transport in a one-channel Luttinger liquid. *Physical Review Letters*, **68**, 1220–1223.

35 Chang, A.M. (2003) Chiral Luttinger liquids at the fractional quantum Hall edge. *Reviews of Modern Physics*, **75**, 1449–1505.

36 Altomare, F., Chang, A.M., Melloch, M.R., Hong, Y., and Tu, C.W. (2006) Evidence for macroscopic quantum tunneling of phase slips in long one-dimensional superconducting Al wires. *Physical Review Letters*, **97** (1), 017001.

37 Bezryadin, A., Lau, C.N., and Tinkham, M. (2000) Quantum suppression of superconductivity in ultrathin nanowires. *Nature*, **404** (6781), 971–974.

38 Bollinger, A.T., Dinsmore, III, R.C., Rogachev, A., and Bezryadin, A. (2008) Determination of the superconductor–insulator phase diagram for one-dimensional wires. *Physical Review Letters*, **101** (22), 227003.

39 Lau, C.N., Markovic, N., Bockrath, M., Bezryadin, A., and Tinkham, M. (2001) Quantum phase slips in superconducting nanowires. *Physical Review Letters*, **87** (21), 217003.

40 Schmid, A. (1983) Diffusion and localization in a dissipative quantum system. *Physical Review Letters*, **51**, 1506–1509.

41 Refael, G., Demler, E., and Oreg, Y. (2009) Superconductor to normal-metal transition in finite-length nanowires: Phenomenological model. *Physical Review B*, **79** (9), 094524.

42 Meidan, D., Oreg, Y., Refael, G., and Smith, R.A. (2007) *Sharp Superconductor-Insulator Transition in Short Wires*, Contrib. Proc. "Fluctuations and phase transitions in superconductors", Nazareth Ilit, Israel.
To be published in *Phys. C*, **468**.

43 Chakravarty, S., Ingold, G.L., Kivelson, S., and Luther, A. (1986) Onset of global phase coherence in Josephson-junction arrays A dissipative phase transition. *Physical Review Letters*, **56**, 2303–2306.

44 Fisher, M.P.A. (1986) Quantum phase slips and superconductivity in granular films. *Physical Review Letters*, **57**, 885–888.

45 Chakravarty, S., Ingold, G.L., Kivelson, S., and Zimanyi, G. (1988) Quantum statistical mechanics of an array of resistively shunted Josephson junctions. *Physical Review B*, **37**, 3283–3294.

46 Bobbert, P.A., Fazio, R., Schön, G., and Zimanyi, G.T. (1990) Phase transitions in dissipative Josephson chains. *Physical Review B*, **41**, 4009–4016.

47 Bobbert, P.A., Fazio, R., Schön, G., and Zaikin, A.D. (1992) Phase transitions in dissipative Josephson chains: Monte Carlo results and response functions. *Physical Review B*, **45**, 2294–2304.

48 Korshunov, S.E. (1989) Phase diagram of a chain of dissipative Josephson junctions. *EPL (Europhysics Letters)*, **9**, 107.

49 Refael, G., Demler, E., Oreg, Y., and Fisher, D.S. (2003) Dissipation and quantum phase transitions of a pair of Josephson junctions. *Physical Review B*, **68** (21), 214515.

50 Villain, J. (1975) Theory of one- and two-dimensional magnets with an easy magnetization plane. II. The planar, classical, two-dimensional magnet. *J. Phys. France*, **36** (6), 581–590.

51 Refael, G. (2001) Randomness, dissipation, and quantum fluctuations in spin chains and mesoscopic superconductor arrays. PhD Thesis, Harvard University.

52 Chakravarty, S. (1982) Quantum fluctuations in the tunneling between superconductors. *Physical Review Letters*, **49**, 681–684.

53 Penttilä, J.S., Parts, Ü., Hakonen, P.J., Paalanen, M.A., and Sonin, E.B. (1999) "Superconductor–insulator transition" in a single Josephson junction. *Physical Review Letters*, **82**, 1004–1007.

54 Khlebnikov, S. (2004) Quasiparticle scattering by quantum phase slips in one-dimensional superfluids. *Physical Review Letters*, **93** (9), 090403.

55 Khlebnikov, S. (2005) Tunneling in a uniform one-dimensional superfluid: Emergence of a complex instanton. *Physical Review A*, **71** (1), 013602.

56 Khlebnikov, S.Y., Rubakov, V.A., and Tinyakov, P.G. (1991) Periodic instantons and scattering amplitudes. *Nuclear Physics B*, **367**, 334–358.

57 Mooij, J.E. and Nazarov, Y.V. (2006) Superconducting nanowires as quantum phase-slip junctions. *Nature Physics*, **2**, 169–172.

58 Feigelman, M. and Larkin, A.I. (1998) Quantum superconductor metal transition in a 2D proximity-coupled array. *Chemical Physics*, **235**, 107–114.

59 Feigelman, M.V., Larkin, A.I., and Skvortsov, M.A. (2001) Quantum superconductor–metal transition in a proximity array. *Physical Review Letters*, **86**, 1869–1872.

60 Spivak, B., Zyuzin, A., and Hruska, M. (2001) Quantum superconductor–metal transition. *Physical Review B*, **64** (13), 132502.

61 Oreg, Y. and Finkel'Stein, A.M. (1999) Suppression of T_c in superconducting amorphous wires. *Physical Review Letters*, **83**, 191–194.

62 Herbut, I.F. (2000) Zero-temperature *d*-wave superconducting phase transition. *Physical Review Letters*, **85**, 1532–1535.

63 Dalidovich, D. and Phillips, P. (2000) Fluctuation conductivity in insulator–superconductor transitions with dissipation. *Physical Review Letters*, **84**, 737–740.

64 Dalidovich, D. and Phillips, P. (2002) Hall conductivity near the $z = 2$ superconductor–insulator transition in two dimensions. *Physical Review B*, **66** (7), 073308.

65 Abrikosov, A.A. and Gor'kov, L. (1960) Contribution to the theory of superconducting alloys with paramagnetic impurities. *Soviet-Physics JETP – USSR*, **12** (6), 1243–1253.

66 Del Maestro, A., Rosenow, B., Shah, N., and Sachdev, S. (2008) Universal thermal and electrical transport near the superconductor–metal quantum phase transition in nanowires. *Physical Review B*, **77** (18), 180501.

67 Shah, N. and Lopatin, A. (2007) Microscopic analysis of the superconducting quantum critical point: Finite-temperature crossovers in transport near a pair-breaking quantum phase transition. *Physical Review B*, **76** (9), 094511.

68 Lopatin, A.V., Shah, N., and Vinokur, V.M. (2005) Fluctuation conductivity of thin films and nanowires near a parallel-field-tuned superconducting quantum phase transition. *Physical Review Letters*, **94** (3), 037003.

69 Anderson, P. (1959) Theory of dirty superconductors. *Journal of Physics and Chemistry of Solids*, **11**, 26–30.

70 Rogachev, A. and Bezryadin, A. (2003) Superconducting properties of polycrystalline Nb nanowires templated by carbon nanotubes. *Applied Physics Letters*, **83**, 512.

71 Rao, S.S., Stesmans, A., Kosynkin, D.V., Higginbotham, A., and Tour, J.M. (2011) Paramagnetic centers in graphene nanoribbons prepared from longitudinal unzipping of carbon nanotubes, *New Journal of Physics*, **13**, 113004, *http://iopscience.iop.org/1367-2630/13*.

72 Yeh, S.S. and Lin, J.J. (2009) Two-channel Kondo effects in $Al/AlO_x/Sc$ planar tunnel junctions. *Physical Review B*, **79** (1), 012411.

73 Del Maestro, A.G. (2008) The superconductor–metal quantum phase transition in ultra-narrow wires, PhD Thesis, Harvard University.

74 Orr, B.G., Jaeger, H.M., Goldman, A.M., and Kuper, C.G. (1986) Global phase coherence in two-dimensional granular superconductors. *Physical Review Letters*, **56**, 378–381.

75 Hertz, J.A. (1976) Quantum critical phenomena. *Physical Review B*, **14**, 1165–1184.

76 Millis, A.J. (1993) Effect of a nonzero temperature on quantum critical points in itinerant fermion systems. *Physical Review B*, **48**, 7183–7196.

77 Moriya, T. and Kawabata, A. (1973) Effect of spin fluctuations on itinerant electron ferromagnetism. *Journal of the Physical Society of Japan*, **34**, 639.

78 Pankov, S., Florens, S., Georges, A., Kotliar, G., and Sachdev, S. (2004) Non-Fermi-liquid behavior from two-dimensional antiferromagnetic fluctuations: A renormalization-group and large-N analysis. *Physical Review B*, **69** (5), 054426.

79 Ma, S.K., Dasgupta, C., and Hu, C.K. (1979) Random antiferromagnetic chain. *Physical Review Letters*, **43**, 1434–1437.

80 Khlebnikov, S. (2012) Critical current of a superconducting wire via gauge/gravity duality. *Physics Letters B*, **715** (1–3), 271–274. *arXiv:1201.5103*.

81 Maldacena, J.M. (1998) The large N limit of superconformal field theories and supergravity. *Advances in Theoretical and Mathematical Physics*, **2**, 231.

4
Duality

4.1
Introduction

Duality in quantum field theory and in many-body systems is a powerful concept which enables one to understand a strongly-coupled regime by mapping into a weakly-coupled regime. Technically, this is done by going from a given representation (of fields) to a dual representation (of fields), where in the simplest scenario, the Lagrangian density (or action) in terms of the dual fields takes the same form as the original formulation, that is, the so-called self-dual. The parameters associated with the terms are mapped from the original parameters. In particular, the coupling constant characterizing the strength of interaction is mapped from a large value (strong coupling) to a small value (weak coupling), and vice versa. Often, in weak-coupling theory, perturbative methods can be employed to calculate physical quantities of interest.

In the context of superconductivity in 2D, a duality between a description based on the Cooper pair, and an alternative description based on the vortex enables one to study the dynamics of the superconducting system, including Cooper pair and vortex dynamics, as well as the superconductor–insulator transition. In single Josephson junctions, the Schmid-type KTB transition between a phase of localized order parameter phase (superconducting), to a phase of large phase fluctuation (insulating), exhibits a duality in the imaginary-time formulation in which the dual description is given in terms of the instantons of the phase-only theory [1, 2].

In the context of 1D superconducting nanowires, in Chapter 3, several examples of duality transformation have already been encountered. These include the dual transform employed by Khlebnikov (Section 3.5.3) to study the behavior of short, narrow wires connected to superconducting leads, where tunneling of a vortex-antivortex pair couples to the Mooij–Schön plasmon mode, and in the dual transform of Refael *et al.* in their treatment of the 1D Josephson junction chain in the strong Josephson-coupling limit (Section 3.4.1), which enabled the derivation of an effective sine-Gordon action.

In this chapter, we explore the implications of such duality in the context of short wires in a much fuller way, particularly regarding new behaviors, devices, and concepts based on such behaviors. In terms of device concept, the existence of a duality

One-Dimensional Superconductivity in Nanowires, First Edition. F. Altomare and A.M. Chang

between a QPS junction and a Cooper pair junction were independently noted by Mooij and Nazarov [3], and by Khlebnikov [4, 5]. From the duality, novel behaviors emerged. Particularly noteworthy are the predictions of the Josephson-type voltage-"phase" relationship, that is, $V = V_c \sin \Phi$ in QPS junctions, in contrast to the conventional current–phase relationship, $I = I_c \sin \varphi$ in Josephson junctions, where Φ is a dual field to φ and represents the dipole moment of the wire (up to a constant of proportionality) [5], the existence of Shapiro current steps under microwave irradiation (rather than voltage steps), and the implementation of 1D superconducting nanowire-based qubits. These predictions are among the most consequential regarding new phenomena associated with ultranarrow superconducting nanowires, and their potential applications.

The analysis of short wires provided a theoretical underpinning for the experimentally observed SIT, and revealed that the effective shunt resistance R_s across the entire short wire controls the quantum phase transition with the transition between superconducting and insulating phases occurring at $R_Q/R_s = 1$. This occurrence, rather than the parameter for long wires determined solely by the plasmon propagation impedance, $\mu = R_Q/2Z$, where $Z = \sqrt{L_{kin}/C}$ with L_{kin} the kinetic inductance and C the geometric capacitance per unit length, is a consequence of the constraints imposed by the boundary. The connection to the large and stiff superconducting leads at the boundary constraints the phase, in such a way that in the Büchler *et al.* [6] and Meidan *et al.* [7] theories, the most probable configuration for QPS–anti-QPS pairs occurs at the boundary, so that in effect only the $k \to 0$ Fourier component provides dissipation (Sections 3.3 and 3.5.2, respectively), while in the Khlebnikov theory, the most likely tunneling path occurs in the middle of the short wire, and the apparent SIT is controlled by the phase stiffness K relative to the QPS fugacity f_{QPS} (Sections 3.5.3 and 4.3). Moreover, Khlebnikov derived a tunneling rate and linear resistance of the form $\sim \exp[-\pi^2 K_s w/(2 L k_B T)]$ while Meidan *et al.* used a self-consistent treatment to produce a sharp drop-off in the resistance below T_c, on the superconducting side of the transition.

Despite these differences between the resistively-shunted theories of Büchler *et al.* [6] and Meidan *et al.* [7], and the plasmon-only theory due to Khlebnikov, which does not explicitly include dissipation via core normal electron [4, 5], the conclusion is the same in that wires that are short – in the sense of $L \ll c_{pl}\hbar\beta$ – act as a unit, in a manner analogous to a single resistively shunted Josephson junction, for example, with the phase-difference across the entire unit (nanowire) as the important field variable, rather than its dependence along the wire interior. Moreover, there is a regime where the QPS interaction is linear in time (imaginary-time) separation, rather than logarithmic. The linear interaction is analogous to an often-used model for the quark confinement potential. This enables one to connect the physics of QPS in short wires that are bounded by stiff superconducting leads with the physics of quark confinement in high-energy physics. In the two approaches, effective shunt-resistor versus only plasmon, the regime where this linear potential becomes relevant differs. In the latter, it is an integral part of the low energy physics that leads to the predictions emerging from the duality. Thus, the deconfined state, equivalent to a quark plasma, is identifiable as the asymptotically free

QPS state, and represents the insulating phase of the superconductor–insulator quantum phase transition, while the confined state, within which the QPS and anti-QPSs form a bound pair, represents the superconducting state.

In this analysis, duality shows up as a mapping between a description based on the phase of the Cooper pair, φ, to a description based on the instantons of the phase-only theory, in other words, based on the QPS. From a historical perspective, such a duality may be anticipated from existing theories of 2D superconducting systems, for disordered films and islands, and in regular or disordered Josephson junction arrays [8–10]. There, the two-dimensionality naturally gives rise to the topological defects in real space, namely, the vortices. In this 2D case, a classical topological phase transition of the KTB variety occurs at a finite temperature determined by the vortex-antivortex binding energy-scale. This occurs provided that Coulomb effects are not so dominating as to cause quantum fluctuations in the phase to destabilize the superconducting state and drive the transition temperature to zero. In the limit of strong Coulomb repulsion, quantum fluctuations dominate, and a superconductor-insulation transition of the quantum variety takes place at $T = 0$.

The existence of a duality between a Cooper pair and a vortex in 2D films, islands, and Josephson arrays is a well-established concept in the theoretical literature, where the duality holds approximately [8–12]. The theories with duality sought to explain experimental observations of the SIT first observed in 2D disorder superconducting films [13, 14]. Nevertheless, in this 2D case, there is evidence that the approximations leading to the dual theory do not capture the full extent of the fluctuations introduced by various processes, such as the Aslamazov–Larkin fluctuations, and thus yield an oversimplified picture of the SIT, as pointed out by Feigel'man, Larkin and Skvortsov [15] and others.

Here, in 1D nanowires, the QPS is a topological excitation of the quantum system in Euclidean $(1 + 1)$ dimension containing one spatial dimension plus one imaginary-time dimension. By mapping the short-wire system into an effective systems analogous to a single Josephson junction, the spatial dimension is effectively integrated out in the low energy effective theory, and only the imaginary-time dimension remains. Specifically, the resulting behavior is related intimately to the Schmid-type transition for dissipative systems [1], for example, the resistively-shunted single Josephson junction, a system which lends itself to a dual transformation between a Cooper pair Bose particle formulation, and an instanton/QPS formulation [1, 2]. In a related context, Khlebnikov utilized a dual transformation [4, 5] to study the influence of the boundary conditions imposed by large superconducting leads on the behavior of a short wire within a phase-only model in the absence of core normal electron dissipation.

The reduction to an effectively $(1 + 0)$ dimensional problem renders the duality to be more exact than in the 2D case. In the context of the 1D SC nanowire, several groups have proposed nearly exact duality: This has led to the discovery of the new phenomena, dual to the well-known Josephson phenomena in JJ, termed QPS junctions, new device concepts, and so on as mentioned above.

From a technical perspective, starting from a 2D or analogously, a (1 + 1) Euclidean action, the dual transformation from a phase-based representation starts with the Villain transformation, which replaces the Josephson coupling term $E_J \cos(\varphi - \varphi')$ by a quadratic term, in a form that preserves the periodicity [9–11, 16]:

$$\exp\left(\int \frac{1}{\hbar} d\tau\, E_J \left\{\cos\left[\varphi(\tau) - \varphi'(\tau)\right]\right\}\right)$$
$$\to \sum_m \exp\left\{-\frac{E_J}{2\hbar}\int \frac{d\tau}{\tau_0}\left[\varphi(\tau) - \varphi'(\tau) + 2\pi m_i(\tau)\right]^2\right\} . \tag{4.1}$$

Here, $m_i(\tau)$ is an integer valued function. This is followed by the introduction of an auxiliary integer field J

$$\sum_m \exp\left\{-\frac{E_J}{2\hbar}\int \frac{d\tau}{\tau_0}\left[\varphi(\tau) - \varphi'(\tau) + 2\pi m_i(\tau)\right]^2\right\}$$
$$\to \sum_{|J_i(\tau)|} \exp\left(\int \frac{d\tau}{\hbar\tau_0}\sum_i\left\{-\frac{1}{2E_J}|J_i(\tau)|^2 - i J_i(\tau)\left[\varphi(\tau) - \varphi'(\tau)\right]\right\}\right) . \tag{4.2}$$

The quadratic term in the J field now has a coefficient $1/E_J$, which is the inverse of the original coupling E_J; thus, a strongly-coupled system (when E_J is large) is mapped into a weakly-coupled one (with a small $1/E_J$). The duality transform for strongly Josephson-coupled chains in Section 3.4.1 is of this variety, for instance. In the case of a short 1D nanowire, due to the linear interaction between QPSs, the Villain approximation is not necessary, and an exact expansion of the $\cos \Phi(\tau)$ term takes place [1, 2, 5], where Φ is a dual field to the phase φ.

4.2
Mooij–Nazarov Theory of Duality – QPS Junctions

The Mooij–Nazarov theory of duality between a Josephson junction and a QPS junction is in part founded on the implications of the Büchler *et al.* analysis (Section 3.3) of the environmental influence on the QPS rate in short wires. For a sufficiently large QPS fugacity f_{QPS}, compared to a dissipation scale parameterized by R_Q/R_s, where R_s is the shunt resistance, QPSs proliferate, and successive QPS events become coherent. Under this condition, the dynamics of QPS needs to be treated quantum-mechanically. The quantization of the QPS dynamics in short 1D nanowires reveals a duality with Cooper pair dynamics in a Josephson junction. The duality mapping is motivated by drawing a parallel between the energy level structure of the Josephson junction-based Cooper box in the presence of Josephson tunneling of Cooper pairs, with that of a superconducting nanowire connected to a wider superconductor in a ring geometry, in the presence of coherent QPS transitions connecting equivalent ground states.

Mooij and Nazarov put forth a Hamiltonian formulation of the two devices. The energy diagrams and device geometries are shown, respectively, for the Cooper box and for the QPS junction device in Figures 4.1a,b. In the case of the Cooper box, the charging energy for the addition of a Cooper pair onto the superconducting island, if dominant over the Josephson energy, tends to enforce states of a fixed Cooper pair number. Analogously, in the QPS junction in a ring geometry, flux quantization enforces an integral number of flux quantum to be enclosed by the ring. Thus, for the Cooper box, the starting formulation is cast in the representation diagonal in the dominant energy contribution arising from the Coulomb energy in the charging of the superconducting island by an integer number of Cooper pairs, with an energy scale of $E_C = e^2/C$, where C is the total capacitance of the island to the environment. The number eigenstates are denoted by $|n_{cp}\rangle$. The Josephson coupling enables Cooper pairs to tunnel on and off the central superconducting island, and thus couples states of n_C to $n_C \pm 1$ to lowest order, with an energy scale given by the Josephson energy E_J. The Hamiltonian for the Josephson junction Cooper box in the basis of the number of Cooper pairs is thus

$$H_{JJ,n_{cp}} = E_C(n_{cp} - n_g)^2 - \frac{E_J}{2}\sum_{n_{cp}}(|n_{cp}+1\rangle\langle n_{cp} + \text{h.c.}) , \qquad (4.3)$$

where h.c. is short for the Hermitian conjugate. The quantity n_g represents an offset introduced by a coupling of the center island to a DC biasing voltage. The energy diagram shown in Figure 4.1a is valid when $E_C \gg E_J$ and is periodic in the continuous variable n_g. At half integer values of n_g, an anticrossing takes place between the Cooper pair branches with $n_{cp,-} = n_g - \frac{1}{2}$ and $n_{cp,+} = n_g + \frac{1}{2} = n_{cp,-} + 1$, respectively. At this symmetry point, the anticrossing splitting equals $2E_J$. In the vicinity of such special points, the Cooper box may be operated as a qubit, with an effective Hamiltonian [17]

$$H_{qubit} = -\frac{E_J}{2}\sigma_z + E_C\left(n_g - \frac{1}{2}\right)\sigma_x . \qquad (4.4)$$

This is a pseudospin formalism, where the σ_is represent Pauli matrices in the i-direction, with $i = (x, y, z)$. These Pauli matrices operate on the pseudospin 1/2 space in which the spin-down projection corresponds to the symmetric combination of $|n_{cp,-}\rangle$ and $|n_{cp,+}\rangle$ states, while the antisymmetric combination corresponds to the pseudospin-up projection along the z-direction.

In direct parallel with the Josephson junction Cooper box Hamiltonian of (4.3), a Hamiltonian with an identical form may be written down for the QPS qubit, based on the number of flux quanta in units of $\phi_0 = hc/(2e)$ (or $\phi_0 = h/(2e)$ in S.I. units). The Hamiltonian in this case is written in terms of an eigenstate of an integer number of flux quanta, $|n_{flux}\rangle$. An external magnetic field can be applied to the QPS qubit in the shape of a ring to produce a flux bias ϕ, which, in units of the flux quantum, is denoted by $f \equiv \phi/(-\phi_0)$. The inductive energy $E_L = \phi_0^2/2L_{kin}$ plays an analogous role as the charging energy in the Cooper box case. Here, the kinetic inductance $L_{kin} = m/(e^2 n_s \mathrm{A}) = \hbar/(\pi e^2 D(E_F) D\Delta_0 \mathrm{A})$ (3.44)

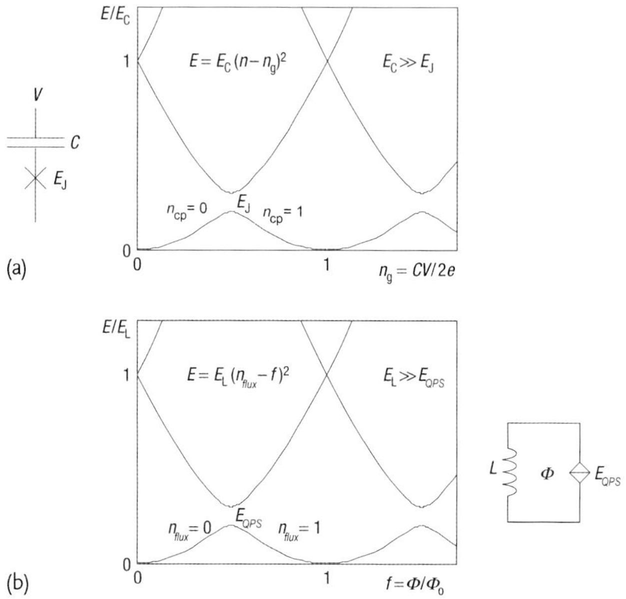

Figure 4.1 Circuit and energy dependence: (a) A Cooper pair box and (b) a QPS qubit. The diamond-shaped symbol in the QPS qubit circuit represents the quantum phase-slip process. The capacitive energy is $E = E_c(n_{cp} - n_g)^2$, while the inductive energy is $E = E_L(n_{flux} - f)^2$. From [3].

dominates over the geometric inductance [18], where m is the electron mass, n_s is the superconducting electron density, and A is the nanowire cross section. The energy scale, which plays the role equivalent to E_J, is driven by the rate of phase slips, that is, $E_{QPS} = \hbar\Gamma_{QPS}$. The Hamiltonian in the $|n_{flux}\rangle$ basis is thus [18]

$$H_{QPS,n_{flux}} = E_L(n_{flux} - f)^2 - \frac{E_J}{2}\sum_{n_{flux}}(|n_{flux} + 1\rangle\langle n_{flux} + \text{h.c.}) \,. \tag{4.5}$$

The exact duality between the two devices can now be obtained by writing down the general Hamiltonian for a Josephson junction and for a QPS junction, through the introduction of the conjugate pair operators of charge $\hat{q}$ (in units of $(-2e)$) and phase φ:

$$[\hat{q}, \hat{\varphi}] = -i \,. \tag{4.6}$$

Furthermore, the influence of the environment, such as those leading to dissipation, may be incorporated using the Caldeira–Leggett approach [19], and in addition, the external bias of the voltage and current varieties are also considered. The Hamiltonian for the Josephson junction is now cast in the form

$$H_{JJ} = E_C\hat{q}^2 - E_J\cos\hat{\varphi} + H_{env} + H_{bias} \,. \tag{4.7}$$

The Hamiltonian of the environment contains the boson modes of the environment which are coupled to the system via terms contained in H_{bias}. The biasing

Hamiltonian comes in two varieties, current-biasing and voltage-biasing, given respectively by

$$H_{I,\text{bias}} = \frac{-\phi_0}{2\pi}(I - \hat{I}_\text{R})\hat{\varphi} \ , \tag{4.8}$$

and

$$H_{V,\text{bias}} = (-2e)(V - \hat{V}_\text{R})\hat{q} \ . \tag{4.9}$$

I and V are externally applied values, and the operators $\hat{I}_\text{R}$ and $\hat{V}_\text{R}$ account for fluctuations in the effective resistor due to the influence of the boson modes of the environment. The effect of the coupling to such modes is modeled by choosing coefficients in the linear combination of the modes to reproduce the response function of the environment, with

$$\hat{I}_\text{R} = \frac{\hbar}{2e}(-i\omega)Y(\omega)\hat{\varphi}(\omega) \ , \tag{4.10}$$

and

$$\hat{V}_\text{R} = (2e)(-i\omega)Z(\omega)\hat{q} \ , \tag{4.11}$$

where the admittance $Y(\omega)$ characterizes a resistor connected in parallel, and the impedance $Z(\omega)$ for a resistor in series with the qubit.

Similarly, the Hamiltonian for the QPS junction reads

$$H_\text{QPS} = E_\text{L}\frac{\hat{\varphi}^2}{(2\pi)^2} - E_\text{QPS}\cos 2\pi\hat{q} + H_\text{env} + H_\text{bias} \ . \tag{4.12}$$

A duality mapping can now be accomplished by making the canonical transformation:

$$(\hat{q}, \hat{\varphi}) \rightarrow \left(-\frac{\hat{\varphi}}{2\pi}, 2\pi\hat{q}\right) \ , \tag{4.13}$$

in concert with the mapping of the parameters:

$$\begin{aligned} E_\text{QPS} &\rightarrow E_\text{J} \ , \\ E_\text{L} &\rightarrow E_\text{C} \ , \\ I &\leftrightarrow \frac{V}{R_\text{Q}} \ , \\ Y(\omega)R_\text{Q} &\leftrightarrow \frac{Z(\omega)}{R_\text{Q}} \ . \end{aligned} \tag{4.14}$$

This exact duality mapping enables one to deduce the transport properties of a QPS junction based on the behavior of Josephson junctions. Below, several outstanding dual properties, such as the dual to the Josephson junction current–phase

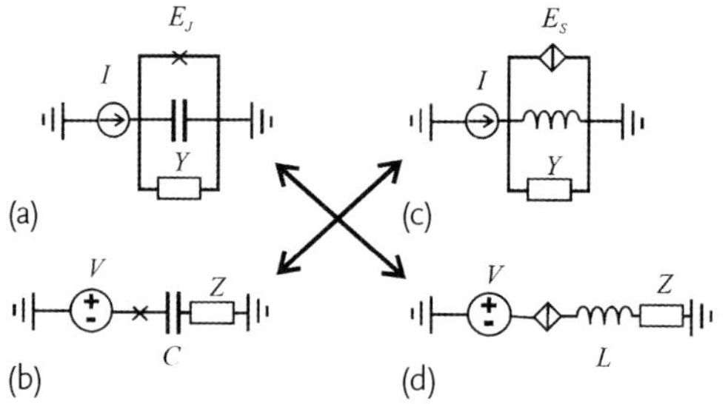

Figure 4.2 Dual equivalence of Josephson and QPS junctions in circuits. (a) A current-biased Josephson junction; (b) a voltage-biased Josephson junction; (c) a current-biased QPS junction, and (d) a voltage-biased QPS junction. Circuit (a) is the exact dual of circuit (d), and circuit (b) is the exact dual of circuit (c). From [3].

relationship and the dual to the Shapiro voltage steps, are discussed. The predictions of these notable dual relations, and in addition, the future development of various QPS junction-based qubits, dual to known Josephson junction-based qubits, such as the Cooper box, flux, and phase qubits, present opportunities for the realization and exploration of new phenomena and devices in this exciting field of 1D superconducting nanowires. In Figure 4.2, we reproduce two types of QPS devices, including the QPS qubit and junctions and their Josephson counterparts.

4.2.1
QPS Junction Voltage-Charge Relationship and Shapiro Current Steps

To illustrate some of the new behaviors of the new QPS devices, we consider a QPS junction in the opposite limit of the QPS qubit already introduced. Here, we are interested in the opposite regime of parameters, where $E_{\text{L}} \ll E_{\text{QPS}}$ instead. For this regime, the charge is localized and the phase exhibits large fluctuations, and thus the nanowire is in a regime toward insulating. The corresponding situation for the dual Josephson circuit then requires $E_{\text{C}} \ll E_{\text{J}}$, and phase is localized. Under current-bias, this situation is well-described by modeling the junction as a classical, resistively-shunted Josephson junction, in which the dynamics of the phase in a tilted washboard potential obeys the familiar differential equation (for electrons, with e the fundamental charge)

$$-\frac{\phi_0}{2\pi}\left(C\frac{d^2\varphi}{dt^2}+\frac{1}{R}\frac{d\varphi}{dt}\right)+I_{\text{c}}\sin\varphi=I(t)\,. \tag{4.15}$$

For current drive below the critical current $I_{\text{c}} = 2\pi E_{\text{J}}/(-\phi_0)$, the phase φ is localized within a local minimum of the washboard potential, and the junction is superconducting.

The corresponding behavior of the dual QPS junction is obtained by making the duality mapping, giving

$$(-2e)\left(L\frac{d^2q}{dt^2}+R\frac{dq}{dt}\right)+V_{\text{c}}\sin 2\pi q=V(t)\,. \tag{4.16}$$

Thus, in the absence of time-dependence, with the application of a DC voltage, we have the relation

$$V_{DC} = V_c \sin 2\pi q \,, \quad V_c = \frac{2\pi}{(-2e)} E_{QPS} \,. \tag{4.17}$$

This is dual to the usual current–phase relation, $I = I_c \sin\varphi$, for a Josephson junction. In addition, the "plasma" frequency of oscillation about a minimum for small voltage drives is

$$\omega_{pl} = \sqrt{2E_L E_{QPS}} \,. \tag{4.18}$$

Aside from the dual version of the Josephson current–phase relationship, other novel effects are anticipated under RF drive. The high-frequency environmental impedance of a submicrometer long QPS junction is determined by the geometric capacitance and inductance of nearby wiring, and is typically around 300 Ω. This impedance leads to a high-quality factor of the plasma oscillations, that is,

$$Q^2_{QPS} \equiv \beta_L = 2\pi \frac{V_c}{2e} \frac{L}{R_N^2} = 2\pi^2 E_{QPS} \frac{1}{E_L} \left(\frac{R_Q}{R_N}\right)^2 \,. \tag{4.19}$$

The parameter β_L is the analog of the McCumber parameter β_C for Josephson junctions. Duality suggests a strong nonlinear hysteretic response to a small RF AC voltage drive, V_{AC}, on the order of $V_{AC} \sim V_c/Q_{QPS}$.

Even more dramatically, and of potential significance, is the dual version of the Josephson voltage steps with

$$I_n = n(-2e)\nu \,, \tag{4.20}$$

with ν as the RF frequency. At $\nu = 50\,\text{GHz}$, the resulting DC current for $n = 1$ is approximately 16 nA. Here, the steps are in the current rather than in the voltage, as shown in Figure 4.3b.

4.2.2 QPS Qubits

Mooij and Harmans proposed the phase-slip flux qubit [18] based on a ring geometry similar to the device in Figure 4.1b. The ring comprises a narrow portion within a wider ring, in which the narrow, weak portion could be implemented via a 1D nanowire, as indicated in Figure 4.4a. The energy diagram in the absence of any QPS, that is, QPS rate $\Gamma_{QPS} = f_{QPS}/\tau_0 = 0$, and the level diagram versus different fluxoid number, $n = n_{flux}$ are shown in Figure 4.4b, while the level repulsion is shown in Figure 4.4c, with a gap equal to $E_{QPS} = \hbar\Gamma_{QPS} = \hbar f_{QPS}/\tau_0$. Here, τ_0 is the (imaginary)-timescale of the QPS core, given in Section 3.2.2. Typically, for a qubit to be operational, the anticrossing gap E_{QPS} must exceed the thermal energy $k_B T$ for typically low temperatures $T \sim 100\,\text{mK} \approx 8.7\,\mu\text{eV}$. This corresponds to a

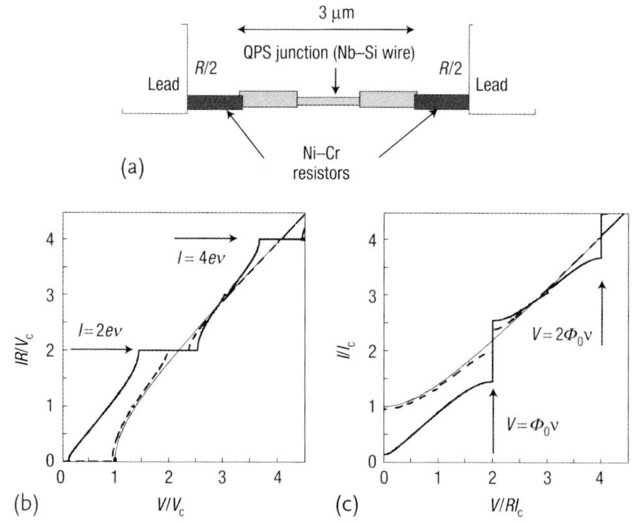

Figure 4.3 Shapiro steps at constant current from a QPS junction. (a) The junction must be embedded in a highly resistive environment. Parasitic capacitances are sufficiently small for the design presented. (b) QPS junction – The parameters correspond to $Q = 0.5$, $2\pi\nu = \omega_{\rm pl}$, $V_{\rm AC} = V_{\rm c} = 0, 1, 5$ for the thick solid, dashed, and thin solid curves, respectively. With the values: $E_{\rm S}/\hbar \equiv E_{\rm QPS}/\hbar = 120\,\rm GHz$, and $E_{\rm L}/\hbar = 30\,\rm GHz$, the plasma frequency $\omega_{\rm pl}/2\pi = 85\,\rm GHz$, and the critical voltage is $V_{\rm c} = 1.56\,\rm mV$, and $R = 115\,\rm k\Omega$. (c) Voltage Shapiro steps in the dual Josephson circuit with equivalent parameters. From [3].

rate $\Gamma_{\rm QPS} \sim 2.1\,\rm GHz$, using (4.21) as a rough guide,

$$\begin{aligned} \Gamma_{\rm QPS} &= \frac{b}{\tau_{\rm GL}}\frac{\mathsf{L}}{\xi(T)}\left(\frac{\Delta F_0}{E_{\rm GL}}\right)^{1/2}\exp\left(-\frac{\Delta F_0}{E_{\rm GL}}\right) \\ &\approx 1.5\,c\frac{\mathsf{L}}{\xi(T)}\sqrt{\frac{R_{\rm Q}}{R_\xi}\frac{k_{\rm B}T}{\hbar}}\exp\left(-0.3\,d\frac{R_{\rm Q}}{R_\xi}\right)\,, \\ E_{\rm GL} &= \frac{\hbar}{a\tau_{\rm GL}} = \frac{8}{a\pi}(T_{\rm c}-T)\,, \quad R_\xi = \frac{\xi}{\mathsf{L}}R_{\rm N}\,. \end{aligned} \tag{4.21}$$

Figure 4.5 shows the QPS rate $\Gamma_{\rm QPS}$ as a function of resistance per unit length for different values of the superconducting coherence length ξ. The constants *a*–*d* are of order unity, and $R_{\rm N}$ is the normal resistance of the wire. The values here correspond to Nb-Si, MoGe, and Nb nanowires. For Al, the coherence length is typically a factor of 10 larger, while the resistance per unit length is a factor of 10 smaller, while length may be as long as 100 μm.

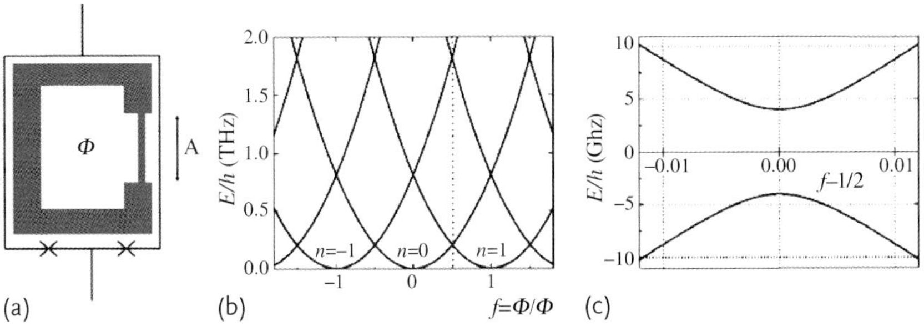

Figure 4.4 (a) Schematic diagram of a narrow wire in a close superconducting loop. A SQUID (superconducting-quantum-interference device) with two Josephson junctions (crosses) is used to measure the qubit state. (b) Energy levels as a function of the applied flux for different fluxoid numbers, for a loop with $L_{\text{kin}} = 4nH$. A phase-slip event changes the fluxoid number n. The arrow indicates the operating point at $f = 1/2$. (c) Lowest two energy levels of the phase-slip qubit with $I_p = 0.25$ A and $\Gamma_{QPS}/2 = 4$ GHz near $f = 1/2$. From [18].

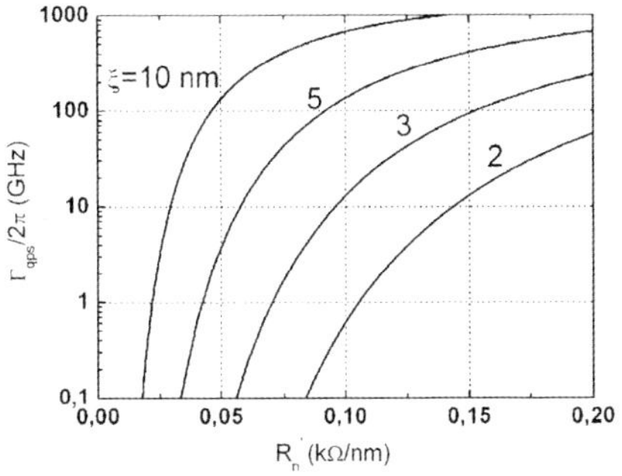

Figure 4.5 Quantum phase-slip rate as a function of resistance per unit length for four values, according to (4.12). Parameters: L = 50 nm, $T_c = 1.2$ K and $c = d = 1$. From [18].

4.3 Khlebnikov Theory of Interacting Phase Slips in Short Wires: Quark Confinement Physics

Much of the result and prediction for Mooij–Nazarov theory for QPS qubit and junctions, derived from a duality with the corresponding Josephson junction-based devices, can be obtained in the Khlebnikov theory of short wires introduced in Section 3.5.3. There, a duality transform was introduced to relate the phase-only theory expressed in terms of the vortex charge, q_ρ, to a dual theory based on a photon field, which represents the plasmon modes of the low amplitude fluctuations of the original phase field.

Considering a short wire, Khlebnikov derived an effective interaction between phase slips, which varies linearly with the imaginary-time temporal separation between them. Focusing on the tunneling of an individual QPS–anti-QPS (instanton–

anti-instanton) pair, Khlebnikov showed that the most likely spatial point for a vortex to tunnel across a narrow wire occurs at the midpoint of the wire along the length. The tunneling is governed by a rate, which, due to the boundary conditions imposed by the stiff superconducting leads connected to the ends of the narrow wire, becomes exponentially suppressed with a functional form (3.181):

$$\Gamma_{\text{sw}} \sim \exp\left(-\frac{\pi^2 K_s w}{2 L k_B T}\right) . \tag{4.22}$$

The coefficient in the exponent to $\beta = 1/(k_B T)$, however, arises from quantum-mechanics, and not classical activation, and is significantly smaller than the free energy barrier between adjacent free energy minima differing in phase by 2π. Here, the coefficients are determined by the energy configurations of the initial and final tunneling states.

Although derived for narrow wires not necessarily narrower than the superconducting coherence length, ξ, importantly, the solution relevant to the most singular behavior responsible for the linear QPS–anti-QPS interaction, and for the above exponential suppression of the tunneling rate in the superconducting phase, was argued to be relevant to superconducting nanowires in the 1D limit as well. This is due to the observation that the $l = 0$ Fourier component across the wire within a narrow wire is dominant. In other words, in the vortex tunneling process, the Fourier component corresponding to an absence of any spatial dependence across the width is responsible for the most notable behavior. This is precisely the relevant scenario for a 1D superconducting nanowire, for which the order parameter is largely uniform across its ultranarrow width dimension since $\xi \gg w$ here.

To obtain the results and the striking predictions, the interaction between multiple QPSs and anti-QPSs, rather than just a single pair, must be considered. In analogy with the representation introduced in Chapter 3 (see (3.59), (3.80), (3.106), and (3.147)), the following partition function is written for the 1D nanowire containing a Gaussian term plus the topological term associated with the phase slips, that is,

$$Z_{1D} = \int \mathcal{D}\Phi(\tau) \exp\left[-\frac{\hbar}{2Kc}\int \frac{d\tau}{\tau_0}\left(\frac{\partial \Phi}{\partial \tau}\right)^2\right] \sum_{N_+=0}^{\infty}\sum_{N_-=0}^{\infty} \frac{\left(\frac{1}{2} f_{\text{QPS}}\right)^{N_+ + N_-}}{N_+! N_-!}$$
$$\times \int_0^{\hbar\beta} \prod_{l=1}^{N_+} \frac{d\tau_l}{\tau_0} e^{i\Phi(\tau_l)} \prod_{m=1}^{N_-} \frac{d\tau'_m}{\tau_0} e^{-i\Phi(\tau'_m)} , \tag{4.23}$$

where Φ represents a dipole moment of the short wire. The fugacity $f_{\text{QPS}} \propto \exp(-S_{\text{core}}/\hbar)$, where S_{core} is the action of the QPS core (3.60), and N_+ is the number of QPSs, and N_- the number of anti-QPSs.

The physical interpretation of the $\Phi(\tau)$ field comes from the duality relationship between the phase field of the order parameter, $\varphi(x, \tau)$, and the spatially-dependent field, $\Phi(x,\tau)$, to be related to $\Phi(\tau)$ below:

$$\partial_t\varphi(x,\tau) \propto \partial_x\Phi(x,\tau)\,, \quad \partial_t\varphi(x,\tau) \propto \partial_x\Phi(x,\tau)\,. \tag{4.24}$$

Let us recall the definition of the quantities $q_\rho \equiv \partial_\rho\varphi$ introduced in Section 3.5.3 in association with the vortices and phase fluctuations for narrow and short wires in (1 + 2) dimensional time-space (3.162). There, the components of q_ρ may be thought of as being proportional to a three-current density. Here, we have a (1 + 1)-dimensional space-time, with x denoting the sole spatial direction. In real time, for the present situation, the two-current density is then

$$(J^0; J^1) = (J^0; J^x) \equiv \frac{-(2e)\hbar n_s}{4m}\left(\frac{\partial\varphi}{c_{\rm pl}\partial t}, \frac{\partial\varphi}{\partial x}\right)\,. \tag{4.25}$$

Here, J^0 represents a charge density multiplied by $c_{\rm pl}$. Moreover, in 1D nanowires, we have found that the effective action is obtained by an expansion about the value $A = 0$ in (3.30). Thus, the spatial term J^x represents the supercurrent density. Making use of the duality relations above (4.24), we thus have

$$\frac{\partial\Phi(x,t)}{\partial t} = J^x\,; \quad \frac{\partial\Phi(x,t)}{\partial x} = J^0 c_{\rm pl}^2. \tag{4.26}$$

In Section 3.5.3, it was pointed out that the large superconducting leads connected to the ends of the nanowire imposed Dirichlet boundary conditions on the time-derivative of the phase φ. Thus, the duality condition indicates a corresponding Neumann condition for $\Phi(x,t)$ at the ends.

For large charging energy $E_{\rm C}$ and low temperature, to avoid excessive energy costs associated with the time-derivative of the φ field, which in turn is proportional to the spatial-derivative of $\Phi(x,t)$, the spatially uniform component of $\Phi(x,t)$ is relevant, in other words, the length normalized, $k = 0$ Fourier component, that is,

$$\Phi(t) \equiv \frac{1}{\mathsf{L}}\int_0^{\mathsf{L}} \Phi(x,t)dx\,. \tag{4.27}$$

This $\Phi(\tau)$ is the quantity introduced in (4.23). Alternatively, one may think in the following way: The QPS–anti-QPS pair action has a shallow minimum at L/2, reflecting the initial and final tunneling state energy. This is the configuration relevant for the QPS. Thus, the x-spatial dependence drops out, leaving only a dependence on τ. $\Phi(\tau)$ may be interpreted as a dipole moment of the entire wire since according to (4.26), $\Phi(x,t)$ may be viewed as being proportional to a dipole density associated with a charge density J^0.

The above forms the basis of the theory of linearly interacting QPSs, from which one can obtain an apparent superconductor-insulator transition (SIT) for short wires connected to large superconducting leads and deduce the consequences of duality, such as the duality Josephson relation $V = V_{\rm c}\sin\Phi$ in a QPS junction.

From the partition function in (4.23), one can demonstrate the linear form of the interaction by examining the contribution of a single QPS–anti-QPS pair

$$\begin{aligned} Z_{\text{pair}} &= \frac{\left(\frac{1}{2} f_{\text{QPS}}\right)^2}{1!1!} \int_0^{\hbar\beta} d\tau_1 d\tau_1' \int \mathcal{D}\Phi(\tau) \\ &\quad \times \exp\left[-\frac{\hbar}{2Kc} \int \frac{d\tau}{\tau_0} \left(\frac{\partial \Phi}{\partial \tau}\right)^2 + i\Phi(\tau_1) - i\Phi(\tau_1')\right] \\ &= \frac{\left(\frac{1}{2} f_{\text{QPS}}\right)^2}{1!1!} \int_0^{\hbar\beta} d\tau_1 d\tau_1' \exp\left[-\left(\frac{1}{2\hbar}\right) K|\tau_1 - \tau_2|\right], \end{aligned} \tag{4.28}$$

where the second line is readily obtained from the first by going into a frequency representation for Φ and performing the $\mathcal{D}\Phi$ integral over the Gaussian function. Note that if the pair were either both QPSs or both anti-QPSs, the relative sign would be positive in the initial expression, and the action would be extremely large and positive, diverging in the limit of zero temperature. Thus, in effect, only terms for which $N_+ = N_-$ in the partition sum contribute.

The string tension K may be related to the superfluid stiffness of a 1D nanowire K_s via the relation

$$\frac{1}{K} = \frac{1}{4\pi^2} \int \frac{dx}{K_s}. \tag{4.29}$$

For an ideal, uniform wire, this yields $K = 4\pi^2 K_s/L$. It should be adequate in the presence of small variations in wire width and height, such that the amplitude of the order parameter does not vary substantially, to permit a treatment based on phase-fluctuations only.

This presence of the linear interaction suggests a similarity to the physics of quark confinement. Khlebnikov draws a parallel in the analysis of this problem with the deconfinement and confinement transition in the quark problem, where the insulating state is equivalent to a deconfined quark plasma with free QPSs, and the superconducting state parallels a confined state with bound QPS–anti-QPS pairs. The analysis is cast in terms of the screening property of the quark system, regarding the ability to screen integer and fractional charges. The technical details are contained in Khlebnikov's original paper [5] and will not be reproduced here. Below, we sketch the scenario which yields the important dual relationship: $V = V_c \sin \Phi$.

The form of the partition in (4.23) can readily be cast into the form of a (1 + 0) dimension sine-Gordon theory by performing the sums and making use of the identity $\cos\Phi = 1/2(e^{i\Phi} + e^{-i\Phi})$:

$$\begin{aligned} &\sum_{N_+=0}^{\infty} \sum_{N_-=0}^{\infty} \frac{\left(\frac{1}{2} f_{\text{QPS}}\right)^{N_+ + N_-}}{N_+! N_-!} \int_0^{\hbar\beta} \prod_{l=1}^{N_+} \frac{d\tau_l}{\tau_0} e^{i\Phi(\tau_l)} \prod_{m=1}^{N_-} \frac{d\tau_m'}{\tau_0} e^{-i\Phi(\tau_m')} \\ &= \exp\left[+f_{\text{QPS}} \int d\tau \cos\Phi(\tau)\right]. \end{aligned} \tag{4.30}$$

This yields an action for the short wire, namely,

$$\frac{S_{1D}}{\hbar} = \int \frac{d\tau}{\tau_0} \left[\frac{\hbar}{2Kc} \left(\frac{\partial \Phi}{\partial \tau} \right)^2 - f_{QPS} \cos \Phi(\tau) \right] . \tag{4.31}$$

This is a periodic potential, for which there are Bloch states and a band structure.

Whereas in the limit of large QPS fugacity, such that $f_{QPS} \gg (c/\hbar) K\tau_0^2$, the system tends to localize with a minimum of the $\cos \Phi(\tau)$ potential, with occasional tunneling between minima, where the tunneling is associated with the instanton solutions of (4.31) of the form [1, 5]:

$$\Phi_{inst} = 4 \arctan e^{\omega_{pl}|\tau|} , \tag{4.32}$$

with an action

$$\frac{S_{inst}}{\hbar} = 8 \sqrt{\frac{\hbar f_{QPS}}{K\tau_0^2 c}} , \tag{4.33}$$

where ω_{pl} is the "plasma" frequency

$$\omega_{pl} = \sqrt{\frac{f_{QPS} Kc}{\hbar}} . \tag{4.34}$$

The instantons here correspond to bound QPS–anti-QPS pairs, and thus rare instantons indicate an insulating-like phase.

In this limit, the system spends most of its time near the minimum of the potential of $f_{QPS} \cos \Phi$, with a nearly constant value of Φ, and with fluctuations of small amplitude about the minimum with a frequency of oscillation given by the ω_{pl}. On rare occasions, large fluctuations occur as the system tunnels between minima via an instanton. Small fluctuations in Φ correspond to large fluctuations in the order parameter phase φ, and the nanowire behaves much like an insulator. Nevertheless, the periodic potential gives rise to Bloch states, and in principle, for a finite ratio $f_{QPS}\hbar/(K\tau_0^2 c) \gg 1$, a supercurrent can still pass through the nanowire due to the instanton tunneling process. The rate of tunneling behaves as

$$\Gamma_{inst} \sim \omega_{pl} e^{-S_{inst}/\hbar} . \tag{4.35}$$

This current is exponentially small, and the wire behaves much like an insulator. The instantons form a dilute gas with an average instanton density in imaginary time:

$$\bar{n}_{inst} \sim \omega_{pl} \sqrt{\frac{S_{inst}}{\hbar}} e^{-S_{inst}/\hbar} . \tag{4.36}$$

In the band structure, there must be a "crystal-momentum" in the path integral formalism, associated with the instanton number Q:

$$Q = \frac{1}{2\pi} \int d\tau \frac{\partial \Phi}{\partial \tau} . \tag{4.37}$$

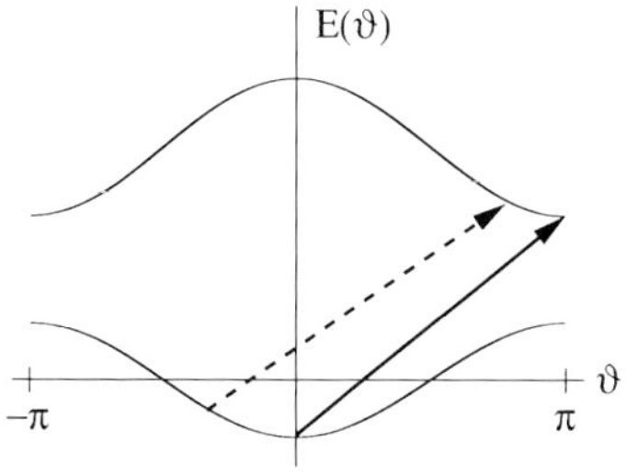

Figure 4.6 A schematic of the band structure. The arrows indicate the exciton transitions for two different values of the biasing current I. In general, the exciton frequency is dependent on the bias current. From [5].

Denoting this "crystal-momentum" by ϑ, the action in (4.23) is generalized by including a "source" term, namely,

$$S_{\text{QPS}} \to S_{\text{QPS}} + i\vartheta Q\,. \tag{4.38}$$

This means that in the partition sum for $Z_{1\text{D}}$, each QPS (anti-QPS) gains an extra factor of $e^{i\vartheta}$ $(e^{-i\vartheta})$, yielding

$$\frac{Z(\vartheta)}{Z(\vartheta = 0)} = \exp\left[2\int \frac{d\tau}{\tau_0}\bar{n}(\cos\vartheta - 1)\right]\,. \tag{4.39}$$

To find the average supercurrent in the ground state requires differentiation of $\ln Z(\vartheta)$ with respect to ϑ in order to give

$$I = \frac{(-2e)}{2\pi} i \left\langle \frac{\partial \Phi}{\partial \tau} \right\rangle = 2(-2e)\bar{n}\sin\vartheta\,. \tag{4.40}$$

Matching this expression with the usual supercurrent in a Josephson junction enables one to identify a critical current $I_c = 2(-2e)\bar{n}$, and importantly, ϑ as the phase difference between the ends of the nanowire $\Delta\varphi$.

An exact solution of the sine-Gordon action in (4.31) should give a band structure that has additional excited state bands periodic in the Brillouin zone in the crystal-momentum ϑ. A transition between bands, or branches of $E(\vartheta)$, where E is the energy of a band, occurs for the same value of $dE/d\vartheta$, which gives an identical current I for the initial and final branches (Figure 4.6). This multivalued behavior in the energy vs. ϑ is responsible for the occurrence of fractional Shapiro steps which were observed by Dinsmore *et al.* [20]. Furthermore, such an excitation changes ϑ and thus gives a voltage pulse in accordance with the relationship:

$$\frac{d\vartheta}{dt} = \frac{d\Delta\varphi}{dt} = (-2e)\frac{V}{\hbar}\,. \tag{4.41}$$

The application of a voltage couples to Φ, the nanowire dipole moment, via a term $(-2e)\Phi V$. Within the deeply "insulating" regime where $f_{\text{QPS}} \gg (c/\hbar)K\tau_0^2$, the system will reside near a minimum of the cos Φ potential for a long time, with small fluctuations about a mean value. The relation between the mean Φ value and an applied DC voltage V is obtained by minimizing the total potential

$$\frac{\hbar f_{\mathrm{QPS}}}{\tau_0}\cos\Phi-\left(-\frac{2e}{2\pi}\right)\Phi V$$

with respect to Φ, giving

$$V=V_{\mathrm{c}}\sin\Phi\,,\quad V_{\mathrm{c}}=\frac{2\pi\hbar f_{\mathrm{QPS}}}{-2e\tau_0}=\frac{2\pi E_{\mathrm{QPS}}}{-2e}\,. \tag{4.42}$$

This relationship dual to the current relationship $I = I_{\mathrm{c}}\sin\varphi$ in a Josephson junction is identical to what was deduced in the Mooij–Nazarov theory in Section 4.2. Recently, several exciting experimental development has indicated evidence for the type of novel behaviors predicted by these duality theories. These experiments are discussed in Chapter 8. On a different note, a recent theoretical preprint is attempting to derive the Mooij–Nazarov and Khlebnikov forms of duality from a different starting point [21].

References

1 Schmid, A. (1983) Diffusion and localization in a dissipative quantum system. *Physical Review Letters*, **51**, 1506–1509.

2 Schön, G. and Zaikin, A.D. (1990) Quantum coherent effects, phase transitions, and the dissipative dynamics of ultra small tunnel junctions. *Physics Reports*, **198**, 237–412.

3 Mooij, J.E. and Nazarov, Y.V. (2006) Superconducting nanowires as quantum phase-slip junctions. *Nature Physics*, **2**, 169–172.

4 Khlebnikov, S. (2008) Quantum phase slips in a confined geometry. *Physical Review B*, **77** (1), 014505.

5 Khlebnikov, S. (2008) Quantum mechanics of superconducting nanowires. *Physical Review B*, **78** (1), 014512.

6 Büchler, H.P., Geshkenbein, V.B., and Blatter, G. (2004) Quantum fluctuations in thin superconducting wires of finite length. *Physical Review Letters*, **92** (6), 067007.

7 Meidan, D., Oreg, Y., and Refael, G. (2007) Sharp superconductor–insulator transition in short wires. *Physical Review Letters*, **98** (18), 187001.

8 Fisher, M.P.A. (1990) Quantum phase transitions in disordered two-dimensional superconductors. *Physical Review Letters*, **65**, 923–926.

9 Fisher, M.P.A. and Lee, D.H. (1989) Correspondence between two-dimensional bosons and a bulk superconductor in a magnetic field. *Physical Review B*, **39**, 2756–2759.

10 Fazio, R. and Schön, G. (1991) Charge and vortex dynamics in arrays of tunnel junctions. *Physical Review B*, **43**, 5307–5320.

11 José, J.V., Kadanoff, L.P., Kirkpatrick, S., and Nelson, D.R. (1977) Renormalization, vortices, and symmetry-breaking perturbations in the two-dimensional planar model. *Physical Review B*, **16**, 1217–1241.

12 van Wees, B.J. (1991) Duality between Cooper-pair and vortex dynamics in two-dimensional Josephson-junction arrays. *Physical Review B*, **44**, 2264–2267.

13 Haviland, D.B., Liu, Y., and Goldman, A.M. (1989) Onset of superconductivity in the two-dimensional limit. *Physical Review Letters*, **62**, 2180–2183.

14 Hebard, A.F. and Paalanen, M.A. (1984) Pair-breaking model for disorder in two-dimensional superconductors. *Physical Review B*, **30**, 4063–4066.

15 Feigel'man, M.V., Larkin, A.I., and Skvortsov, M.A. (2001) 5. Superconductor–metal–insulator transitions: Quantum superconductor–metal transition in a proximity array. *Physics Uspekhi*, **44**, 99–104.

16 Villain, J. (1975) Theory of one- and two-dimensional magnets with an easy magnetization plane. II. The planar, classical, two-dimensional magnet. *Journal de Physique France*, **36** (6), 581–590.

17 Devoret, M.H., Wallraff, A., and Martinis, J.M. (2004) Superconducting qubits: A short review. *arXiv:cond-mat/0411174*, p. 41.

18 Mooij, J.E. and Harmans, C.J.P.M. (2005) Phase-slip flux qubits. *New Journal of Physics*, **7**, 219.

19 Caldeira, A.O. and Leggett, A.J. (1981) Influence of dissipation on quantum tunneling in macroscopic systems. *Physical Review Letters*, **46**, 211–214.

20 Dinsmore, R.C., Bae, M.H., and Bezryadin, A. (2008) Fractional order Shapiro steps in superconducting nanowires. *Applied Physics Letters*, **93** (19), 192505.

21 Kerman, A.J. (2012) Flux-charge duality and quantum phase fluctuations in one-dimensional superconductors. *arXiv:1201.1859v4*.

5
Proximity Related Phenomena

5.1
Introduction

Our discussion up to this point has focused mainly on homogeneous systems of a uniform, or nearly uniform, wire of narrow lateral dimensions. In several instances, influence of environment [1] or large connecting electrical leads [2–4] were accounted for by the imposition of boundary conditions on the order parameter field at the wire ends. On the other hand, there are situations, as well as more complex systems, for which such a treatment is inadequate; for instance, a highly nonlinear or nonequilibrium system under large current drive, or proximity induced superconductivity in a ballistic, normal metal nanowire. In this chapter, we will focus on the heterogeneous system of a superconductor connected to a normal metal. The configurations of our discussion are of two varieties: (1) a superconducting nanowire connected to normal metal leads, and (2) a ballistic, few-quantum channel nanowire connected to large, superconducting leads.

The first case pertains to the behavior of a 1D superconducting nanowire (SCNW) in the dirty limit, when it is connected to a large normal metal lead (N) on its ends, and in others, in a N-SCNW-N configuration. For this system, which represents a special case of the general problem of a superconductor-normal metal interface, much is already known from a long history of experimental and theoretical investigations.

Regarding the second case, we are interested in the superconducting proximity effect between large superconducting leads, and a ballistic normal metal wire containing a few quantum channels sandwiched in between. Specifically, we are interested in the superconductor-normal-superconductor (SNS) bridge configuration.

Let us now look at the two cases in detail. In the first case of a 1D SCNW connected to large normal-metal pads, when a current is applied through the normal leads, normal electrons are injected into the superconductor across the interface. For the current to pass through the superconductor, a conversion process must take place to form Cooper pairs. Close to T_c, this conversion occurs on a length scale termed the charge imbalance length [5, 6]. Depending on the coupling or degree of transparency at the interface, the electronic density of states on the two sides,

One-Dimensional Superconductivity in Nanowires, First Edition. F. Altomare and A.M. Chang

and other factors, a proximity region with a varying degree of superconducting correlation develops on the normal metal side of the interface. At the same time, an electric field penetrates into the superconductor. The proximity region will carry the supercurrent, while the penetration of the electric field will produce a region within the superconductor, in which most of the current is carried by quasiparticles, resulting in a voltage drop. This voltage drop is indicative of a nonequilibrium distribution of quasiparticles. This region adjacent to the interface is characterized by a size-scale on the order of the charge imbalance length at higher temperature, but of order of the superconducting coherence length at temperatures well below T_c [7–9].

Moreover, nonlinear effects can arise as the current is increased. The natural theoretical framework for analyzing this type of behavior for dirty superconductors is via the Usadel equations introduced in Section 1.5. In the case of interest here, where the interface occurs between a 1D superconducting nanowire and a normal metal lead, what is found experimentally is that a voltage drop occurs, leading to dissipation in this interface region [10].

Understanding issues of nonequilibrium quasiparticle distribution, voltage drop induced by the normal electron to Cooper pair conversion process, nonlinear effects, and so on, will help shed light on how these issues can influence the current carrying capability of a 1D superconducting nanowire. For instance, Altomare, Li, and Chang [11] found that the current at which a nanowire switches from a superconducting state to a normal state drops by roughly 50% when the large superconducting leads are driven normal by the application of a small magnetic field. The issues put forth thus far are generic even for wider wires, with a diameter of $\sim 0.2-1\,\mu m$, as they have been extensively investigated experimentally, and analyzed via the Usadel equations, by Klapwijk and coworkers [7–9].

Thus, these issues arise even in the absence of the prevalence of phase slips. Understanding them enables one to differentiate between the effects caused by interface processes and those that arise from the intrinsic properties of one-dimensionality of the system, and in particular, those caused by phase slips of both the thermal (TAP) and quantum varieties (QPS). Detailed analysis of the systems with interfaces will also form the basis for understanding the limitations and constraints, if superconducting nanowires were to be used as low-dissipation interconnects in future generation ULSI (ultra-large-scale integration). Unless all active device elements become of the superconducting variety, the properties at the interface of normal metallic semiconductor devices and a superconducting nanowire conduit will strongly influence what is feasible and set limits on the minimum amount power dissipation caused by the interface resistance, as well as the maximum amount of current which may be carried without driving the superconducting interconnects into the normal state.

In the second case of an SNS bridge, the normal metal may be a metallic atomic chain, of either normal or superconducting material, or a semiconductor (Sm). The latter type of S-Sm-S devices is currently of great interest due to the fact that their properties can be tuned by tuning the semiconductor via coupling to a nearby electrostatic gate. Our focus is on the induced superconducting correlation within the normal metallic nanowire (NW) and the supercurrent which can be passed through the nanowire. For example, a universal quantized critical value of the supercurrent has been predicted in the limit of a short NW [12, 13], as well as nonuniversal current steps in S-SmNW-S with a SmNW region longer than the coherence length ξ_0 [14]. In addition, novel phenomena can arise, such as the predicted existence of a Majorana fermion zero energy mode when the nanowire electronic system has a strong spin–orbit coupling [15–19]. A Majorana fermion is distinct from a Dirac fermion as it is its own antiparticle [20, 21] and is widely speculated to be a possibility for describing the neutrino elementary particle. In condensed matter, recent experimental advances [22–25] and the possibility of implementing a topological, fault tolerant quantum computer [26] have generated a tremendous amount of excitement in the Majorana mode.

5.2 Transport Properties of Normal-Superconducting Nanowire-Normal (N-SCNW-N) Junctions

The Usadel equations are ideal for analyzing the behavior of a 1D superconducting nanowire connected at its two ends to normal metal leads, such as the N-SCNW-N aluminum nanowire system investigated by Altomare *et al.* [10], where the superconductor is in the dirty limit. The Usadel equations in their original form presented in Section 1.5 were derived from the equilibrium quasiclassical, energy-averaged Green's function introduced by Eilenberger and by Larkin and Ovchinnikov. While these equations are suitable for describing response in the linear and nonlinear regime, in many experimental situations, one is interested in the behavior near the switching current I_s, the current at which the superconducting nanowire is driven into the normal state by an excessive amount of bias current. As this is a highly nonequilibrium situation, a nonequilibrium version of the Usadel equations must be used. This will be the subject of the next section.

5.2.1 Nonequilibrium Usadel Equations

In Section 1.5, we introduced the Usadel equations [27], which describe the equation of motion of the quasiclassical Green's functions (both normal and anomalous) in the dirty limit. Those equations were based on an equilibrium formulation of the propagators averaged over the energy variable due to Eilenberger and to Larkin and Ovchinnikov [28, 29], in which the equation of motion for the Gorkov Green's function is treated in the quasiclassical approximation, with a neglect of terms containing second order spatial derivatives. These terms are small compared to first order derivative terms due to the fact that the relevant length scales of the superconducting state, of order ξ and λ, far exceed the Fermi wavelength λ_F. A subsequent expansion of the solutions in terms of an isotropic term, plus a small component dependent on the direction of the momentum (on or near the Fermi surface) yielded the Usadel equations. The Usadel equations have proven to be extremely useful for understanding a large number of the properties of dirty superconductors, including transport. Moreover, inhomogeneous systems, specifically, those involving one or more interfaces between a superconductor and a normal metal, can readily be accounted for.

To treat the nonequilibrium situations under voltage-bias or current-bias, a fully nonequilibrium formulation, extending the earlier equilibrium treatment, was developed by extending the earlier equilibrium treatment by Larkin and Ovchinnikov in the mid 1970s [30] by using the Keldysh formalism (Section 3.2.1.1) for the Green's functions. The resulting generalized Usadel equations for the dirty limit are extremely powerful and are able to account for the nonequilibrium and nonlinear regime, beyond the linear regime.

Because the equations are highly nonlinear and involve a self-consistent solution of coupled integral/differential equations, they are seldom amenable to analytic solutions, and numerical methods are usually employed. Here, we start with a summary of the nonequilibrium equations [31–33]. The formulation is in matrix form and is an extension of the form introduced in the equilibrium formulation in Section 1.5. There, the Green function matrix was defined in Nambu space, where the spinors in the Hilbert space are direct products of the Nambu, $[\psi_\uparrow, \psi_\downarrow^\dagger]$, and spin, $[\psi_\uparrow, \psi_\downarrow]$, parts, with $\Psi^{\mathrm{T}} = 1/\sqrt{2}[\psi_\uparrow, \psi_\downarrow, \psi_\downarrow^\dagger, -\psi_\uparrow^\dagger]$ [34].

In the Keldysh formalism, developed for superconductivity by Larkin and Ovchinnikov [30], the Green's function is given by a (4×4) matrix:

$$\check{G} = \begin{pmatrix} \hat{G}^{\mathrm{R}} & \hat{G}^{\mathrm{K}} \\ 0 & \hat{G}^{\mathrm{A}} \end{pmatrix}, \tag{5.1}$$

where $\hat{G}^{\mathrm{A}}$ is the advanced Green's function, and $\hat{G}^{\mathrm{R}}$ is the retarded Green's function, which together describe the equilibrium properties, and $\hat{G}^{\mathrm{K}}$ the Keldysh Green's function describing the nonequilibrium properties, yielding information on the distribution of quasiparticles. These Green's functions are represented by (2×2) matrices in the Nambu particle-hole space, with the field vectors $(\psi_\uparrow, \psi_\downarrow^\dagger)$. They are given by (with $\alpha = \{\mathrm{R,A,K}\}$)

$$\hat{G}^{\alpha}(t, \boldsymbol{r}; t', \boldsymbol{r}') = \begin{pmatrix} G^{\alpha}(t, \boldsymbol{r}; t', \boldsymbol{r}') & F_1^{\alpha}(t, \boldsymbol{r}; t', \boldsymbol{r}') \\ F_2^{\alpha}(t, \boldsymbol{r}; t', \boldsymbol{r}') & -G^{\alpha}(t, \boldsymbol{r}; t', \boldsymbol{r}') \end{pmatrix}. \tag{5.2}$$

Using the abbreviated notation $x = (t, \boldsymbol{r})$, the elements of the Nambu-space matrix, consisting of the normal Green's functions G^{α} and anomalous Green's functions

F^{α}, are

$$\begin{aligned}
G^{\mathrm{R}}(x,x') &= -i\Theta(t-t')\langle\{\psi(x),\psi^{\dagger}(x')\}\rangle\,,\\
G^{\mathrm{A}}(x,x') &= i\Theta(t'-t)\langle\{\psi(x),\psi^{\dagger}(x')\}\rangle\,,\\
G^{\mathrm{K}}(x,x') &= i\langle[\psi(x),\psi^{\dagger}(x')]\rangle\,,\\
F_1^{\mathrm{R}}(x,x') &= -i\Theta(t-t')\langle\{\psi(x),\psi(x')\}\rangle\,,\\
F_2^{\mathrm{R}}(x,x') &= -i\Theta(t-t')\langle\{\psi^{\dagger}(x),\psi^{\dagger}(x')\}\rangle\,,\\
F_1^{\mathrm{A}}(x,x') &= i\Theta(t'-t)\langle\{\psi(x),\psi(x')\}\rangle\,,\\
F_2^{\mathrm{A}}(x,x') &= +i\Theta(t'-t)\langle\{\psi^{\dagger}(x),\psi^{\dagger}(x')\}\rangle\,,\\
F_1^{\mathrm{K}}(x,x') &= i\langle[\psi(x),\psi(x')]\rangle\,,\\
F_2^{\mathrm{K}}(x,x') &= i\langle[\psi^{\dagger}(x),\psi^{\dagger}(x')]\rangle\,,\,.
\end{aligned} \tag{5.3}$$

with $[\ldots,\ldots]$ and $\{\ldots,\ldots\}$ denoting the commutator and the anticommutator, respectively. F_1 and F_2 are as defined in (1.45).

These Green's functions can be related to the Keldysh Green's functions $\hat{G}_{ij}$, $i,j = 1$ or 2, defined on the forward and back branches of the Keldysh contour c^{K}, where the forward contour branch, c^{F}, runs from $t = -\infty$ to $+\infty$, and the backward branch, c^{B}, from $t = \infty$ back to $-\infty$ [31]. Denoting the forward branch by the index 1 and the backward branch by 2, the relations are

$$\begin{aligned}
\hat{G}^{\mathrm{R}}(x,x') &= \hat{G}_{11}(x,x') - \hat{G}_{12}(x,x') = \hat{G}_{21}(x,x') - \hat{G}_{22}(x,x')\,,\\
\hat{G}^{\mathrm{A}}(x,x') &= \hat{G}_{11}(x,x') - \hat{G}_{21}(x,x') = \hat{G}_{12}(x,x') - \hat{G}_{22}(x,x')\,,\\
\hat{G}^{\mathrm{K}}(x,x') &= \hat{G}_{21}(x,x') + \hat{G}_{12}(x,x') = \hat{G}_{11}(x,x') + \hat{G}_{22}(x,x')\,.
\end{aligned} \tag{5.4}$$

Note that only three of the four G_{ij}s are independent.

To apply the quasiclassical approximation utilized by Eilenberger [28] and Larkin and Ovchinnikov [29], we pass to the energy integrated Green's function in the Fourier transformed representation with respect to the relative coordinate introduced in Section 1.5 ((1.49) and (1.55)), $\check{g}(\boldsymbol{R},\boldsymbol{p}_{\mathrm{F}},t,t')$,

$$\begin{aligned}
\check{g}(\boldsymbol{R},\boldsymbol{p},t,t') &= \frac{i}{2\pi}\int d\epsilon_p\,\check{G}(\boldsymbol{R},\boldsymbol{p},t,t')\,,\\
\check{G}(\boldsymbol{R},\boldsymbol{p},t,t') &= \int d\Delta\boldsymbol{r}' e^{-i\boldsymbol{p}\cdot\Delta\boldsymbol{r}'}\,\check{G}\left(\boldsymbol{R}+\frac{1}{2}\Delta\boldsymbol{r}',\boldsymbol{R}-\frac{1}{2}\Delta\boldsymbol{r}',t,t'\right)\,,
\end{aligned} \tag{5.5}$$

where the center of mass coordinate $\boldsymbol{R} = \frac{1}{2}(\boldsymbol{r}+\boldsymbol{r}')$, and $\Delta\boldsymbol{r} = \boldsymbol{r}-\boldsymbol{r}'$. The quasiclassical Green's function $\check{g}$ only depends on the direction on the Fermi surface via $\boldsymbol{p}_{\mathrm{F}}$, with the magnitude of the momentum fixed at $\boldsymbol{p}_{\mathrm{F}}$. The quasiclassical Green's function satisfies the normalization condition

$$\int dt''\check{g}(t,t'')\check{g}(t'',t')dt'' = \check{1}\delta(t-t')\,. \tag{5.6}$$

We are interested in the situation where the system is stationary and the dependence on the time coordinates is on the time difference $\Delta t = t - t'$. The time difference Fourier transform is given by

$$\check{g}(\boldsymbol{R}, \boldsymbol{p}, \omega) = \int d\Delta t \check{g}(\boldsymbol{R}, \boldsymbol{p}, t - t') e^{i\omega\Delta t} ,$$
$$[\check{g}(\boldsymbol{R}, \boldsymbol{p}, \omega)]^2 = \check{1} . \tag{5.7}$$

In the diffusive case considered by Usadel [27], the Green's functions are nearly isotropic, and are expanded as an isotropic term, g_0, plus a velocity-dependent term in the form

$$\check{g} = \check{g}_0 + \boldsymbol{v} \cdot \check{\boldsymbol{g}}_1 . \tag{5.8}$$

The normalization condition enables one to eliminate the velocity-dependent term, leading to the nonequilibrium Usadel equations on the isotropic term in the space-time Fourier transformed Green's functions (subscript "0" will be omitted from here on), that is,

$$\hbar D \nabla(\check{g} \nabla \check{g}) + i[\check{H}, \check{g}] + [\check{\Sigma}, \check{g}] = 0 . \tag{5.9}$$

Here, $D = 1/3 v_{\text{F}} l_{\text{mfp}}$ is the diffusion constant, and

$$\check{H} = -e\phi\check{1} + \hbar\omega\check{\sigma}_z - \check{\Delta} ,$$
$$\check{\Delta} = \begin{pmatrix} \hat{\Delta} & 0 \\ 0 & \hat{\Delta} \end{pmatrix} ,$$
$$\hat{\Delta} = \begin{pmatrix} 0 & \Delta \\ -\Delta^* & 0 \end{pmatrix} ,$$
$$\check{\sigma}_z = \begin{pmatrix} \hat{\sigma}_z & 0 \\ 0 & \hat{\sigma}_z \end{pmatrix} ,$$
$$\hat{\sigma}_z = \begin{pmatrix} 1 & 0 \\ 0 & -1 \end{pmatrix} , \tag{5.10}$$

with Δ denoting the pair potential. The self-energy $\check{\Sigma}$ has the same structure as the Green's function. In the absence of spin-flip scattering, its main contributions come from inelastic scattering processes, for example, from thermal phonons, which are negligible at low temperatures.

Therefore, at low temperatures, the Usadel equations, along with the normalization condition, can be written in component form

$$\hbar D \nabla(\hat{g}^{\text{R}} \nabla \hat{g}^{\text{R}}) = -i[\check{H}, \hat{g}^{\text{R}}] ,$$
$$\hbar D \nabla(\hat{g}^{\text{R}} \nabla \hat{g}^{\text{K}} + \hat{g}^{\text{K}} \nabla \hat{g}^{\text{A}}) = -i[\check{H}, \hat{g}^{\text{K}}] ,$$
$$\hat{g}^{\text{R}} = -\hat{\sigma}_z \hat{g}^{\text{A}\dagger} \hat{\sigma}_z . \tag{5.11}$$

The electrostatic potential ϕ is determined by the condition of charge neutrality [9]

$$e\phi(\boldsymbol{R}) = \int_{-\infty}^{\infty} N(\epsilon) f_{\mathrm{T}}(\epsilon, \boldsymbol{R}) d\epsilon \ . \tag{5.12}$$

To relate the Keldysh Green's function to the quasiparticle distribution, we first introduce the distribution function matrix $\hat{f}$ in Nambu space, that is,

$$\hat{f} = f_{\mathrm{L}} \hat{1} + f_{\mathrm{T}} \hat{\sigma}_z \ , \tag{5.13}$$

where f_{L} represents the even part (energy or longitudinal mode), and f_{T} represents the odd part (charge or transverse mode) in particle–hole space [35], with the full distribution function $f(\boldsymbol{R}, \epsilon)$ given by

$$2 f(\boldsymbol{R}, \epsilon) = 1 - f_{\mathrm{L}}(\boldsymbol{R}, \epsilon) - f_{\mathrm{T}}(\boldsymbol{R}, \epsilon) \ . \tag{5.14}$$

The Keldysh Green's function can now be expressed in terms of $\hat{f}$ as

$$\begin{aligned} \hat{g}^{\mathrm{K}} &= \hat{g}^{\mathrm{R}} \hat{f} - \hat{f} \hat{g}^{\mathrm{A}} \\ &= \begin{pmatrix} (g^{\mathrm{R}} + g^{\mathrm{R}\dagger})(f_{\mathrm{L}} + f_{\mathrm{T}}) & f_1^{\mathrm{R}}(f_{\mathrm{L}} - f_{\mathrm{T}}) - f_2^{\mathrm{R}\dagger}(f_{\mathrm{L}} + f_{\mathrm{T}}) \\ f_2^{\mathrm{R}}(f_{\mathrm{L}} + f_{\mathrm{T}}) - f_1^{\mathrm{R}\dagger}(f_{\mathrm{L}} - f_{\mathrm{T}}) & -(g^{\mathrm{R}} + g^{\mathrm{R}\dagger})(f_{\mathrm{L}} - f_{\mathrm{T}}) \end{pmatrix} \end{aligned} \tag{5.15}$$

making use of the last expression in (5.11). Specializing to the 1D case, where the spatial variable is along the x-direction only, we find the following coupled equations for the Green's functions g^{R}, f_1^{R} and f_2^{R} from the Usadel equations [8, 9] with $\epsilon = \hbar\omega$

$$\begin{aligned} \hbar D \left(g^{\mathrm{R}} \frac{\partial^2 f_1^{\mathrm{R}}}{\partial x^2} - f_1 \frac{\partial^2 g^{\mathrm{R}}}{\partial x^2} \right) &= -2i\Delta g^{\mathrm{R}} - 2i\epsilon f_1^{\mathrm{R}} \ , \\ \hbar D \left(f_1^{\mathrm{R}} \frac{\partial^2 f_2^{\mathrm{R}}}{\partial x^2} - f_2 \frac{\partial^2 f_1^{\mathrm{R}}}{\partial x^2} \right) &= 2i\Delta f_2^{\mathrm{R}} + 2i\Delta^* f_1^{\mathrm{R}} \ . \end{aligned} \tag{5.16}$$

The coupled equations for the distribution function arising from the Usadel equations for the Keldysh Green's functions reads

$$\begin{aligned} \hbar D \frac{d j_{\mathrm{energy}}}{dx} &= 0 \ , \\ \hbar D \frac{d j_{\mathrm{charge}}}{dx} &= 2 R_{\mathrm{L}} f_{\mathrm{L}} + 2 R_{\mathrm{T}} f_{\mathrm{T}} \ . \end{aligned} \tag{5.17}$$

The terms in these equations are defined as

$$
\begin{aligned}
j_{\text{energy}} &= \Pi_{\text{L}} \frac{d f_{\text{L}}}{dx} + \Pi_X \frac{d f_{\text{T}}}{dx} + j_\epsilon f_{\text{T}} \,, \\
j_{\text{charge}} &= \Pi_{\text{T}} \frac{d f_{\text{T}}}{dx} - \Pi_X \frac{d f_{\text{L}}}{dx} + j_\epsilon f_{\text{L}} \,, \\
\Pi_{\text{L}} &= \frac{1}{4}\left(2 + 2|g^{\text{R}}|^2 - |f_1^{\text{R}}|^2 - |F_2^{\text{R}}|^2\right) \,, \\
\Pi_{\text{T}} &= \frac{1}{4}\left(2 + 2|g^{\text{R}}|^2 + |f_1^{\text{R}}|^2 + |f_2^{\text{R}}|^2\right) \,, \\
\Pi_X &= \frac{1}{4}(|f_1^{\text{R}}|^2 - |f_2^{\text{R}}|^2) \,, \\
j_\epsilon &= \frac{1}{2}\text{Re}\left\{ f_1^{\text{R}} \frac{d f_2^{\text{R}}}{dx} - f_2^{\text{R}} \frac{d f_1^{\text{R}}}{dx} \right\} \,, \\
R_{\text{L}} &= -\frac{1}{2}\text{Im}\left\{ \frac{d^2 f_2^{\text{R}}}{dx^2} + \frac{d^2 f_1^{\text{R}\dagger}}{dx^2} \right\} \,, \\
R_{\text{T}} &= -\frac{1}{2}\text{Im}\left\{ \frac{d^2 f_2^{\text{R}}}{dx^2} - \frac{d^2 f_1^{\text{R}\dagger}}{dx^2} \right\} \,.
\end{aligned} \tag{5.18}
$$

However, as can be observed from the equations above, the formulation is quite technical and cumbersome, and solutions for experimentally relevant situations are obtained in a self-consistent way using numerical methods. Below, we sketch the general ideas and parameterization leading to the numerical solutions based on the work of Klapwijk and collaborators. We are particularly interested in the behavior of the N-SCNW-N configuration, in order to help interpret the experimental findings.

5.2.2
Parameterization of the Usadel Equations

Because of the normalization condition, $(g^{\text{R}})^2 + f_1^{\text{R}} f_2^{\text{R}} = 1$, it is straightforward to parameterize the Green's functions in terms of two angles, θ and ϕ [9, 33], that is,

$$
\hat{g}^{\text{R}} = \begin{pmatrix} \cos\theta & \sin\theta\, e^{i\phi} \\ \sin\theta\, e^{-i\phi} & -\cos\theta \end{pmatrix} \,. \tag{5.19}
$$

The Usadel equations for $\hat{g}^{\text{R}}$ in (5.16) above now read

$$
\begin{aligned}
\hbar D \left[\frac{\partial^2 \theta}{\partial x^2} - \sin\theta \cos\theta \left(\frac{\partial \phi}{\partial x}\right)^2 \right] &= -2i\epsilon \sin\theta - \cos\theta\,(\Delta e^{-i\phi} + \Delta^* e^{i\phi}) \,, \\
\hbar D \frac{\partial}{\partial x}\left(\sin^2\theta \frac{\partial \phi}{\partial x}\right) &= i \sin\theta\,(\Delta e^{-i\phi} - \Delta^* e^{i\phi}) \,.
\end{aligned} \tag{5.20}
$$

For (5.18), the various terms are (with the subscript 2 denoting the quantum component)

$$\begin{aligned}
\Pi_{\mathrm{L,T}} &= 1 + |\cos\theta|^2 \mp |\sin\theta|^2 \cosh(2\phi_2) \,, \\
\Pi_X &= -|\sin\theta|^2 \sinh(2\phi_2) \,, \\
j_\epsilon &= 2\mathrm{Im}\left(\sin^2\theta \frac{\partial\phi}{\partial x}\right) \,, \\
R_{\mathrm{L,T}} &= \mathrm{Re}\left[\sin\theta\,(\Delta e^{-i\phi} \mp \Delta^* e^{i\phi})\right] \,.
\end{aligned} \tag{5.21}$$

The local density of states $N(\epsilon, x)$ and electrostatic potential are

$$N(\epsilon, x) = \mathrm{Re}[g^{\mathrm{R}}] = \mathrm{Re}\{\cos\theta\} \,, \quad -e\phi(x) = \int_{-\infty}^{\infty} N(\epsilon, x) f_{\mathrm{T}}(\epsilon, x) dx \,, \tag{5.22}$$

and the self-consistent equation for the gap $\Delta(x)$ is

$$\begin{aligned}
\Delta(x) = \frac{N(0)V}{4i} \int_{-\hbar\omega_{\mathrm{D}}}^{\hbar\omega_{\mathrm{D}}} d\epsilon \Big[&\left(\sin\theta\, e^{i\phi} - \sin\theta^* e^{i\phi^*}\right) f_{\mathrm{L}} \\
&-(\sin\theta\, e^{i\phi} - \sin\theta^* e^{i\phi^*}) f_{\mathrm{T}} \Big] \,.
\end{aligned} \tag{5.23}$$

5.2.3
Numerical Results

Klapwijk and co-workers have performed a series of experimental and theoretical (numerical) studies of the N-SCNW-N system for aluminum nanowires with lateral width in the range of 0.2–1 μm, and thickness below 100 nm. The normal leads are made of copper. The initial study was performed with a small driving current, and the system was in the linear regime [7]. A contact resistance is produced by the normal electron to Cooper pair conversion process; this resistance is associated with a finite voltage, which penetrates into the aluminum superconducting nanowire. The numerical calculation using the equilibrium Usadel equations [27, 35] produced the profile shown in Figure 5.1a. The density of states as a function of position along the length of the NW is shown in Figure 5.1b. Extending this into the nonlinear regime [9], two stable configurations were found, as can be seen in Figure 5.2. The first is one in which the entire nanowire (aluminum) is in a single, coherent superconducting state. However, near the interface with the normal metal bilayer (thick copper on top of thin aluminum), the resistance due to the normal electron to Cooper pair conversion is clearly present. The second is a bimodal configuration where in two small and independent superconducting regions reside close to the normal leads, and the center of the nanowire is occupied by a normal state of large spatial extent. This state arises as a result of a strong energy mode f_{L},

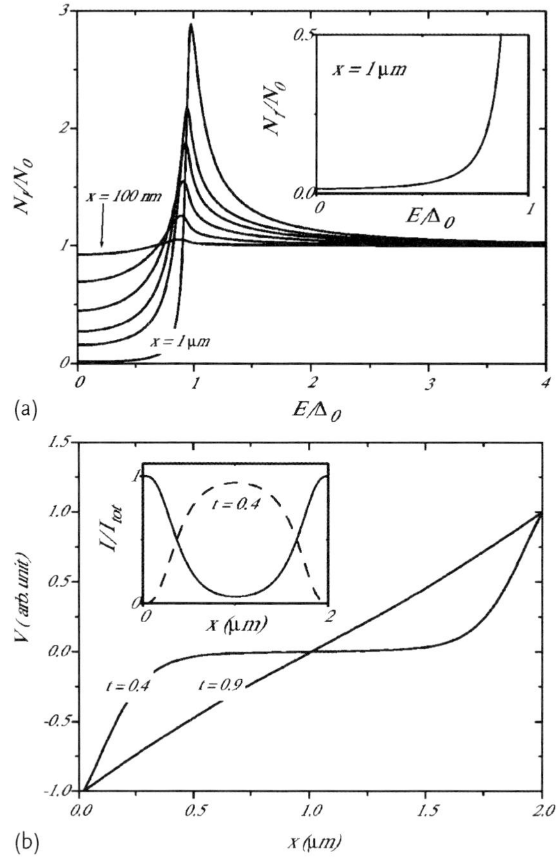

Figure 5.1 (a) The calculated density-of-states N_1 at various distances from the reservoirs ($x = 100, 200, 300, 400, 500, 1000$ nm) for a 2 μm long aluminum wire ($t = 0.4$, $D = 160\,\text{cm}^2/\text{s}$, and $\Delta_0 = 192$ μeV). Note the exponentially small but finite subgap density-of-states in the middle (at $x = 1$ μm; see inset). (b) The voltage in the superconducting wire as a function of position for two different temperatures ($t = 0.4$ and 0.9). At $t = 0.9$, the wire behaves as a normal metal and for $t = 0.4$, the voltage is clearly present to a depth ξ (wire length 2 μm with $D = 160\,\text{cm}^2/\text{s}$ and $\Delta_0 = 192$ μeV). The inset shows the position-dependent normal currents and supercurrents. From [7].

which suppresses superconductivity in the middle of the wire, as the nonequilibrium distribution f_L approaches that of a thermal distribution, but at an elevated temperature. At the ends near the cold, normal leads, the emergence of a finite superconducting gap is favored.

In Figure 5.3, the two-terminal differential resistance (dV/dI) is presented for a Cu/Al-Al-Cu/Al N-SCNW-N device, as a function of temperature in Figure 5.3a, for a fixed biased current, versus bias current in Figure 5.3b, and versus the magnetic field at a fixed bias current in Figure 5.3c. The numerical solutions of the Usadel equation are able to reproduce the observed results and are presented in Figure 5.3a for comparison. Particularly noteworthy is the finding that the nonlin-

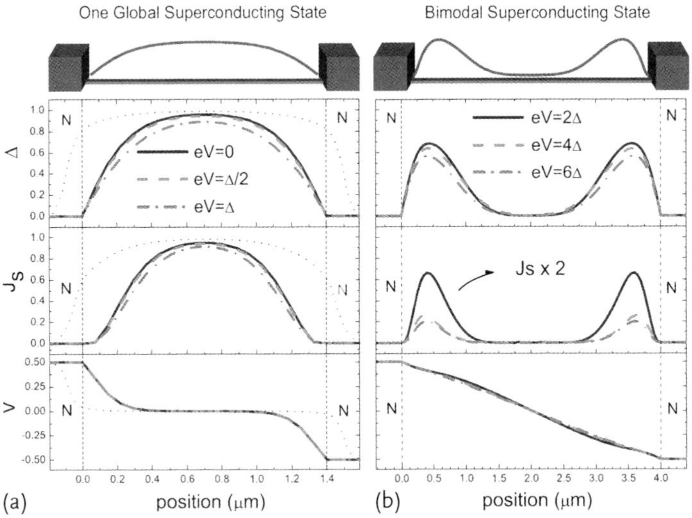

Figure 5.2 (a) The complete wire is in a single superconducting state with order parameter $\Delta(x)$. However, near the normal reservoirs, the condensate only carries a small fraction J_S of the current as a supercurrent, which results in a resistance and a voltage drop at the ends of the wire, over roughly a coherence length. At the lowest temperatures, a small proximity effect can occur at the connection of the bilayer reservoirs to the wire (schematically illustrated by dotted black lines). (b) Two distinct superconducting domains at the ends of the wire are separated by a normal region in the center of the wire. Due to the small supercurrent, the voltage profile is almost equal to the normal state. From [9].

ear differential conductance reflecting the interface resistance in Figure 5.3b only shows a minimal change, that is, only increases by $\sim$ 50%, before the nanowire is driven into the normal state at $\sim$ 120 μA.

The behavior of the contact resistance in this nonlinear regime is relevant to interpreting the results of Altomare *et al.*, where they measured the nonlinear voltage–current characteristics of a 100 μm long and $\sim$ 7 nm diameter aluminum nanowire [10], connected to leads which were driven normal by the application of a weak magnetic field, that is, in an N-SCNW-N configuration. There, in order to separate out the contributions of the "contact" resistance at the N-SCNW interface, from the intrinsic nonlinearity of the 1D superconducting nanowire, it is crucial to have a reliable model and an understanding of the behavior of the interface resistance in the nonlinear, nonequilibrium regime, brought about by the sizable driving current passing through the 1D superconducting nanowire.

Based on the results of the Klapwijk work, it is clear that just below the switching current, where the superconducting nanowire is driven normal by a large current, the so-called switching current I_s, the differential resistance, dV/dI only approximately doubles, and is not a large effect. This will enable one to rule out the contact resistance at the interface as the source of the observed nonlinearity in the work of Altomare *et al.* The presentation and discussion of this experimental work is found in Section 7.4.2.

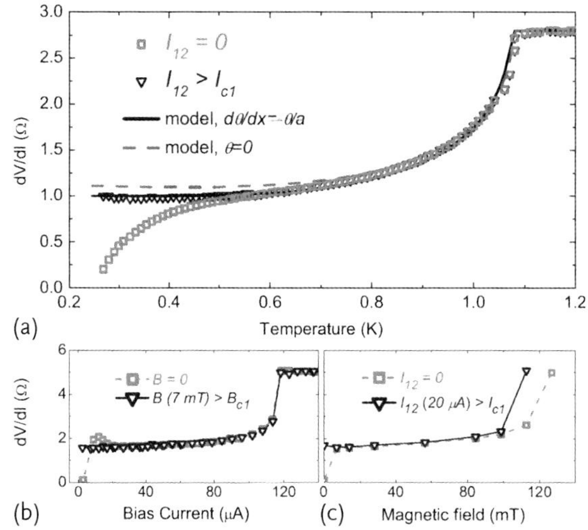

Figure 5.3 (a) The two-terminal resistance versus temperature of a 1.4 μm long-wire (sample 1a). Due to the proximity effect of the wire on the normal reservoirs, the resistance becomes negligible at low temperatures. This weak proximity effect can be suppressed by applying a small bias current (b) or small magnetic field (c) (sample 4, 200 mK). This "corrected" wire resistance is constant down to the lowest temperatures (magenta squares of (a)). A model (dashed line) with rigid normal boundary conditions for the pairing angle $\theta = 0$ slightly overestimates the observations. A weaker boundary condition (full line), in which θ decays gradually to zero over a characteristic length shows excellent agreement with the experiment. From [9].

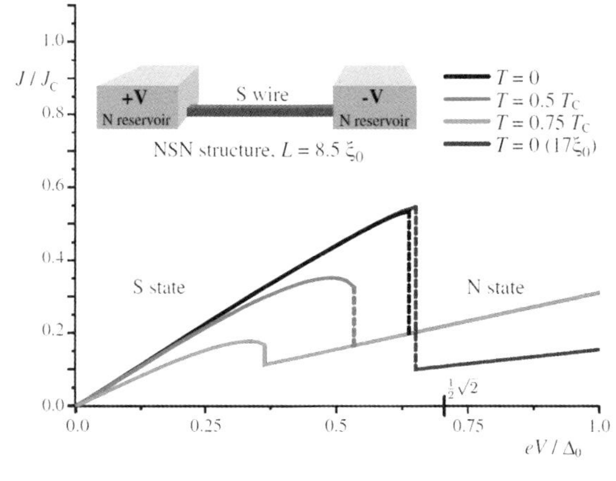

Figure 5.4 The calculated current (J)–voltage (V) relation of a superconducting wire of length L $= 8.5\xi_0$ between normal metallic reservoirs (see inset) at several temperatures, and for a wire of length $17\xi_0$. J_c is the critical current density, and Δ_0 the bulk gap energy. From [8]. For a color version of this figure, please see the color plates at the beginning of the book.

In [8], Klapwijk and coworkers found another important feature which arises in the N-SCNW-N system: the reduction of the switching current, I_s, to well below the theoretical depairing current (I_c) – the absolute maximum supercurrent sustainable due to velocity induced Cooper pair-breaking. This is a consequence of the nonequilibrium distribution of quasiparticles. In Figure 5.4, we show the numerically computed switching current, which is roughly 60% of the depairing current I_c at $T = 0$. This result is consistent with the behavior observed in Chang's group [11], in which the application of a weak magnetic field to an S-SCNW-S system drives the electrodes normal (and thus the system becomes N-SCNW-N) and reduces the switching current in the aluminum nanowire by almost 50%.

5.3 Superconductor–Semiconductor Nanowire–Superconductor Junctions

In the previous section, we examined the inhomogeneous system of a 1D superconductor connected to normal metal leads on each end. In this section, we address the reverse configuration of superconducting leads connected to normal metal narrow wire, as well as a weak link such as an atomic chain, which may be superconducting as well as being normal. Moreover, up to this point, we have focused on metallic systems. An active area of research on superconducting nanowires is the superconductor–semiconductor nanowire–superconductor (S-SmNW-S) system. A typical semiconductor nanowire wire or diameter $\sim$ 100–200 nm can readily be tuned into the few transverse quantum channel (mode) limit by the application of a voltage on a gate electrode, capacitively coupled to the SmNW. The nanowires are often quite pure, and ballistic transport can take place through them. If the nanowire is sufficiently short, that is, not long compared to the coherence length, proximity induced superconductivity can readily occur within the nanowire.

Outstanding new behaviors include the prediction and observation of universal critical values of the supercurrent in the short-wire limit for ballistic wires [12, 13], nonuniversal critical current steps in longer ballistic wires [14, 36], and subgap features under finite voltage bias, in both the ballistic and diffusive regimes [37–40].

Moreover, this is relevant to the recent interest in topologically protected quantum computing [26] in topological insulator-related [41–43] materials, in which a Majorana fermion zero-energy mode is predicted to exist [18]. In the past few years, theorists have proposed a S-SmNW-S devices, where the semiconductor has strong spin–orbit interaction, such as InAs or InSb, in which Majorana fermions are expected to be observed [15, 16]. This has brought on an intense search worldwide, culminating in recent announcements of the first evidence for the existence of such a mode in the condensed matter context [22–25]: these experiments are briefly reported on in Chapter 9.

In Section 2.3, we briefly discussed the behavior of uniform, disorder-free nanowires. Here, we will extend the discussion to inhomogeneous systems of S-SmNW-S and S-SQPC-S (superconductor-superconducting quantum point contact-superconductor). There are two regimes of interest regarding the amount of disorder within the normal region between the superconducting leads, that of the clean, ballistic limit, and the opposite dirty limit, where electron transport occurs via a diffusive motion.

The behaviors of such systems are readily treated via the methods of the Bogoliubov–de Gennes equations (BdG), which enables one to determine the electron-like and hole-like quasiparticle excitations above the ground state, and the Usadel equations, which apply in the dirty limit. The inhomogeneous system is readily handled by the imposition of appropriate boundary conditions at the interfaces.

The BdG equations are given by (1.26)

$$E_n u_n(\boldsymbol{r}) = \left[\frac{\boldsymbol{p}^2}{2m} + U(\boldsymbol{r})\right] u_n(\boldsymbol{r}) + \Delta(\boldsymbol{r}) v_n(\boldsymbol{r}) ,$$

$$E_n v_n(\boldsymbol{r}) = -\left[\frac{\boldsymbol{p}^2}{2m} + U(\boldsymbol{r})\right] v_n(\boldsymbol{r}) + \Delta(\boldsymbol{r})^* u_n(\boldsymbol{r}) . \tag{5.24}$$

Within the superconducting leads, the energy gap takes on a spatially uniform value Δ_0, with a phase φ_1 on the left lead (SC1), and a phase φ_2 on the right lead (SC2). Within the NW, the gap is taken to be identically zero ($\Delta_{\text{NW}} = 0$). This simplified model of the gap is assumed to be close to the self-consistently determined value [44, 45].

The Josephson effect in a S-SmNW-S junction may be conceptualized as a combination of the penetration of Cooper pairs through the S-SmNW or SmNW-S interface, and the propagation between the interfaces within the semiconductor. In order for a supercurrent to pass through the entire device, phase coherence must be maintained within the semiconductor. The supercurrent is driven by a difference in phase between the two superconductors connected to the ends of the SmNW. Adreev reflections at the S-SmNW interface is the key process, that enables a supercurrent to pass through the normal (SmNW) region.

To make the problem tractable and conceptually transparent, in the ballistic limit of a clean SmNW, one may imagine a situation in which the width of the nanowire gradually narrows away from the S-SmNW interface, into a constriction (or point contact), within which a few modes (or quantum channels) are occupied below the Fermi level at the narrowest point. For a sufficiently gradual narrowing, an adiabatic approximation may be used [12, 14], and the wavefunctions of the quasiparticles which solve the BdG equation in the SmNW take the form with a WKB approximation [14]

$$\psi_j(x, y) = \left[\frac{2}{\mathrm{w}(x)}\right]^{1/2} \sin\left\{j\pi\left[\frac{y}{\mathrm{w}(x)} + \frac{1}{2}\right]\right\} \times \boldsymbol{\phi}_j(x) , \tag{5.25}$$

where $\boldsymbol{\phi}_j(x)$ is a two-component vector wavefunction. Inserting the form into BdG equations yield the probability amplitude for the Andreev reflection of an electron-like quasiparticle of energy ε, injected from the left superconducting lead in the jth quantum channel. The reflection amplitude is given by

$$a_j(\delta\varphi, \epsilon) = \frac{\Delta(e^{i\Phi_j} - e^{i\delta\varphi})}{(\epsilon + \hbar\Omega)e^{i\delta\varphi} - (\epsilon - \hbar\Omega)e^{i\Phi_j}} . \tag{5.26}$$

Here, $\Phi_j(\epsilon)$ is the phase accumulated in a round trip from the left S-SmNW interface, to the right SmNW-S interface, ε is the energy referenced to the Fermi energy

E_F, and $\delta\varphi \equiv \varphi_2 - \varphi_1$ is the order parameter phase difference between the two superconducting leads. A quasiparticle moving from the left toward the right interface gains a phase of $\int_0^L k_j^+(x)dx$, and the retro-reflected quasihole returns and "gains" the phase of $-\int_0^L k_j^-(x)dx$, yielding

$$\begin{aligned}
\Phi_j\left(\frac{\epsilon}{\hbar}\right) &= \int_0^L k_j^+(x)dx - \int_0^L k_j^-(x)dx\,,\\
k_j^{\pm} &= \sqrt{\frac{2m}{\hbar^2}(E_F \pm \epsilon) - \left[\frac{j\pi}{\mathrm{w}(x)}\right]^2}\,,\\
\hbar\Omega &= \sqrt{\epsilon^2 - \Delta^2}\,.
\end{aligned} \tag{5.27}$$

Note that the direction of propagation of the electron-like or hole-like quasiparticle is given by its group velocity $1/\hbar\ d\epsilon/dk$.

The DC critical supercurrent is expressible in terms of the Andreev reflection amplitudes as [44, 46]

$$\begin{aligned}
I_c &= \frac{(-e)\Delta}{\hbar\beta}\sum_j\sum_{\omega_n}\frac{1}{\sqrt{\hbar^2\omega_n^2 + \Delta^2}}[a_j(\varphi, i\omega_n) - a_j(-\varphi, i\omega_n)]\\
&= \frac{2e\Delta^2}{\hbar\beta}\sum_j\sum_{\omega_n}\frac{\sin\varphi}{\left[\begin{array}{l}(2\hbar^2\omega_n^2 + \Delta^2)\cos\Phi_j(i\omega_n) - 2i\hbar\omega_n\sqrt{\hbar^2\omega^2 + \Delta^2}\\ \times\sin\Phi_j(i\omega_n) + \Delta^2\cos\varphi\end{array}\right]}\,,\\
\Phi(i\omega_n) &\equiv \Phi\left(\frac{\epsilon}{\hbar} \to i\omega_n\right)\,,
\end{aligned} \tag{5.28}$$

where $\omega_n = (2n+1)\pi k_B T = (2n+1)\pi/\beta$ is the Matsubara frequency.

For long SmNWs with length $\mathrm{L} \gg \xi_0 = \hbar v_F/(\pi\Delta)$, the zero temperature critical current is approximated by [14]

$$I_c \approx \frac{-2e}{\pi \mathrm{L}}\sum_j^N v_j\,, \quad \frac{1}{v_j} \equiv \int_0^{\mathrm{L}}\frac{dx}{\mathrm{L}}\frac{m}{\hbar k_j(x)}\,, \quad k_j(x) \equiv k_j^{\pm}(x, \epsilon = 0)\,. \tag{5.29}$$

For each quantum channel labeled by index j, the contribution to the current can be thought of as the product of $(-2e)\Delta/\hbar$ and the phase-coherence factor $\xi_j/\mathrm{L} = \hbar v_j/(\pi\Delta\mathrm{L})$.

The main contribution to the current comes from energies, at which the denominator vanishes. The values of the energy correspond to bound Andreev states in the SmNW [45, 47]. For instance, in the case of a single channel, they occur at [47, 48]

$$\epsilon_n^{\pm} = \frac{\hbar v_F}{2(L + \pi\xi_0)}\left[2\pi\left(n + \frac{1}{2}\right) \pm \delta\varphi\right]\,, \tag{5.30}$$

where the $\pm$ label corresponds to the positive and negative energy states, respectively.

The above expressions are adequate in the limit of smooth change in the width $w(x)$. In the case of abrupt changes on the scale of the Fermi wavelength λ_F, it is necessary to expand the wavefunctions in terms of a basis set, and the reflection coefficients must be summed over the basis set, and the reflection and transmission coefficients which mix the different quantum channels must be determined by matching the boundary conditions at the S-SmNW (left) and SmNW-S (right) interfaces [14, 44].

In the long wire limit, the critical current exhibits clear steps as the width of the constriction is decreased, as shown in Figure 5.5. The step height is not universal, however. For the opposite limit of smooth change in width $w(x)$, on the other hand, the steps are rounded and can almost be nonexistent. For a single channel, the current–phase relationship is linear at $T = 0$, starting at $I_c = 0$ at $\delta\varphi = 0$, and reaching $(-e)v_F/(L + \hbar v_F/\Delta)$ just below π, and suffers a jump of $2ev_F(L + \hbar v_F/\Delta)$ just above π [47], becoming sinusoidal in $\delta\phi$ near T_c. This behavior as a function of $\delta\varphi$ is reminiscent of what was found in 3D ballistic S-N-S junctions by Kulik [49], Ishii [50], and Bardeen and Johnson [51].

In the limit of very short SmNWs and weak links, with $L \ll \xi_0$ and with an adiabatic change in the width, Beenakker and van Houten derived a universal value for the critical current [12], namely,

$$I_c = N\frac{(-e)}{\hbar}\Delta_0(T)\sin\frac{\delta\varphi}{2}\tanh\left[\frac{\Delta_0(T)}{2k_BT}\cos\frac{\delta\varphi}{2}\right]. \tag{5.31}$$

As $T \to 0$, this expression approaches

$$I_c = N\frac{(-e)}{\hbar}\Delta_0\sin\frac{\delta\varphi}{2}. \tag{5.32}$$

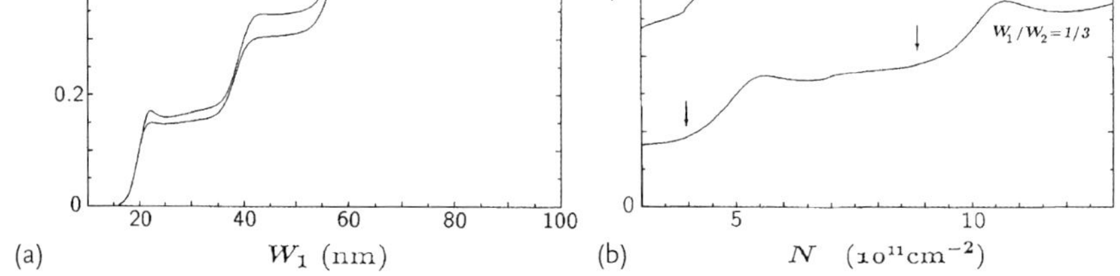

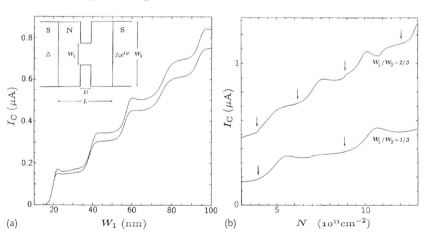

Figure 5.5 (a) Critical current as a function of the width of the constriction: The upper curve is for $T = 0.5$ K and the lower curve is for $T = 1.0$ K. Inset: Schematic of the model used in the numerical calculations. (b) Critical current as a function of the carrier density. The arrows indicate the densities at which another channel opens. From [14].

Here, N is the total number of quantum channels in the short constriction, which may be an SmNW, provided the S-SmNW interface is transparent, or a superconducting weak link. Note that the critical current per quantum channel is given by the universal number $(-e)\Delta_0/\hbar$. This arises due to the fact that in the short channel, adiabatic limit where a WKB approximation for the wavefunction holds, only one bound state per mode (one for each direction of propagation). The energy $\epsilon^{\pm}$ is given by

$$\arccos\left(\frac{\epsilon^{\pm}}{\Delta_0(T)}\right) = \pm\frac{1}{2}\delta\varphi \,. \tag{5.33}$$

There is only a single bound state per mode, independent of the quantum mode (channel) index! Since I_c is determined by the Andreev bound states, the identical energy position, independent of channel (or mode) index guarantees the universality. This predicted quantized current has now been verified in experiments on superconducting atomic point contacts [13].

We have focused our presentation on the DC critical value of the supercurrent, and its dependence on the phase difference $\delta\varphi$ on the two superconducting leads. A host of interesting and striking behaviors are also present at finite voltage bias and AC response. Particularly striking are subharmonic gap structures at finite voltage bias across the S-SmNW-S junction. The treatment of AC and subharmonic structures for ballistic SmNW devices is readily carried out by extending the BdG treatment to the time-dependent case [37–39].

In the dirty limit, two approaches are available. The first makes use of the quantum channel quasi-1D energy structure in the SmNW, and averages over the disorder to account for backscattering of quasiparticles in the SmNW [39]. In the fully diffusive limit, where channel structures become less well-defined, the time-dependent Usadel equations are solved to yield the transport behavior under finite

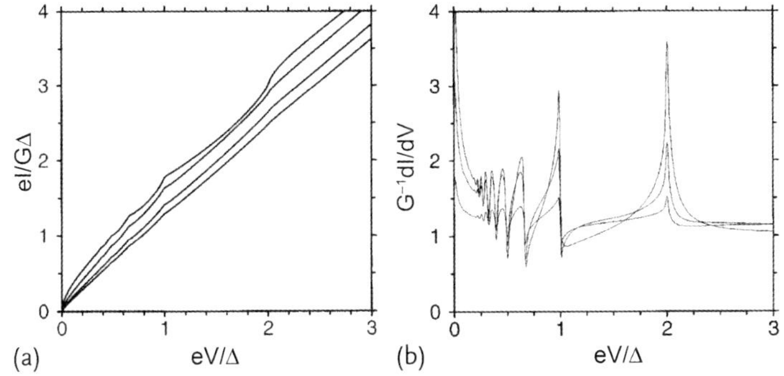

Figure 5.6 The quasiparticle current (a) and the differential conductance (b) versus voltage at various temperatures for a disordered SNS junction. From top to bottom, $T = 0, 1, 2, 3\Delta$. At low voltages, the DC current (a) has a square root dependence on voltage, while at high voltages, it exhibits excess current. The conductance (b) possesses subharmonic singularities, diverges at low voltages, and at high voltages, asymptotically tends to the normal state conductance of the disordered region. From [39].

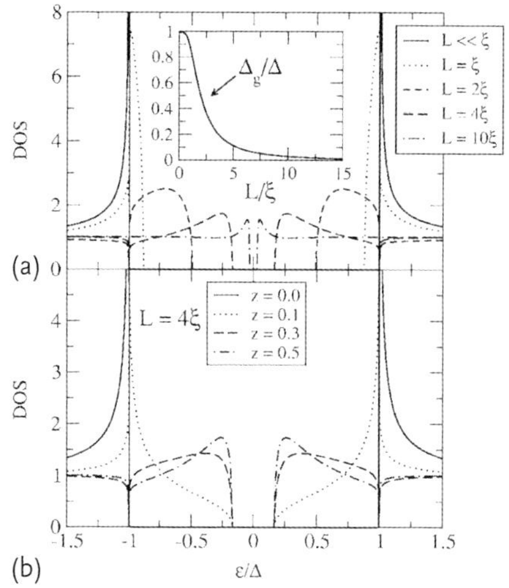

Figure 5.7 (a) Normalized density of states in the middle of the wire ($z = 0.5$) as a function of energy for different wire lengths. The inset shows the minigap Δ_g as a function of the length. (b) Normalized density of states as a function of the energy in different positions (z) along a wire of length $L = 4\xi$. From [40].

voltage bias [40]. These approaches are able to reproduce the most salient features observed experimentally.

In Figure 5.6, we show an example of the subharmonic structure in the differential resistance dV/dI under finite voltage bias for a ballistic SmNW in the single quantum channel limit. In Figure 5.7, the behavior in a fully diffusive SmNW is presented. Note that within the nanowire, the proximity induced gap is independent of the position along the wire length and approaches the value given by the Thouless energy associated with electron diffusion: $\Delta_{\text{SmNW}} \sim 3.5E_{\text{T}} = \hbar D/L^2$, when the wire length is much longer than the superconducting coherence length.

5.4 Majorana Fermion in S-SmNW-S Systems with Strong Spin–Orbit Interaction in the Semiconductor

A new major development in condensed matter systems within the past 10 years is the prediction and experimental discovery of the topological insulator [41–43, 52–55]. This new type of insulator is found in novel materials systems that possess a strong spin–orbit interaction, which leads to a gapped excitation spectrum within the bulk, but gapless edge states on the boundary for 2D materials such as $HgTe/HgCd_xTe_{1-x}$ quantum wells [52] with an inverted band structure, or surface states in bismuth (Bi), antimony (Sb), and tellurium (Te)-based compounds [53]. The states of the edges are protected from imperfections and disorder by their topological nature. Moreover, Fu and Kane [18, 19] noted that the surface of a topological insulator, when proximity-coupled to an ordinary s-wave superconductor, can support a nondegenerate, zero-energy, Majorana mode – a mode for which the excitation is its own antiparticle, and is thus composed of an equal amount of electron and hole components. A great deal of excitement abounds regarding its potential application as a platform for fault-tolerant quantum computing [26].

Following the initial proposal of the existence by Fu and Kane, in 2010, a group led by Das Sarma [15], and independently by Oreg, Refael, and von Oppen [16] proposed an elegant S-SmNW-S system in which a zero energy, Majorana fermion excitation was predicted to exist. In place of a topological insulator, it was recognized that a semiconductor with strong spin–orbit, such as InSb or InAs, can also produce a Majorana mode, provided that the spin degeneracy is lifted via proximity to a ferromagnetic material [56], or by the application of a sufficiently strong magnetic field, as was recognized by Lutchyn *et al.* [15], Oreg *et al.* [16], and Alicea [57], breaking Kramer's degeneracy, but taking care not to quench the superconductivity.

The geometry is one in which an SmNW with strong spin–orbit interaction, usually InSb or InAs, is proximity-coupled to a conventional s-wave superconductor, for example Nb, at its two ends, as shown in Figure 5.8. The proximity induces a nonzero superconducting gap $\Delta_0 < \Delta_{\mathrm{Nb}}$ within the SmNW, which is modeled as

$$H_{\mathrm{SC}} = \int_{-\infty}^{\infty} \left[\Delta(x)\psi_{\uparrow}^{\dagger}(x)\psi_{\downarrow}^{\dagger}(x)] + \mathrm{h.c.}\right] ,$$
$$\Delta(x) = \Delta_0 \Theta(x - \mathrm{L}) + \Delta_0 e^{i\varphi} \Theta(-x - \mathrm{L}) . \tag{5.34}$$

Here, L is the length of the SmNW between superconducting leads, and is much shorter than the coherence length of Nb ($\mathrm{L} \ll \xi_{\mathrm{Nb}}$). The SmNW extends over and is in strong contact with the superconducting leads, with a length $\mathrm{L}_1 \gg \xi_{\mathrm{Nb}}$ to ensure good proximity-coupling.

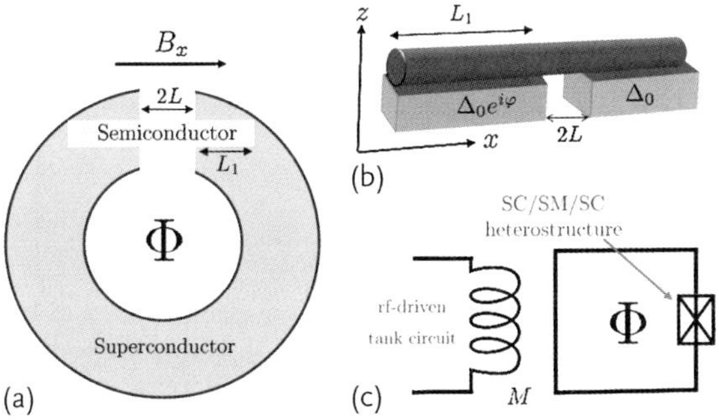

Figure 5.8 (a) Top view of an SM/SC heterostructure embedded in a small-inductance SC loop. (b) Side view of the SM/SC heterostructure. The nanowire can be top-gated to control chemical potential. Here, we assume $L \ll \xi$ and $L_1 \gg \xi$ with ξ being the SC coherence length. (c) Proposed readout scheme for the Andreev energy levels. Inductively coupled RF-driven tank circuit allows time-resolved measuring of the effective state-dependent Josephson inductance [19]. From [15].

The spin–orbit interaction, which in this type of compound semiconductors is well-modeled by a Rasha spin–orbit interaction, is given by

$$H_{\text{Rasha}} = \alpha \hat{\boldsymbol{z}} \cdot [\boldsymbol{p} \times \boldsymbol{\sigma}] = i\hbar\alpha\sigma_y \frac{\partial}{\partial x} , \tag{5.35}$$

where $\boldsymbol{\sigma}$ is the Pauli matrix operating on the electron spin, and α characterizes the strength of the Rashba interaction.

The Hamiltonian of the SmNW also includes a magnetic field B_x along the wire which can be tuned to remove the double-degeneracy of the spin states by opening a Zeeman gap. The complete Hamiltonian now reads

$$H_{\text{SmNW}} = \sum_{\sigma,\sigma'} \int_{-\infty}^{\infty} dx\, \psi_\sigma \left(-\frac{\hbar^2}{2m^*} \frac{\partial^2}{\partial x^2} - \mu + i\hbar\alpha\sigma_y \frac{\partial}{\partial x} + \frac{g_{\text{SmNW}}}{2} \mu_B B \sigma_x \right)_{\sigma\sigma'} \times \psi_{\sigma'}(x) + H_{\text{SC}} . \tag{5.36}$$

Here, m^* is the carrier effective mass and μ is the chemical potential.

Since the nanowires are assumed to be ballistic, and in the most favorable case, only one quantum mode is present, the BdG equations are naturally suited to carry out the analysis. The presence of the spin–orbit term requires the explicit labeling of the spin in the electron and hole-like components, u and v, leading to a four-component vector, which in the four-component Nambu space, is given by [15]

$$\Psi(x)^{\text{T}} = [u_\uparrow(x), u_\downarrow(x), v_\downarrow(x), -v_\uparrow(x)] . \tag{5.37}$$

Ψ satisfies the BdG equation

$$\hat{H}_{\text{BdG}} \Psi(x) = E\Psi(x) , \tag{5.38}$$

with the BdG full Hamiltonian given in matrix form ($\partial_x \equiv \partial/\partial x$ and $\partial_x^2 \equiv \partial^2/\partial x^2$):

$$\hat{H}_{\text{BdG}} = \begin{pmatrix} -\frac{\hbar^2}{2m}\partial_x^2 - \mu & \frac{g_{\text{SmNW}}}{2}\mu_B B + \hbar\alpha\partial_x & \Delta(x)^* & 0 \\ \frac{g_{\text{SmNW}}}{2}\mu_B B - \hbar\alpha\partial_x & -\frac{\hbar^2}{2m}\partial_x^2 - \mu & 0 & \Delta(x)^* \\ \Delta(x) & 0 & \frac{\hbar^2}{2m}\partial_x^2 + \mu & \frac{g_{\text{SmNW}}}{2}\mu_B B - \hbar\alpha\partial_x \\ 0 & \Delta(x) & \frac{g_{\text{SmNW}}}{2}\mu_B B + \hbar\alpha\partial_x & \frac{\hbar^2}{2m}\partial_x^2 + \mu \end{pmatrix} . \tag{5.39}$$

This equation for the SmNW is supplemented by the boundary conditions at the interface with the superconducting leads. The solution then yields the Andreev bound state, as before in the absence of spin–orbit interaction.

We now look for solutions for $\varphi = \pi$, that is, the left and right superconducting leads are 180° out of phase in their complex order parameter, Δ, and more specifically, for a mode at zero energy ($E = 0$), which could result in a Majorana mode: $\hat{H}_{\text{BdG}} \Psi_0(x) = 0$. The BdG equation above simplifies as the Hamiltonian becomes purely real. With the introduction of the length scale $x_{\mathrm{NW}} = \hbar/(m^*\alpha)$, energy scale $E_{\mathrm{NW}} = m^*\alpha^2$, and the corresponding dimensionless distance $s = x/x_{\mathrm{NW}}$, chemical potential $\tilde{\mu} = \mu/E_{\mathrm{NW}}$, gap $\tilde{\Delta}(x) = \Delta/E_{\mathrm{NW}}$, and Zeeman coupling $\tilde{Z}_x \equiv g_{\mathrm{SmNW}}\mu_B B/(2E_{\mathrm{NW}})$, the BdG equation becomes

$$\hat{H}_{\mathrm{BdG}} \equiv \begin{pmatrix} -\frac{\partial^2}{\partial s^2} - \tilde{\mu} & \tilde{Z}_x + \frac{\partial}{\partial s} & \tilde{\Delta}(x) & 0 \\ \tilde{Z}_x - \frac{\partial}{\partial s} & -\frac{\partial^2}{\partial s^2} - \tilde{\mu} & 0 & \tilde{\Delta}(x) \\ \tilde{\Delta}(x) & 0 & \frac{\partial^2}{\partial s^2} + \tilde{\mu} & \tilde{Z}_x - \frac{\partial}{\partial s} \\ 0 & \tilde{\Delta}(x) & \tilde{Z}_x + \frac{\partial}{\partial s} & \frac{\partial^2}{\partial s^2} + \tilde{\mu} \end{pmatrix} . \tag{5.40}$$

One can now seek a solution, $\Psi_0(x)$, which is entirely real. Particle–hole symmetry of the BdG dictates that if

$$\Psi_0^{\mathrm{T}} = [u_\uparrow(x), u_\downarrow(x), v_\downarrow(x), -v_\uparrow(x)] \tag{5.41}$$

is a solution, then

$$\Psi_0^{\mathrm{T}} = [v_\uparrow(x), v_\downarrow(x), u_\downarrow(x), -u_\uparrow(x)] \tag{5.42}$$

is also a solution, leading to the condition that only

$$v_\uparrow(x) = \lambda u_\uparrow(x)\,, \quad v_\downarrow(x) = \lambda u_\downarrow(x)\,, \quad \lambda = \pm 1\,, \tag{5.43}$$

such that the 4×4 BdG matrix is reduced to a 2×2 matrix.

Seeking an Andreev solution bound to the SmNW region, in the form

$$u_{\uparrow,\downarrow} \propto e^{zs}\,, \tag{5.44}$$

leads to the secular equation

$$z^4 + 4(\tilde{\mu} + 1)z^2 + 8\lambda\tilde{\Delta}_0 z + 4C_0 = 0\,, \quad C_0 = \tilde{\mu}^2 + \tilde{\Delta}^2 - \tilde{Z}_x^2\,. \tag{5.45}$$

For $C_0 > 0$, the equations, in conjunction with the boundary conditions, are overconstrained, and no solutions exist. This corresponds to the situation for the ordinary, nontopological insulator phase. For $C_0 < 0$, however, a solution does exist, for either sign of λ. This is the topological phase, with a zero-energy Majorana mode given by the solution. The phase boundary occurs at $C_0 = 0$.

The BdG equation can be solved numerically for the geometry in Figure 5.8 as a function of the phase φ, in the limit of a very short SmNW, where $\mathsf{L} \ll \xi$. The results for both phases at selected values of the parameters are shown in Figure 5.9.

These predictions have now led to major experimental efforts to search for them. In Chapter 8, the latest development in the experimental evidence for the existence of the Majorana fermion zero-energy mode is discussed in detail.

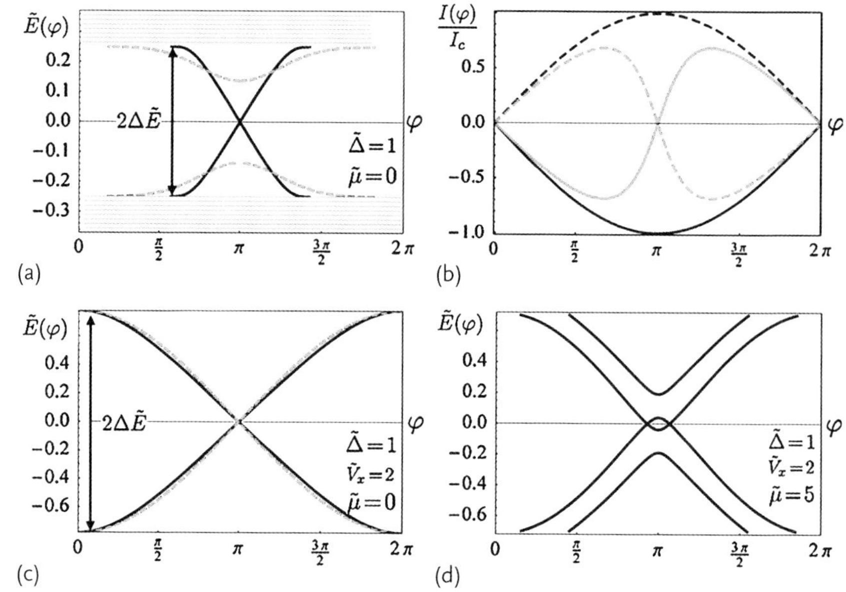

Figure 5.9 Andreev energy spectrum in SmNW/SC heterostructure for the junction with L → 0. (a) Energy spectrum in TP (topological) trivial (dashed line: $V_x = 0.75$) and nontrivial (solid line: $V_x = 1.25$) states. The two TP distinct phases differ by having an even and odd number of crossings, respectively. (b) Schematic plot of the Josephson current through the junction carried by Andreev states: gray and black lines describe Josephson current in TP trivial and nontrivial phases, respectively. (c) and (d) describe the evolution of the Andreev energy spectrum with chemical potential. (c) The spectrum in TP nontrivial phase. The dashed line is a fit to $\pm\cos(\varphi/2)$ function. (d) The spectrum in TP trivial phase. There is no crossing at $\varphi = \pi$. From [15].

References

1 Büchler, H.P., Geshkenbein, V.B., and Blatter, G. (2004) Quantum fluctuations in thin superconducting wires of finite length. *Physical Review Letters*, **92** (6), 067007.

2 Khlebnikov, S. (2008) Quantum phase slips in a confined geometry. *Physical Review B*, **77** (1), 014505.

3 Khlebnikov, S. (2008) Quantum mechanics of superconducting nanowires. *Physical Review B*, **78** (1), 014512.

4 Sachdev, S., Werner, P., and Troyer, M. (2004) Universal conductance of nanowires near the superconductor–metal quantum transition. *Physical Review Letters*, **92** (23), 237003.

5 Clarke, J. (1972) Experimental observation of pair-quasiparticle potential difference in nonequilibrium superconductors. *Physical Review Letters*, **28**, 1363–1366.

6 Tinkham, M. and Clarke, J. (1972) Theory of pair-quasiparticle potential difference in nonequilibrium superconductors. *Physical Review Letters*, **28**, 1366–1369.

7 Boogaard, G.R., Verbruggen, A.H., Belzig, W., and Klapwijk, T.M. (2004) Resistance of superconducting nanowires connected to normal-metal leads. *Physical Review B*, **69** (22), 220503.

8 Keizer, R.S., Flokstra, M.G., Aarts, J., and Klapwijk, T.M. (2006) Critical voltage of a mesoscopic superconductor. *Physical Review Letters*, **96** (14), 147002.

9 Vercruyssen, N., Verhagen, T.G.A., Flokstra, M.G., Pekola, J.P., and Klapwijk, T.M. (2012) Evanescent states and nonequilibrium in driven superconducting nanowires. *Physical Review B*, **85** (22), 224503.

10 Altomare, F., Chang, A.M., Melloch, M.R., Hong, Y., and Tu, C.W. (2006) Evidence for macroscopic quantum tunneling of phase slips in long one-dimensional superconducting Al wires. *Physical Review Letters*, **97** (1), 017001.

11 Altomare, F., Li, P., and Chang, A.M., unpublished.

12 Beenakker, C.W.J. and van Houten, H. (1991) Josephson current through a superconducting quantum point contact shorter than the coherence length. *Physical Review Letters*, **66**, 3056–3059.

13 Muller, C.J., van Ruitenbeek, J.M., and de Jongh, L.J. (1992) Conductance and supercurrent discontinuities in atomic-scale metallic constrictions of variable width. *Physical Review Letters*, **69**, 140–143.

14 Furusaki, A., Tsukada, M., and Takayanagi, H. (1991) Theory of quantum conduction of supercurrent through a constriction. *Physical Review Letters*, **67**, 132–135.

15 Lutchyn, R.M., Sau, J.D., and Das Sarma, S. (2010) Majorana fermions and a topological phase transition in semiconductor–superconductor heterostructures. *Physical Review Letters*, **105** (7), 077001.

16 Oreg, Y., Refael, G., and von Oppen, F. (2010) Helical liquids and Majorana bound states in quantum wires. *Physical Review Letters*, **105** (17), 177002.

17 Kitaev, A.Y. (2001) 6. Quantum computing: Unpaired Majorana fermions in quantum wires. *Physics Uspekhi*, **44**, 131.

18 Fu, L. and Kane, C.L. (2008) Superconducting proximity effect and Majorana fermions at the surface of a topological insulator. *Physical Review Letters*, **100** (9), 096407.

19 Fu, L. and Kane, C.L. (2009) Probing neutral Majorana fermion edge modes with charge transport. *Physical Review Letters*, **102** (21), 216403.

20 Majorana, E. (1937) *Il Nuovo Cimento*, **14**, 171–184.

21 Wilczek, F. (2009) Majorana returns. *Nature Physics*, **5**, 614–618.

22 Nadj-Perge, S., Pribiag, V.S., van den Berg, J.W.G., Zuo, K., Plissard, S.R., Bakkers, E.P.A.M., Frolov, S.M., and Kouwenhoven, L.P. (2012) Spectroscopy of spin–orbit quantum bits in indium antimonide nanowires. *Physical Review Letters*, **108** (16), 166801.

23 Williams, J.R., Bestwick, A.J., Gallagher, P., Hong, S.S., Cui, Y., Bleich, A.S., Analytis, J.G., Fisher, I.R., and Goldhaber-Gordon, D. (2012) Unconventional Josephson effect in hybrid superconductor – Topological insulator devices. *Physical Review Letters*, **109** (5), 056803.

24 Rokhinson, L.P., Liu, X., and Furdyna, J.K. (2012) The fractional a.c. Josephson effect in a semiconductor – superconducting nanowire as a system of Majorana particles. (2012) *Nature Physics*, **8** (11), 795–799.

25 Deng, M.T., Yu, C.L., Huang, G.Y., Larsson, M., Caroff, P., and Xu, H.Q. (2012) Anomalous zero-bias conductance peak in a Nb–InSb nanowire–Nb hybrid device. *Nano Letters*, **12**(12), 6414–6419.

26 Kitaev, A.Y. (2003) Fault-tolerant quantum computation by anyons. *Annals of Physics*, **303**, 2–30.

27 Usadel, K.D. (1970) Generalized diffusion equation for superconducting alloys. *Physical Review Letters*, **25**, 507–509.

28 Eilenberger, G. (1968) Transformation of Gorkov's equation for type II superconductors into transport-like equations. *Zeitschrift für Physik*, **214** (2), 195–213.

29 Larkin, A.I. and Ovchinnikov, Y.N. (1969) Quasiclassical method in the theory of superconductivity. *Soviet Journal of Experimental and Theoretical Physics*, **28**, 1200.

30 Larkin, A.I. and Ovchinnikov, Y.N. (1975) Nonlinear conductivity of superconductors in the mixed state. *Soviet Journal of Experimental and Theoretical Physics*, **41**, 960.

31 Rammer, J. and Smith, H. (1986) Quantum field-theoretical methods in transport theory of metals. *Reviews of Modern Physics*, **58**, 323–359.

32 Nazarov, Y.V. and Stoof, T.H. (1996) Diffusive conductors as Andreev interferometers. *Physical Review Letters*, **76**, 823–826.

33 Stoof, T.H. and Nazarov, Y.V. (1996) Kinetic-equation approach to diffusive superconducting hybrid devices. *Physical Review B*, **53**, 14496–14505.

34 Feigel'man, M.V., Larkin, A.I., and Skvortsov, M.A. (2000) Keldysh action for disordered superconductors. *Physical Review B*, **61**, 12361–12388.

35 Schmid, A. and Schön, G. (1975) Linearized kinetic equations and relaxation processes of a superconductor near T_c. *Journal of Low Temperature Physics*, **20**, 207–227.

36 Schüssler, U. and Kümmel, R. (1993) Andreev scattering, Josephson currents, and coupling energy in clean superconductor–semiconductor–superconductor junctions. *Physical Review B*, **47**, 2754–2759.

37 Averin, D. and Bardas, A. (1995) AC Josephson effect in a single quantum channel. *Physical Review Letters*, **75**, 1831–1834.

38 Bratus', E.N., Shumeiko, V.S., and Wendin, G. (1995) Theory of subharmonic gap structure in superconducting mesoscopic tunnel contacts. *Physical Review Letters*, **74**, 2110–2113.

39 Bardas, A. and Averin, D.V. (1997) Electron transport in mesoscopic disordered superconductor–normal-metal–superconductor junctions. *Physical Review B*, **56**, 8518.

40 Cuevas, J.C., Hammer, J., Kopu, J., Viljas, J.K., and Eschrig, M. (2006) Proximity effect and multiple Andreev reflections in diffusive superconductor–normal-metal–superconductor junctions. *Physical Review B*, **73** (18), 184505.

41 Kane, C.L. and Mele, E.J. (2005) Quantum spin Hall effect in graphene. *Physical Review Letters*, **95** (22), 226801.

42 Kane, C.L. and Mele, E.J. (2005) Z_2 topological order and the quantum spin Hall effect. *Physical Review Letters*, **95** (14), 146802.

43 Bernevig, B.A. and Zhang, S.C. (2006) Quantum spin Hall effect. *Physical Review Letters*, **96** (10), 106802.

44 Furusaki, A. and Tsukada, M. (1991) DC Josephson effect and Andreev reflection. *Solid State Communications*, **78** (4), 299–302.

45 Furusaki, A. and Tsukada, M. (1991) Current-carrying states in Josephson junctions. *Physical Review B*, **43**, 10164–10169.

46 Furusaki, A. and Tsukada, M. (1990) A unified theory of clean Josephson junctions. *Physica B: Condensed Matter*, **165/166**, Part 2 (0), 967–968.

47 van Wees, B.J., Lenssen, K.M.H., and Harmans, C.J.P.M. (1991) Transmission formalism for supercurrent flow in multiprobe superconductor–semiconductor-superconductor devices. *Physical Review B*, **44**, 470–473.

48 Büttiker, M. and Klapwijk, T.M. (1986) Flux sensitivity of a piecewise normal and superconducting metal loop. *Physical Review B*, **33**, 5114–5117.

49 Kulik, I.O. (1970) *Soviet Journal of Experimental and Theoretical Physics*, **30**, 944.

50 Ishii, C. (1970) Josephson currents through junctions with normal metal barriers. *Progress of Theoretical Physics*, **44**, 1525–1547.

51 Bardeen, J. and Johnson, J.L. (1972) Josephson current flow in pure superconducting–normal-superconducting junctions. *Physical Review B*, **5**, 72–78.

52 König, M., Wiedmann, S., Brüne, C., Roth, A., Buhmann, H., Molenkamp, L.W., Qi, X.L., and Zhang, S.C. (2007) Quantum spin Hall insulator state in HgTe quantum wells. *Science*, **318**, 766.

53 Hsieh, D., Qian, D., Wray, L., Xia, Y., Hor, Y.S., Cava, R.J., and Hasan, M.Z. (2008) A topological Dirac insulator in a quantum spin Hall phase. *Nature (London)*, **452**, 970–974.

54 Qi, X.L. and Zhang, S.C. (2010) The quantum spin Hall effect and topological insulators. *Physics Today*, **63** (1), 010000.

55 Hasan, M.Z. and Kane, C.L. (2010) Colloquium: Topological insulators. *Reviews of Modern Physics*, **82**, 3045–3067.

56 Sau, J.D., Lutchyn, R.M., Tewari, S., and Das Sarma, S. (2010) Generic new platform for topological quantum computation using semiconductor heterostructures. *Physical Review Letters*, **104** (4), 040502.

57 Alicea, J. (2010) Majorana fermions in a tunable semiconductor device. *Physical Review B*, **81** (12), 125318.

Part Two Review of Experiments on 1D Superconductivity

6
Experimental Technique for Nanowire Fabrication

6.1
Experimental Technique for the Fabrication of Ultra Narrow Nanowires

The study of superconductivity in 1D was effectively started by Little [1], who in 1967 was the first author to investigate the effect of thermodynamic fluctuations of the order parameter in ring-shaped superconductors: he demonstrated that in one- and two-dimensional samples, no infinitely sharp change of resistivity occurs but, instead, the resistance drops smoothly and rapidly towards zero as $T \to 0$ K. After reading a preprint of this work, Parks and Groff [2] studied the effect of fluctuations in the order parameter on the critical current density in a Sn superconducting microstrip close to T_c. Almost at the same time, Hunt and Mercereau [3] published new results on the observation of a depression in the superconducting transition temperature of thin film with respect to wider films. A few months later, Langer and Ambegaokar [4] published their seminal paper on "Intrinsic Resistive Transition in Narrow Superconducting Channels." Consequently, this encouraged Webb and Warburton [5] to perform experiments on tin whisker crystals. In order to provide a closer correspondence between the LA theory and these experiments, McCumber and Halperin [6] provided a better estimate of the prefactor of the exponential dependence in the LA theory: the result of the combined theoretical effort is known today as the Langer Ambegaokar McCumber Halperin (LAMH) theory (please see Sections 2.4.3–2.4.5).

This theory predicts that the resistance of a 1D superconductor as a function of decreasing temperature has an exponentially activated form, that is,

$$R \propto \Omega \exp\left(-\frac{\Delta F}{k_B T}\right),$$

where ΔF is the free energy barrier that the phase of the condensate has to overcome and

$$\Omega = \left(\frac{L}{\xi}\right)\left(\frac{\Delta F}{k_B T}\right)^{1/2}\frac{\hbar}{\tau_{GL}}$$

is an attempt frequency (2.125), with L the length of the superconductor, ξ the superconducting coherence length and τ_{GL} the Ginzburg–Landau time constant,

One-Dimensional Superconductivity in Nanowires, First Edition. F. Altomare and A.M. Chang

inversely proportional to ΔT. The first experimental verification of LAMH theory was reported by Lukens *et al.* in 1970 [7] and was followed by Newbower *et al.* a couple of years later [8]: in these early experiments, whiskers of tin were measured at a temperature very close to the critical temperature of the superconductor: since the superconducting coherence length diverges as T approaches T_c for $T - T_c \ll T_c$, whiskers which have lateral dimension (w, h) of the order of a few micrometers can be considered 1D as $\xi \gg$ (w, h). The experimental results could be fit by the LAMH expression over six orders of magnitude in the measured resistance. Fast forwarding about 20 years and, thanks to the development of the step-edge technique [9], Giordano was able, for the first time, to fabricate a nanowire with lateral dimensions of the order of 100 nm [10]. This allowed him to explore a regime in which the temperature was much lower than the superconductor critical temperature, but the sample could still be considered 1D. This was the first major advance in several years and provided the initial evidence of a transport mechanism different from thermal activation, opening the door to the investigation of quantum tunneling in nanowires, the main subject of this chapter. Again, the lack of a suitable technique to produce a uniform nanowire of even smaller size hindered progress in the field until the molecular templating technique fabrication was developed by Tinkham and Bezryadin [11]. These results spurred new interest and since then, several groups have invested time and resources into developing new techniques for studying transport in 1D superconductors.

As can be evinced from this very brief history, the interplay of theory and experiments has been crucial in the initial development of the field. However, an extremely important role for reaching our current level of understanding has been played by progress in the fabrication of nanowires with a sufficiently small lateral size with respect to the superconducting coherence length. It is therefore natural to discuss the main techniques used for the fabrication of nanowires, the main subject of this chapter. In the following chapters, we will provide an overview of the major experimental results in 1D superconductivity in an effort to provide a more unified view of the field.

6.2
Introduction to the Techniques

This section will be used to provide a basic understanding of the fabrication techniques and the terminology used in the following sections so that the more advanced fabrication techniques will be grasped more readily by readers with limited experience/knowledge of fabrication techniques, or who are just entering the field. Being well aware that *the devil is in the details,* we will describe the various fabrication techniques, including the basic fabrication steps in quite a general manner. Many steps, repeated in different laboratories, may produce significantly different end results due to uncontrollable factors such as air humidity. Even when a temperature and humidity controlled clean room is used, such problems cannot always be avoided due to other subtle details inherent to the equipment in use. To ensure reproducibility, a fine tuning of the process is often necessary. For instance, in the same lab, an etching solution will require calibration every now and then, as the relative composition of the solution will vary in time due to evaporation or chemical degradation processes.

6.2.1 Lithography

The word *lithography* has a Greek etymology and means "write to a stone." This process was invented at the end of eighteenth century [12] for the publication of theatrical work and is described in great detail in [13]. In our context, lithography is the fundamental tool for sample fabrication. While several forms of lithography exist and are used, we will briefly describe the two most common processes: photolithography and electron-beam (e-beam) lithography, which differ mainly in the type of radiation used for exposure. These two processes are very similar in nature: a substrate (silicon and gallium arsenide are the most common) is covered by a polymer coating: *resist*. The polymer-coated substrate is then irradiated in a predefined pattern. For a *positive tone resist*, the radiation will change the molecular composition of the resist and will make the exposed region soluble in a *developer solution*. As it is easy to imagine, there is also a *negative tone resist*. In this case, the exposed region is rendered harder to be dissolved in the developer and *only* the exposed region will remain.

How can we fabricate a nanowire? A simple way would be to pattern a positive tone resist and deposit a metal. Another alternative would be to deposit a metal on the substrate, pattern the resist as described above, and chemically remove the excess material.

What are the limits of this process? Several factors need to be taken into account. The main ones are:

- Lithography: photolithography (or optical lithography) uses radiation in the optical range, while electron-beam lithography uses an electron beam appropriately driven by coils in an electron microscope.
- The resist, which depends on the type of lithography (optical or e-beam), can be positive or negative.
- The developer depends on the resist used. It must be noted that many developer solutions can also etch metal.

Optical lithography is favored for the fabrication of large structures, typically 0.5 μm or larger in a laboratory setting. In this case, is necessary to prefabricate a mask. The mask is nothing other than a reproduction of the design we want to imprint in the resist. It can be made by printing a drawing on a transparent slide (this process is actually used in many laboratories to prototype small circuit boards) or by etching the chrome plating deposited on a glass plate. In both cases, the image on the mask is projected all at once onto the resist-coated substrate. While there are some tricks that can be employed (at a substantial cost), the main limit to the size is due to the wavelength used to project the image on the resist and light diffraction effects. One of most common is the "i-line" (365 nm) from a mercury lamp.

E-beam lithography relies, instead, on the beam generated in a scanning electron microscope. Low end systems are electron microscopes that are converted by using pattern generators.[1] In this case, the pattern is drawn using computer software and then written using the pattern generator which will move the beam according to the desired pattern, a procedure known as rastering, while taking care of delivering the specified electron dosage to each pixel of the pattern. While this is inherently slower than the optical lithography, the typical size of a properly focused beam is a couple of nanometers. However, the minimal size will be of the order of 10–20 nm, depending on the resist and exact developing conditions. Another factor that limits the minimal size achievable by e-beam lithography is the proximity effect. When an electron collides with the nucleus of a substrate, it can be scattered back in a direction close to the incident direction. In this case, it will have sufficient energy to interact with the resist in a region close to the pattern. For this reason, light and thin substrates are typically used to reduce such multiple-scattering induced proximity effects when trying to achieve the smallest lithographic features. While the backscattering can be beneficial as it provides an undercut in the resist which will allow a clean removal of the unexposed resist (for positive tone resist) after depositing the metal, in some cases, such as a high-density pattern used for a photon detector, it can be detrimental as it will smear the pattern features. As we will discuss later, the exact developer and development conditions employed will also come into play in order to achieve the smallest lithographic size.

6.2.2
Metal Deposition

Once the pattern has been drawn on the substrate, a metal is typically deposited to fill the void. However, evaporated metal tends to grow by forming grains: if the size of the pattern is smaller than the grain size, the metal will not coalesce into a smooth and uniform line. For this reason, especially when attempting to fabricate extremely narrow lines, materials with small grain such as the alloy gold/palladium ($Au_{1-x}Pd_x$) are favored. Metal deposition is predominantly carried out by evaporation or by sputtering. In evaporation, the material to be deposited is heated up by means of a current flowing through a holder (boat or crucible), or by an electron beam impinging on the material. Once the melting temperature is reached, the material sublimates and, due to the extremely low ambient pressure and thus low density of background atoms with which to collide, is able to reach the substrate where it condenses. Typically, the source of material is seen as a point source and the directionality of the evaporation can be used to fabricate even smaller features. Alternatively, during sputtering, once the desired vacuum is reached, an amount of argon (noble gas) is admitted to the chamber. The argon is ionized and accelerated

1) http://www.raith.com/ and http://www.jcnabity.com/ (last accessed: December 26, 2012).

toward the source of material to be deposited (target): because of its high velocity, the argon will knock off atoms from the target which will condense on the substrate. Contrary to evaporation, this method is not directional as atoms from the target are accelerated in all directions. While typically better-suited for deposition of films to be etched, it can also be used with some particular "trick" in lieu of evaporation. In both cases, the lower the vacuum the better the film quality as less impurities are incorporated. Also, the quality of the deposited material depends on the rate of deposition, with higher rates typically favoring higher quality. The grain size can depend rather strongly on the rate, and often a higher rate yields smaller grain sizes. After metal deposition, the excess resist, together with the metal evaporated on top, needs to be removed. This process is called *lift-off* and is performed by submersing the device in a liquid that dissolves the resist, often accompanied by agitation in an ultrasonic bath.

6.2.3 Etching

Another important step common to many fabrication methods is *etching*. Etching, which means drawing a figure by removing excess material using some type of corrosion process, can be employed for preparing the substrate used for the device fabrication or for the device itself. Typically, we can distinguish *wet-etching* and *dry-etching*. Wet-etching of deposited metal is typically isotropic, that is, the etchant removes material at the same speed in all directions. In the case of crystalline structures (semiconductors, for example), some etchants act preferably in one direction, typically corresponding to a well-defined crystalline orientation. For example, an aqueous solution of potassium hydroxide etches silicon preferentially in the (111) crystallographic direction: as silicon substrates are typically grown in the (001), the solution will etch at 45° with respect to the direction of growth. Additionally, this solution does not etch silicon dioxide, or more accurately, it etches silicon oxides at a much slower rate than silicon: in this case, the etching is termed *selective* and the ratio of the etching rate is provided. Dry-etching, depending on the exact details, can be isotropic (plasma etching) or highly directional (deep reactive ion etching, for example). These are performed in a vacuum chamber with the low pressure of the reactive gas used as the etchant: once the plasma is ignited, it interacts chemically with the material to be etched. Photoresist or e-beam resist is typically used to protect the areas that need not be etched.

6.2.4 Putting It All Together

What we have just described constitutes the building blocks of many (if not all) fabrication techniques. However, the problem that the experimentalists are facing is to fabricate 1D superconducting nanowires. Again the constraint of being 1D is not an absolute, but rather means 1D with respect to the relevant physics we are trying to study. This being superconductivity, the natural length scale is the

Table 6.1 Common superconducting materials, bulk superconducting length and superconducting length in the dirty regime assuming a mean free path of 1 and 10 nm. Most of the superconductors in the table have 1 nm $< l_{mfp} <$ 10 nm, with the exception of $Mo_{1-x}Ge_x$ (l_{mfp} = 0.3 nm). For $Mo_{1-x}Ge_x$ nanowires, the commonly accepted coherence length is $\xi \simeq$ 10 nm. Assuming a l_{mfp} = 0.3 nm [15], the bulk coherence length of $Mo_{1-x}Ge_x$ would be $\xi_{bulk} >$ 1000 nm. (Data adapted from [16] Table I.)

Superconductor	ξ_{bulk} (nm)	$\xi_{est}^{l_{mfp}=1\,nm}$ (nm)	$\xi_{est}^{l_{mfp}=10\,nm}$ (nm)
Al	1600	34	108
In	360	16	50
Zn	1800	38	120
Sn	230	16	41
Pb	90	8	25
$Mo_{1-x}Ge_{x,film}$	5	2	6
Nb	30		15
Ti	6200[a]	67	248

[a] Data for Ti have been estimated (See [16] Table I).

superconducting coherence length ξ. Following Likharev's approach [14], we can consider a nanowire 1D if the w $\lesssim 4.41\xi$, or we can choose a more strict approach and require w $\ll \xi$. One fact to keep in mind is that typically thin superconducting films and wire are in the so-called *dirty regime*. In this case, the superconducting coherence length is $\xi \sim 0.85\sqrt{\xi_{bulk} l_{mfp}}$, where ξ_{bulk} is the superconducting coherence length of the bulk material and l_{mfp} is the elastic electron mean free path in the disordered thin film. Table 6.1 reproduces the ξ_{bulk} values for the most common superconductors and the value of the dirty-limit superconducting coherence length assuming a mean free path of 10 nm, although, except for aluminum, this might be a gross overestimate in the other materials.

As we can see from Table 6.1, we are faced with the necessity of fabricating nanowires of lateral size smaller than 10–20 nm. Note that even for Al or Zn, which have long coherence lengths ξ, additional constraints based on free energy barrier considerations for the phase slip process will necessitate such small dimensions – see Chapters 2 and 3. Is this feasible with the experimental technique discussed above? Except for recent developments discussed below, the answer would be no. Then, there is the issue of uniformity: the goal is to study a uniform wire, not a wire interrupted by a constriction. Constrictions or, in general, nonuniformity in the nanowire can introduce artifacts which could be misinterpreted. Finally, the formation of natural oxide can be a limiting factor to the nanowire size: in zinc, in fact, which has one of the longest bulk coherence lengths, the oxide formation poses a limit to the minimum lateral size achievable. In order to push these limits, it has been necessary to develop new fabrication techniques which will be discussed below. It turns out that instead of fabricating an extremely narrow nanowire, in some cases, it is easier to fabricate a *stencil*, that is, a support on which to fabricate the nanowire. While this might not seem to be a big advantage, in many cases, this provides better results in terms of uniformity, due to the relative ease of fabricating uniform stencils. Typically, once the stencil is fabricated, the actual wire fabrication becomes (relatively) trivial. In the following, we will discuss some of the newly developed techniques used for the preparation of superconducting nanowires and we will attempt to provide a description in as simple terms as possible (*toy-model*).

6.3 Step-Edge Lithographic Technique

In the early 1980s, the main technique used for submicrometer fabrication of narrow lines was e-beam lithography on thin substrates which were used in order to reduce the backscattering of the electron. The step-edge lithographic technique [9] was developed in Prober's lab at Yale University: this technique allowed the authors to produce wires that were of extremely narrow lateral dimension and could be as long as 0.5 mm. As in all stencil techniques, the most challenging part is the fabrication of the stencil while the fabrication is relatively straightforward. To facilitate conceptualization, imagine a *toy-model* version of this technique which involves filling the vertical corner formed between two steps in a staircase. The difficulty lies in fabricating a uniform vertical step and uniformity in the caulking used to fill the step, which represents the nanowire.

In detail, the fabrication proceeds as follows: A chrome film is deposited on a glass substrate using photolithographic methods. After resist lift-off, the chrome film containing a smooth, straight edge is used as a mask for an ion etch at normal incidence. The ion etching produces a step (the corner between two steps in a staircase), whose depth largely determines the nanowire cross section. In fact, after the step formation, the chrome film is removed and the material of choice for the nanowire formation is deposited in the step filling it (caulking). At this point, two variants are considered for the nanowire fabrication:

Process A Softer metals: the next step is either a directional deposition (thermal or e-beam evaporation) at an angle suitable to fill the step, or conformal deposition (sputtering) in order to fully coat the step. Then, ion etching at a suitable angle (in order to shadow the step in the substrate) is used until all the excess metal is removed, and a wire of triangular cross section is formed at the step edge. This method is suitable for the fabrication of wires of soft materials that can be etched with reactive gas.

Process B Harder metals: it is possible to evaporate the desired metals directly on the vertical edge of the step. As there is always an unintentional light coating of metal on the rest of the glass substrate, a light ion etching will be used to remove the excess metal and make sure that only the metal evaporated on the step is electrically connected.

The fabrication steps are reported in Figure 6.1. The main advantages of this process are that it is self-aligning and can yield extremely long nanowires (0.5 mm);

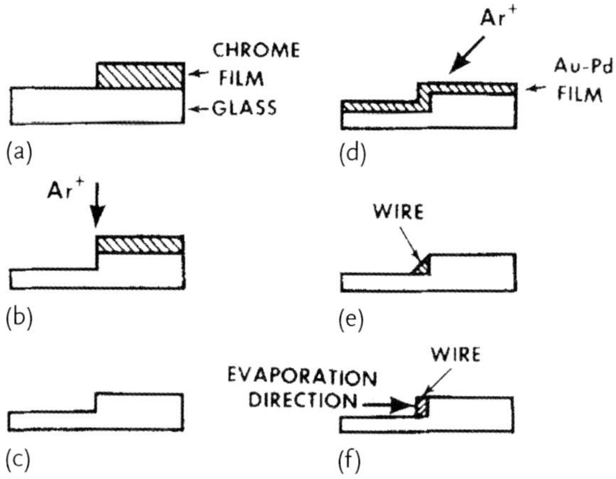

Figure 6.1 Fabrication procedures for production of fine metal wires. The substrate is a microscope cover glass shown in side view. Procedures a–c are the same for both process A and process B. (a) Half the substrate is coated with a thin chrome film. (b) The substrate is ion etched to produce a square step. (c) The chrome film is removed with a chemical etch. Process A: (d) The substrate is coated with the metal film (e.g., Au-Pd), and ion etched at an angle until (e). A nearly triangular wire is formed along the step edge. Process B: (f) The metal film is evaporated parallel to the substrate to coat only the step edge. A subsequent ion etching normal to the substrate may be required to remove the light coating on the rest of the substrate. From [9].

the wires are not subject to any chemical treatment after deposition and most of the wire is protected by an ion milling process. Additionally, during the deposition, the substrate can be heated or cooled, depending on the material to be deposited. The main limit on the nanowire size is represented by the raggedness of the step edge ($\approx$ 10–20 nm) due to the initial lithographic step. A minor problem with this technique is that Cr does not produce completely uniform edges; while NiCr is better from this point of view as it has the disadvantage of lower etching selectivity. Furthermore, while the final ion etching is not extremely long in duration, it could produce structural modification in the outer portion of the nanowire.

6.4
Molecular Templating

A recent technique for the fabrication of even smaller nanowires was developed in 2000 by A. Bezryadin in Tinkham's group at Harvard University [11]. A *toy-model* requires the fabrication of a rope distended across two platforms. During a snowstorm, the snow sticks to the platforms and to the rope, mostly on the top, but partially covers the sidewalls as well. The contacts and nanowire are represented by the snow connecting the two platforms through the snow sticking on the rope: additionally, the thinner the rope, the smaller the diameter of the nanowire.

The role of the "rope" is *realized* by a carbon nanotube (diameter ~ 1–10 nm) which can now be readily bought through various sources. This is a textbook example of how to transfer all the difficulties in the fabrication of a very thin nanowire to the fabrication of the stencil: the carbon nanotube acts only as a mechanical support on which to deposit metal. The nanowire width will be (largely) determined by the width of the underlying nanotube. If the nanotube is suspended above a trench (suspended rope above), and the nanotube is insulating, all the electrical conduction will be confined to the nanowire deposited on top of the nanotube (snow). The other important part in the fabrication of the stencil is the fabrication of the support over which the nanotube is distended: this is accomplished by a combination of standard techniques (lithography and etching).

In detail, the fabrication technique proceeds as follows (Figure 6.2). A silicon substrate is used as a starting point: this can be either insulating or intentionally doped to allow electrical back gating of the sample. On top of the silicon substrate, a layer of silicon oxide (SiO_2) is grown and a low stress silicon nitride layer (SiN) is deposited on top of the SiO_2 using low pressure chemical vapor deposition. After

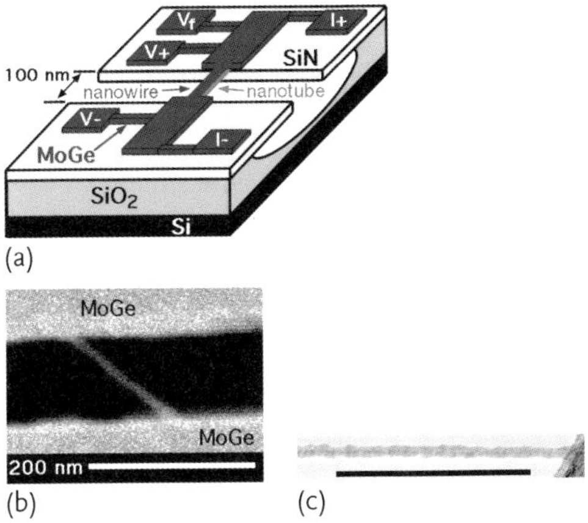

Figure 6.2 Fabrication and imaging of a nanowire. (a) Diagram of the sample. The Si substrate (shown in black) is covered with a 0.5-μm layer of SiO_2 and a 50 nm SiN film. A 100 nm wide slit is patterned in the SiN film using electron beam lithography and reactive ion etching. The SiO_2 layer under the slit is removed using HF to form an undercut (shown white). To make a nanowire, nanotubes (red) were first deposited followed by sputtering of a 5 nm thick amorphous $Mo_{79}Ge_{21}$ film and a 1.5 nm Ge protective layer. Initially, the metal film covers the entire substrate, including the free-standing carbon nanotubes or bundles crossing the slit. Next, a suitable single nanowire was found (using scanning electron microscopy, SEM) with a subsequent patterning of the electrodes (blue) using optical lithography and reactive ion etching. The Mo-Ge film is interrupted by the slit and forms two electrodes connected electrically through a single nanowire. (b) SEM image of a sample, showing a free-standing nanowire which connects two Mo-Ge electrodes. This is one of the thinnest wires found under the SEM. Its apparent width, including blurring in the SEM image, is $W < 10$ nm. The scale bar (white) is 200 nm. (c) One of the thinnest wires found under a higher-resolution transmission electron microscope. The wire width is $W_{TEM} < 5.5 \pm 1$ nm. The scale bar (black) is 100 nm. From [11].

this step, the entire wafer is coated with an e-beam resist (PMMA, or polymethyl methacrylate). E-beam lithography is then performed to write several lines with different widths (which will determine the distance between the two platforms and therefore the nanowire length) and markers. The former will be used to create the trenches: the nanotube will be suspended across the trenches so that the metal covering the nanotube does not short to the metal in the trench to either side. The latter are used to easily locate the slit in a scanning electron microscope (SEM) to find the most suitable nanotubes to be targeted. After developing the exposed resist, the wafer is exposed to reactive ion etching using a SF_6 plasma. This plasma etches the exposed SiN, thus transferring its pattern into the substrate. After removing the PMMA, the wafer is coated (for protection) and diced in small chips. HF is then used to produce an undercut in the SiO_2 due to the higher etching rate of SiO_2 with respect to SiN. After rinsing, a nanotube suspension is deposited on the chip: when the solvent dries, the nanotubes are left on it. A blanket evaporation covers the chip with the desired superconducting material. To find a suitable sample, SEM inspection is used to search for nanotubes which lie across the trenches. Once their position is recorded, in a final step, optical lithography is performed to define a quasi-4-terminal configuration[2)] to perform electrical measurement.

The main advantage of this technique lies in its ability to create extremely narrow and short nanowires. However, selection of superconducting material is limited: only a few superconducting elements (most notably Nb, $Mo_{1-x}Ge_x$) adhere to the nanotube without the need for an adhesion layer. One minor problem with this technique is that the nanowire length is determined by the size of the slit and the angle of the nanotube with respect to it. It is not possible to decide the length of the nanowire beforehand. Additionally, nanowires longer than 200–300 nm invariably appear to be inhomogeneous when characterized in a transport measurement, exhibiting multiple resistive transitions, resistive tails and multiple critical currents [17]. It is, in fact, important to notice that the SiN etching produces sloped edges and long nanotubes, which tend to follow the slope of SiN, and may not be properly metalized, causing the formation of weak links [18]. Notice that carbon nanotube [11], fluorinated carbon nanotube [17], DNA molecules [19] and WS2 nanorods [20], and scaffolds [21] have all been used as suspension bridges. However, in order to eliminate possible misinterpretation of the experimental data, an insulating stencil is preferable. Despite these drawbacks, this method has proven to be extremely successful in producing short and narrow $Mo_{1-x}Ge_x$, and Nb nanowires. Also, a newly developed *in-situ* annealing process [22] has been applied to the study of single-crystalline Mo_3Ge nanowires.

2) This is termed a quasi-4-terminal measurement as a portion of the film is typically measured in series with the nanowire.

6.5 Semiconducting Stencils

Carbon nanotubes are not the only manufactured stencil that can be used as a stencil for nanowires. In fact, other techniques use stencils created by Molecular Beam Epitaxy (MBE). This crystal growth technique is widely used in the semiconducting industry for producing compound semiconductor heterostructures – structures composed of layers of different chemical compositions but matching lattice constants – and has the capability to create such structures by depositing one atomic layer after another [23]. A semiconducting layer of only a few nanometers, when viewed edge-on, would be an extremely convenient stencil for the fabrication of nanowires. These stencils would not be limited in length or in width and offer a wider choice of superconducting material. In the following, we will describe two techniques in detail which have been used for the fabrication of superconducting nanowires. First, a variant of these techniques developed by Natelson *et al.* [24] will also be briefly discussed. This method has been used only to study transport in nonsuperconducting AuPd nanowires [25, 26].

6.6 Natelson and Willet

The first use of a semiconducting stencil for the fabrication of sub-10 nm nanowires was developed by D. Natelson *et al.* [24] while working in R. Willet's group at Bell Laboratories in 2002.

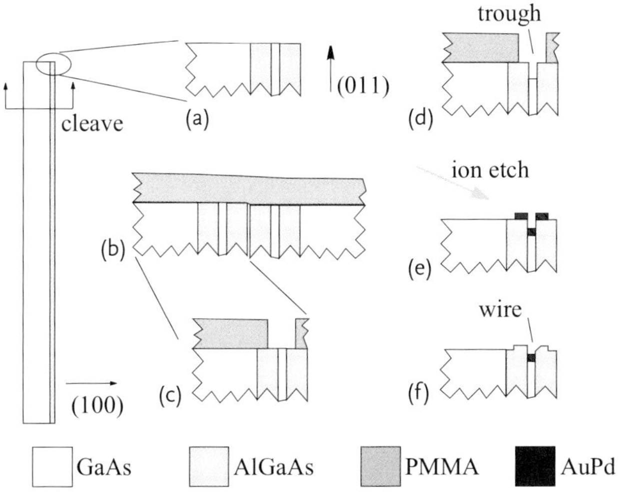

Figure 6.3 A schematic representation of the wire fabrication process, showing cross-sectional views of a sample. The individual steps are described in the text. From [24].

The *toy-model* of this technique is similar to creating a very precise mold in the form of a trench which is later filled by the desired material. If any extra material were to overflow, a scalpel can be used to quickly remove the excess.

In this case, the stencil was made of Ga/$Al_xGa_{1-x}As$ ($x \sim 30\%$): these two materials are extremely well lattice-matched, a characteristic that reduces the amount of imperfections at the interface between the two. This is the key to fabricating low stress interfaces where the result is superior mechanical properties. In detail, the fabrication proceeds as follows. The starting point is a commercially available GaAs substrate (Figure 6.3). A 2 μm layer of $Al_xGa_{1-x}As$ is epitaxially grown on the substrate. This is followed by a GaAs layer of thickness w and additional 2 μm of $Al_xGa_{1-x}As$. GaAs is a fcc crystal and substrates grown in the (100) direction are well-known to cleave in the orthogonal (011) direction along the lattice plane. After cleaving a strip, the actual fabrication happens on the (011) surface. PMMA is spun on the sample and after e-beam lithography, a window is defined in the resist. This window exposes the epitaxial grown GaAs (edge-wise) so that it can be selectively etched by a solution of NH_4OH and H_2O_2 diluted in water, without removing the $Al_xGa_{1-x}As$. This etch step creates a trench of width w and arbitrarily long length: this is the *mold* of our toy-model. Without removing the resist, metal is evaporated using either thermal/e-beam evaporation or directional sputtering, thus expanding the range of material that can be utilized. Upon removal of the excess material deposited on the edges of the trench, the metal deposited inside the trench will form the nanowire. The removal step is carried out with a reactive ion etcher and a N_2 plasma (*scalpel*): the sample is oriented in such a way that the bottom of the trench is not exposed to the plasma (reminiscent of the analogous step in Giordano's step-edge fabrication). The final step consists of depositing the lead for forming the electrical contact using an additional e-beam step. The nanowire formed with this process can have widths smaller than 5 nm and can be several micrometers long. This technique has not been applied to the study of superconductivity in nanowires. Among its benefits, we can count the ability to evaporate a large range of materials and to fabricate extremely long nanowires as this is determined only by the stencil characteristics. The drawbacks of this technique are the ion milling step, which can alter the structural properties of the outer layer of the superconducting material, and the necessity to perform a second e-beam lithography step which requires resist curing at elevated temperature $\approx 150\,^{\circ}C$: this step can impact the quality of the superconducting material.

6.7
SNAP Technique

Similar to the technique described in the previous section, the SNAP technique, developed in the laboratories of P.M. Petroff and J.R. Heath at the University of California at Santa Barbara and Caltech relies on the excellent lattice matching between GaAs and $Al_xGa_{1-x}As$. In its first incarnation [27], metal would be evaporated directly on the GaAs$Al_xGa_{1-x}As$ superlattice grown by MBE; the nanowire thus formed would be carefully stamped onto another substrate with predeposited contact. Here, we will describe a variation of this technique that has been applied to the fabrication of superconducting nanowires [28].

A *toy model* of this technique can be represented by a comb. If we dip the tip of the comb's teeth in ink, we can paint with the comb on a stack of paper: the marks will be left in correspondence with the teeth of the comb. And, if we move the comb in the direction perpendicular to its length, we can obtain very long lines. After letting the ink dry, we can cut the entire stack of paper around the ink: we will then project the lines on the entire stack of paper.

The starting point of this technique entails producing the comb, and commences with the MBE growth of a $GaAs/Al_xGa_{1-x}As$ superlattice, that is, several tens of alternating GaAs and $Al_xGa_{1-x}As$ layers. After cleaving the substrate in strips, fabrication takes place on the (011) surface. A 20–30 nm deep etch with a dilute solution of buffered hydrofluoric acid in H_2O is used to selectively etch the $Al_xGa_{1-x}As$ with respect to the GaAs, leaving a comb-like structure (in this case, the comb is very long in the direction perpendicular to the teeth and it is not necessary to physically move the comb here). A directional evaporation is then used to deposit platinum (Pt) with the superlattice oriented with an angle of $\approx 36^\circ$ with respect to the incoming evaporation plume. The angle chosen and the large distance between the superlattice and the material source ensure that only the top of the GaAs layers is coated with metal: this step corresponds to dipping the comb in ink. Seperately, in the meantime, a thin film of Nb (11 nm) is sputtered on a clean Si substrate; without breaking the vacuum, the film is protected from oxidation by sputtering a thin (4 nm) layer of SiO_2 (stack of paper). The as-prepared substrate is covered with a resin of epoxy and PMMA and undercured so that it is still somewhat soft. The comb-like structures are then pressed on the resin, in a manner analogous to a nanoprinting process. After curing, the sample is immersed in a solution of KI (4 g)/I_2 (1 g)/H_2O (100 ml) to remove the GaAs oxide layer at the interface between the semiconductor heterostructure and the metal. This step is designed to release the Pt wires on top of the resin and is equivalent to drawing on the stack of paper and letting the ink dry. The resulting structure is shown in Figure 6.4a. Then, e-beam lithography is used to design the leads and transfer the pattern onto the SiO_2 protective layer: this opening is then protected by sputtering Al_2O_3 (Figure 6.4b). Finally, the combined nanowire array/lead pattern is transferred to the superconducting film by highly directional etching corresponding to cutting around the dried ink (Figure 6.4c,d).

Heath's group at Caltech has, so far, mainly concentrated on the study of arrays of nanowires and, due to the versatility of this technique, has been able to systematically study arrays of Nb nanowires as a function of length and width. Up to now, individual nanowires have not been studied. The main advantage of this technique is: its ability to use a vast array of materials (as long as they can be deposited in thin films); additionally, once the master (etched superlattice) is created, it can be reused several times. One disadvantage of this technique is that the superconducting film is subject to additional processing after deposition which can somewhat

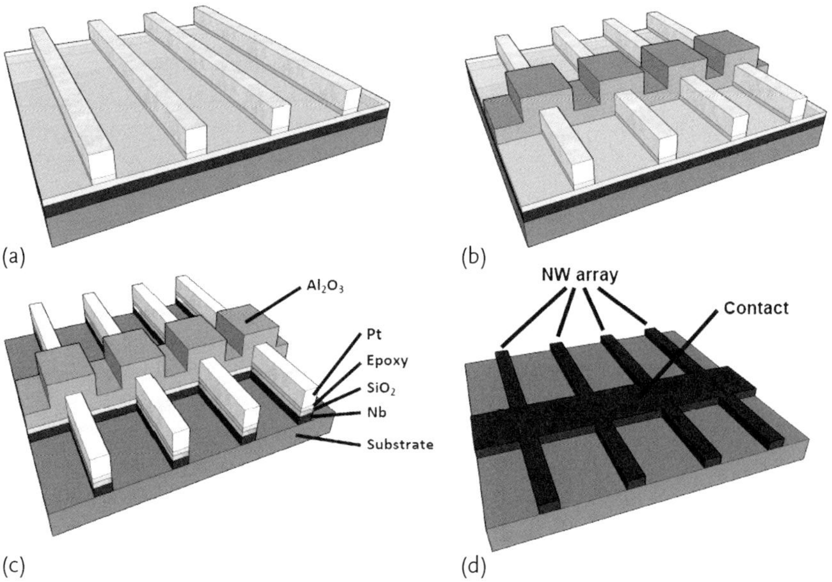

Figure 6.4 Process flow for the fabrication of superconducting NW array circuits. (a) An array of Pt SNAP NWs is placed onto a SiO_2-covered superconducting Nb film. (b) Al_2O_3 is deposited as a pattern for the contact electrodes. (c) A monolithic (all Nb) NW array circuit is obtained after the pattern is transferred with a directional dry etch. (d) The resultant NW array circuit is drawn here. In practice, it is protected by the SiO_2 cover layer. From [28].

affect its characteristics. A recent account of the many possibilities offered by this technique, including the ability to create 2D interconnects can be found in [29].

6.8 Chang and Altomare

Another stencil technique that relies on the MBE growth of a suitable heterostructure was proposed by A.M. Chang and developed by F. Altomare and A.M. Chang [30]. The idea is somewhat complementary to the one implemented by Natelson *et al.* [24]: instead of creating a trench, they create a ridge on which they deposit the metal forming the nanowire.

A *toy-model* equivalent to this technique could be represented by the rope connecting two platforms, as in the molecular templating technique. However, in this case, instead of having a rope, we have a very thin wall with vertical edges. The snow in this case will fall down vertically and therefore will sit on top of the platform and on top of the wall and below, but not on the sides. The snow below is disconnected from that on top.

The starting point of the MBE template is an undoped GaAs substrate. On this substrate, $In_xGa_{1-x}As$ is epitaxially grown with an indium concentration that varies by 0.16/μm. This is done in order to slowly vary the lattice spacing from GaAs to the InP lattice: since the fabrication is done on the cleaved edge, it is important to reduce the structural defects. The final indium concentration is $In_{0.52}Ga_{0.48}As$, which is nearly lattice-matched to InP. A layer of this InP semiconductor is grown for a thickness w; finally, a protective cap layer of $In_{0.52}Ga_{0.48}As$ completes the growth. After cleaving the substrate into strips (Figure 6.5), the fabrication is continued on the (011) surface. A round of e-beam lithography serves the purpose of defining the leads (corresponding to the platform) for a quasi-4-terminal measurement: these are formed by thermal evaporation of titanium (as adhesion layer) and AuPd for the formation of electrical contact; as previously discussed, AuPd is chosen for its small grain size. The last step of the stencil fabrication consists of room temperature wet-etching with a solution of phosphoric acid and hydrogen peroxide which selectively etches both InGaAs and GaAs with respect to InP. The wet-etching attacks all semiconductors not protected by the metal pads, with the exception of InP, while producing an undercut all around the metal pads. What remains is a ridge of InP connecting two consecutive pads with their separation determining the nanowire length (the wall of the toy-model). As a side note, it is worth noticing that this technique was initially developed in a $GaAs/Al_xGa_{1-x}As$ system in which the $Al_xGa_{1-x}As$ had the role of the InP. However, it turned out that all of the available wet-etches tested[3)] attacked the top of the (oxidized) $Al_xGa_{1-x}As$ ridge so that it was no longer flush with the platform [31]. As the electrical contacts to the nanowire are already in place, the semiconducting stencil is now ready to be used with any material compatible with the semiconductor: a directional evaporation, perpendicular to the plane, will form the wire on the InP ridge. The evaporated material will be deposited on the pads as well, and the undercut formed around the contact will prevent the formation of shorting paths to the metal below, so that current will only be carried through the nanowire. The main advantage of this technique lies in its ability to use a large variety of materials, as long as they can be deposited using a directional method. Additionally, once the nanowire has been deposited, there are no additional processing steps, and the nanowire length can be anywhere between 1 and 100 μm. Superconducting electrical contacts have also been used [32] and array of nanowires or ring-like geometries (ring formed by two parallel nanowires together with portion of the pads) can also be studied. The drawback of this technique is that the fabrication can be challenging as the InP ridge is only a few micrometers away from the edge.

6.9
Template Synthesis

Two groups led by M.H.W. Chan at Pennsylvania State University and F.M. Peeters at Universiteit Antwerpen together with S. Michotte at the Universite Catholique de Louvain have focused their attention on using template syntheses (see [33] for a review). A *toy-model* can be represented as a piece of wood in which several cylin-

3) The etching solution had to etch selectively GaAs with respect to $Al_xGa_{1-x}As$

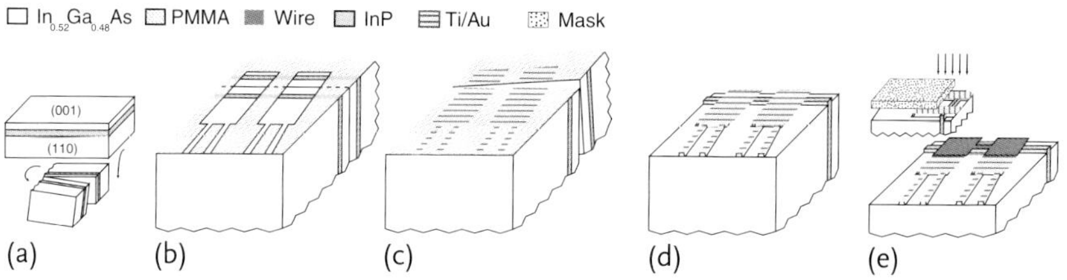

Figure 6.5 Schematic of sample fabrication: (a) The sample is cleaved in small strips which are cut in half and glued together with the two (001) planes facing each other, (b) PMMA is spun on the (110) crystallographic plane of the two pieces and a pattern is written using standard EBL and then developed, (c) thermal evaporation is used to deposit a film of Ti/Au (the portion of the film deposited on the PMMA is not shown) and then the two halves are separated, (d) after lift-off and oxygen plasma etching, wet-etching is used to define the InP ridge, (e) appropriately masking the substrate of the sample, the wire is formed through the final evaporation (top: side view of the evaporation arrangement, bottom: final result). From [30].

drical holes are made through a drill whose size will determine the size of the hole. The holes are then filled with play-dough, representing the superconducting material.

Instead of using a piece of wood, the starting point of this method is a porous membrane. The two types of membranes commonly used are: track-etched membranes and porous alumina membranes. The track-etched membranes, commercially available in a large variety of pore sizes, are formed by bombarding a nonporous sheet of the desired material with nuclear fission fragments. These fragments damage the membrane so that it can be attacked by a chemical etchant: the chemical etch follows the track of the fragments, thus etching pores in the membrane. The porous alumina membrane are electrochemically prepared from aluminum metal. Typically, they have larger pore densities, but only a limited number of sizes are available. Once the desired membrane type has been chosen, the nanowire is prepared by an electrochemical process after deposition of a cathode on one side of the membrane (Figure 6.6). This method is naturally suited to the fabrication of arrays of nanowires [34]. However, single nanowires can be obtained either by dissolving the membrane and releasing the nanowire on a substrate where one of them can be contacted (similarly to nanotubes) [35] or by depositing a thin layer of metal on the opposite side of the cathode [36]. In this case an electrical contact is established through the nanowire, the voltage difference between the cathode and the anode (in the solution) is removed so as to interrupt the growth.

The details of the process, electrolyte, cathode, and anode clearly depend on the desired material: tin [37], lead [36], aluminum [35], and zinc [34, 38] nanowires have all been fabricated. The main advantages of this technique are its ability to fabricate long nanowires and that the nanowires are often made of a collection of single crystals. The main disadvantages of this technique lie in the size of the nanowire (the smallest obtained so far are about 20 nm in diameter) and that the chemical reaction for some material might be not trivial to develop. Additional disadvantages are related to the uniformity of the lateral dimension of the nanowires: differences

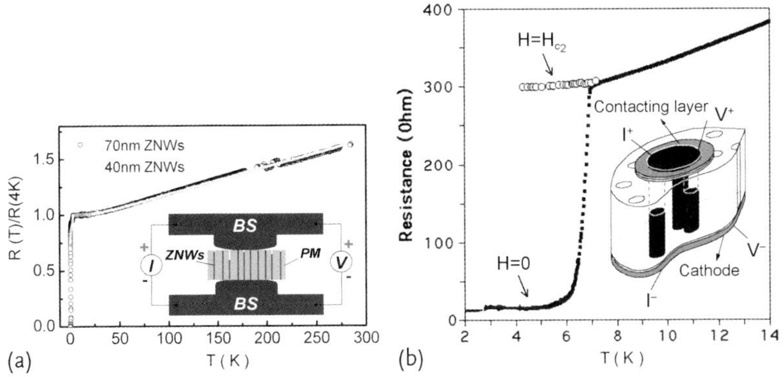

Figure 6.6 (a) $R(T)/R(4\,\text{K})$ vs. T of 70 and 40 nm ZNWs (zinc nanowires) between bulk In electrodes from 0.47 to 300 K, and the schematic of the transport measurement. Adapted from [34]. (b) Resistance of a single 40 nm lead nanowire at low temperature measured by applying a current of 0.1 μA. The critical temperature is around 7.2 K. The value of the resistance when superconductivity is destroyed by a magnetic field applied parallel to the nanowire axis is also shown and the inset depicts a schema of the sample. From [36].

might exist among wires in the same array and in the same nanowire as both the track-etched and porous alumina membrane might suffer from significant nonuniformity.

6.10 Other Methods

The methods that we will briefly touch upon in this section, are not stencil methods. However, through some clever fabrication, the researchers have managed to create nanowires with the lateral dimension of the order of 10 nm nanowires.

6.10.1 Ion Beam Polishing

This technique was developed by M. Zgirski and K.Y. Arutyunov [39] in K.Y. Arutyunov's group at the University of Jyvaskyla.

A *toy model* of this technique can be thought of as making a toothpick out of a much larger piece of wood. By patiently using a file, it is possible to make the piece of wood thinner and thinner until the desired dimensions (toothpick) are reached. Starting from a large nanowire, it is possible to obtain a much thinner one by carefully removing the outer portion until the desired size is obtained.

The first step (Figure 6.7) requires the fabrication of an (aluminum) nanowire several micrometers long, 60 nm wide and 100 nm tall: this is relatively easy to accomplish with e-beam lithography. This wire is then progressively reduced using Ar ion (Ar^+) sputtering, a process known as ion milling. The Ar^+ milling

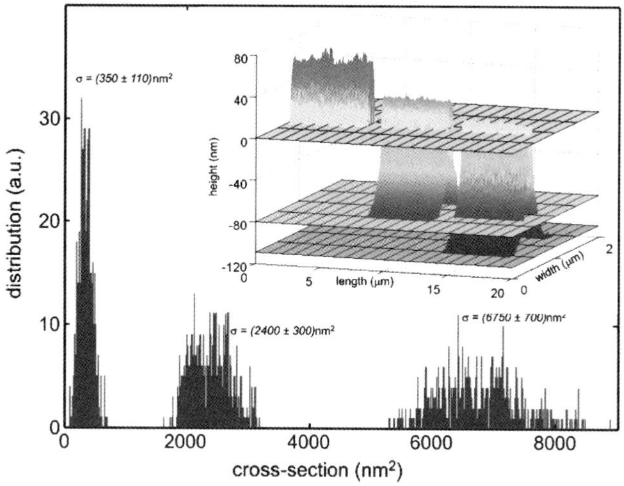

Figure 6.7 Histograms showing the distribution of the wire cross section before and after sputtering. To collect statistics, about 500 SPM (scanning probe microscope) scans were taken across the wire with the step along its axis ~ 12 nm, which is comparable to the radius of curvature of the SPM tip. Narrowing of the histograms is due to the "polishing" effect of ion sputtering. The inset shows the evolution of the sample shape while sputtering measured by SPM. The bright color above the gray plane corresponds to Al and the blue below the plane to Si. Planes (gray, green, orange) indicate Si substrate base levels after successive sessions of sputtering. As Si is sputtered faster than Al, the wire is finally situated at the top of the Si pedestal. Gray plane (height = 0) separates Si from Al. From [39]. For a color version of this figure, please see the color plates at the beginning of the book.

is performed in steps and after each step, the nanowire is investigated and measured using both a scanning probe microscope and an SEM. The authors observed a "polishing" effect which tends to smooth out the wire while its size is reduced. Care must be taken during the ion milling step as the ion mill has little selectivity between the aluminum nanowire and the silicon substrate.

The main advantages of this technique are the minimum nanowire dimension and the possibility of choosing between a large array of materials as the main requirements for its applicability is the ability for the material to be etched through ion milling without considerably damaging the substrate. Additionally, this is so far the only technique which allows one to study the progressive reduction of size on the *same* device. The main drawback of this technique is that almost the entire fabrication is done using ion milling. According to a calculation performed with the SRIM (Stopping and Ranges of Ions in Matter) software, the penetration depth of Ar^+ ions in the aluminum matrix is shorter than 2 nm [40], and thus comparable to the thickness of naturally grown aluminum oxide. However, despite the smoothing effect observed during the size reduction [40], the ion mill is not completely uniform and, in extremely small samples, the standard deviation of the lateral size is quite large compared to the mean which can lead to the creation of very small constrictions in their smallest samples [39]. In any case, the main selling point of this technique is extremely appealing: the ability to measure the same sample as a function of its size. This technique has been recently applied to the study of titanium nanowires [41, 42], but can also be applied to study materials that are more conveniently deposited in film as Nb and $Mo_{1-x}Ge_x$: after sputter deposition of a film, the nanowire could be defined using suitable negative resist and then ion milling can be used to reduce its dimension.

6.10.2
Angled Evaporation

Y. Chen and A.M. Goldman, in A.M. Goldman's group at the University of Minnesota, have implemented another technique that allows the fabrication of nanowires with lateral size of the order of 5 nm [43].

For our *toy-model*, we can again use the analogy of the snow. If we are trying to hit the floor on the bottom of a trench with vertical walls, we will have the largest coverage of snow on the floor if there is no wind, and the snow falls vertically. However, if there is a steady wind across the trench and the snow falls at an angle, we will only be able to cover a smaller part of the floor with snow, with the rest ending up on the vertical wall. In practice, the trench will be a narrow and long opening in the resist with the substrate as the floor and the snow will be the metal. The size of the nanowire is represented by the area on the floor that we are able to cover.

The initial step consists of patterning a wide channel ($\approx$ 50 nm) in PMMA resist on top of a substrate (both SiO_2 and TiO_2 were used). After development of the exposed resist, the sample is placed in a directional evaporator on a stage attached to a liquid nitrogen reservoir. By adjusting the angle between the channel and the directionally evaporated metal so that the substrate is not perpendicular to the direction of evaporation, the width of the nanowire is reduced, as can be seen from geometrical consideration. A schematic of the evaporation geometry, effect of varying the evaporation angle and resultant nanowires are shown in Figure 6.8

In obtaining sub-10 nm nanowires with this technique, a crucial role is played by the evaporation rate and by the cooled evaporator stage: in fact, cooling suppresses the formation of large metallic grains and reduces the lateral diffusion of atoms. Also worth noting is the possibility of using liquid helium to cool the stage, however, there is a possibility that PMMA can crack during the cooling. While the authors have only used gold, this process allows the use of any material which can be directionally evaporated.

6.10.3
Resist Development

While not properly classifiable as a novel fabrication technique, there has been an effort in recent years to better understand the limits of e-beam lithography, with particular focus on the resist. The most common e-beam resist is poly(methylmetha-acrylate), or PMMA. It is both a positive tone resist and a nega-

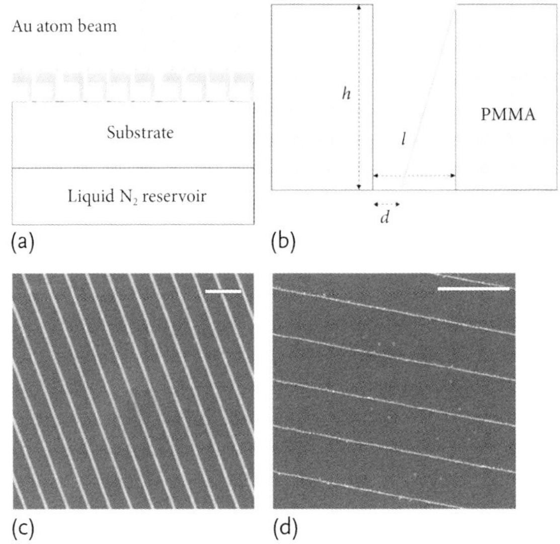

Figure 6.8 (a) Schematic showing the angled deposition onto a substrate with channels prepatterned using e-beam lithography. (b) The width of the wires can, in principle, be expected to be *d* when the deposition angle is set to $\theta = \tan^{-1}[(l - d)/h]$. (c) Scanning Electron Microscope (SEM) images of (c) 50 nm wide wires deposited in the usual way $\theta = 0°$ (the white scale bar is 1 µm). (d) 15 nm wide wires deposited in the usual way using an angled deposition with $\theta = 28°$ (the white scale bar is 1 µm). From [43].

tive tone resist. It has been used since the late sixties [44, 45] for its high resolution and simplicity of use. In 2007, B. Cord *et al.* [46], in K.K. Berggren's group at MIT, showed that with simple enough precautions, it is possible to obtain a sub-10 nm feature (Figure 6.9). In fact, using a Raith-150 SEBL tool at an accelerating voltage of 30 keV to expose the resist, if the pattern is developed with the commonly used solution of solution 3 : 1 of isopropanol : methyl-isobutyl-ketone (IPA : MIBK) at a temperature of about −15 °C, it is possible to obtain wire with a lateral size of ≈ 8 nm at pitches of 40–100 nm. The same group also analyzed HSQ as a resist pattern and found that by using a salty development process [47], it is possible to produce patterns with 12 nm pitch [48] (Figure 6.10). A better understanding of the limit and characteristics of an e-beam resist has great potential for the fabrication of superconducting nanowires: this method has been applied, so far, to the fabrication of nanowire photon detectors [49, 50] and HSQ has recently been used in the fabrication of long $Mo_{1-x}Ge_x$ nanowires [15].

6.11
Future Developments

It is really hard to predict new developments from a fabrication point of view. Certainly, one thing that will benefit the fabrication of narrow nanowires is the ability to reduce grain size; therefore, the use of cooled substrates will definitely be beneficial as the metal grain size is the ultimate limit to fabrication of continuous and uniform nanowires. From the point of view of size reduction, probably the most promising techniques are semiconducting templating, as MBE grown templates are probably the most suitable in terms of creating long and uniform nanowires, as a result of the ability to control growth down to the monolayer. Purely resist-based techniques have made noteworthy progress in the last few years and can now compete with stencil-based techniques ($\sim$ 10 nm wire). The possible improvements in resist technology and e-beam lithography will make those technique even more enticing. The ion beam polishing technique is also quite promising, however, its ability to generate small and uniform nanowires might depend on the material used, as titanium seems to be smoother than aluminum after polishing. In particular, techniques depositing material using evaporation seem to produce polycrystalline nanowires without further processing. In such respects, the technique developed by Chang and Altomare is probably unique in terms of minimum achievable size, (no-)post-processing of the superconductor and uniformity, as the film has a similar resistivity to the nanowire. On the other hand, the SNAP technique, or the ion beam polishing technique, can be used on an epitaxially grown superconductor and that would allow the study of single crystalline nanowire, analogous to nanowires obtained using the molecular templating technique after *in-situ* annealing.

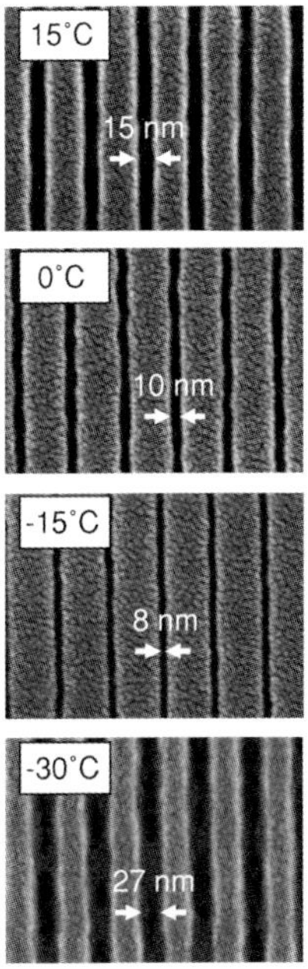

Figure 6.9 SEM images of 60 nm pitch gratings developed in 3 : 1 IPA : MIBK for 30 s at 15, 0, −15, and −30 °C, and etched into a Si substrate, showing the minimum achievable linewidth at each development temperature. As expected from the contrast data, the resolution improves as the temperature is reduced, peaks at −15 °C, and then drops sharply at −30 °C. The poor line-edge definition and bridging in the −30 °C micrograph are characteristic of sloped resist sidewalls, a symptom of poor resist contrast. From [46]

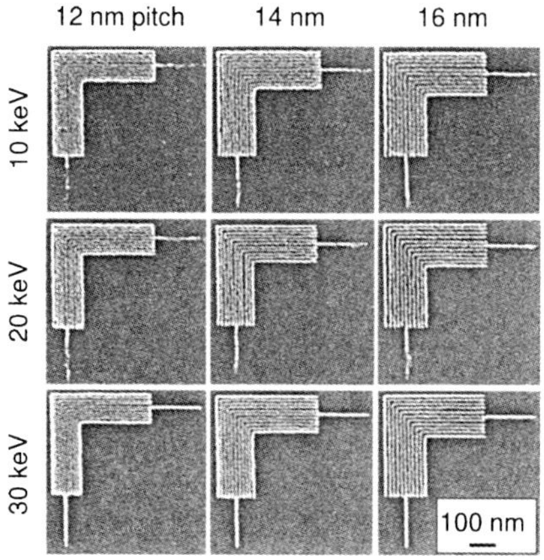

Figure 6.10 Scanning electron micrographs of nested-L structures with pitches ranging from 12 (the smallest yielded at any voltage) to 16 nm, exposed at 10, 20, and 30 keV in 25 nm of HSQ on a Si substrate. The resist was developed for 4 min using 1% NaOH/4% NaCl, and the developed features were coated with 2 nm of Au-Pd prior to imaging. While there is some resolution degradation at lower acceleration voltages for the smaller pitches, it is minimal and, at 16 nm and above, almost completely absent. From [48].

References

1 Little, W.A. (1967) Decay of persistent currents in small superconductors. *Physical Review*, **156**, 396–403.

2 Parks, R.D. and Groff, R.P. (1967) Evidence for thermodynamic fluctuations in a superconductor. *Physical Review Letters*, **18**, 342–345.

3 Hunt, T.K. and Mercereau, J.E. (1967) Quantum phase correlation in small superconductors. *Physical Review Letters*, **18**, 551–553.

4 Langer, J.S. and Ambegaokar, V. (1967) Intrinsic resistive transition in narrow superconducting channels. *Physical Review*, **164**, 498–510.

5 Webb, W.W. and Warburton, R.J. (1968) Intrinsic quantum fluctuations in uniform filamentary superconductors. *Physical Review Letters*, **20**, 461–465.

6 McCumber, D.E. and Halperin, B.I. (1970) Time scale of intrinsic resistive fluctuations in thin superconducting wires. *Physical Review B*, **1**, 1054–1070.

7 Lukens, J.E., Warburton, R.J., and Webb, W.W. (1970) Onset of quantized thermal fluctuations in "one-dimensional" superconductors. *Physical Review Letters*, **25**, 1180–1184.

8 Newbower, R.S., Beasley, M.R., and Tinkham, M. (1972) Fluctuation effects on the superconducting transition of tin whisker crystals. *Physical Review B*, **5**, 864–868.

9 Prober, D.E., Feuer, M.D., and Giordano, N. (1980) Fabrication of 300-Å metal lines with substrate-step techniques. *Applied Physics Letters*, **37**, 94.

10 Giordano, N. (1988) Evidence for macroscopic quantum tunneling in one dimensional superconductors. *Physical Review Letters*, **61**, 2137–2140.

11 Bezryadin, A., Lau, C.N., and Tinkham, M. (2000) Quantum suppression of superconductivity in ultrathin nanowires. *Nature*, **404** (6781), 971–974.

12 Wikipedia (2012), Lithography –, the free encyclopedia. http://en.wikipedia.org/w/index.php?title=Lithography&oldid=509972521, (accessed: 11 September 2012).

13 Senefelder, A. (1911) *The Invention of Lithography*, The Fuchs & Lang Manufacturing Company, New York.

14 Likharev, K.K. (1979) Superconducting weak links. *Reviews of Modern Physics*, **51** 101–161.

15 Kim, H., Jamali, S., and Rogachev, A. (2012) Superconductor–insulator transition in long MoGe nanowires. *Physical Review Letters*, **109** (2), 027002.

16 Arutyunov, K.Y., Golubev, D.S., and Zaikin, A.D. (2008) Superconductivity in one dimension. *Physics Reports*, **464**, 1–70.

17 Bollinger, A.T., Rogachev, A., Remeika, M., and Bezryadin, A. (2004) Effect of morphology on the superconductor–insulator transition in one-dimensional nanowires. *Physical Review B*, **69** (18), 180503.

18 Bezryadin, A. (2008) Topical Review: Quantum suppression of superconductivity in nanowires. *Journal of Physics Condensed Matter*, **20** (4), 043202.

19 Hopkins, D.S., Pekker, D., Goldbart, P.M., and Bezryadin, A. (2005) Quantum interference device made by DNA templating of superconducting nanowires. *Science*, **308**, 1762–1765.

20 Johansson, A., Sambandamurthy, G., Shahar, D., Jacobson, N., and Tenne, R. (2005) Nanowire acting as a superconducting quantum interference device. *Physical Review Letters*, **95** (11), 116805.

21 Bae, M.H., Dinsmore, R.C., Aref, T., Brenner, M., and Bezryadin, A. (2009) Current-phase relationship, thermal and quantum phase slips in superconducting nanowires made on a scaffold created using adhesive tape. *Nano Letters*, **9**, 1889–1896.

22 Aref, T., Levchenko, A., Vakaryuk, V., and Bezryadin, A. (2012) Quantitative analysis of quantum phase slips in superconducting $Mo_{76}Ge_{24}$ nanowires revealed by switching-current statistics. *Physical Review B*, **86**, 024507.

23 Cho, A.Y. and Arthur, J.R. (1975) Molecular beam epitaxy. *Progress in Solid State Chemistry*, **10** (0), 157–191.

24 Natelson, D., Willett, R.L., West, K.W., and Pfeiffer, L.N. (2000) Fabrication of extremely narrow metal wires. *Applied Physics Letters*, **77**, 1991.

25 Natelson, D. (2000) Molecular-scale metal wires. *Solid State Communications*, **115**, 269–274.

26 Natelson, D., Willett, R.L., West, K.W., and Pfeiffer, L.N. (2001) Geometry-dependent dephasing in small metallic wires. *Physical Review Letters*, **86**, 1821–1824.

27 Melosh, N.A., Boukai, A., Diana, F., Gerardot, B., Badolato, A., Petroff, P.M., and Heath, J.R. (2003) Ultrahigh-density nanowire lattices and circuits. *Science*, **300**, 112–115.

28 Xu, K. and Heath, J.R. (2008) Controlled fabrication and electrical properties of long quasi-one-dimensional superconducting nanowire arrays. *Nano Letters*, **8**, 136–141.

29 Heath, J.R. (2008) Superlattice nanowire pattern transfer (snap). *Accounts of Chemical Research*, **41**, 1609–1617.

30 Altomare, F., Chang, A.M., Melloch, M.R., Hong, Y., and Tu, C.W. (2005) Ultranarrow AuPd and Al wires. *Applied Physics Letters*, **86** (17), 172501.

31 Altomare, F. (2004) Superconducting ultra-narrow aluminum nanowires, PhD Thesis, Purdue University.

32 Li, P., Wu, P.M., Bomze, Y., Borzenets, I.V., Finkelstein, G., and Chang, A.M. (2011) Switching currents limited by single phase slips in one-dimensional superconducting Al nanowires. *Physical Review Letters*, **107** (13), 137004.

33 Martin, C.R. (1994) Nanomaterials: A membrane-based synthetic approach. *Science*, **266** (5193), 1961–1966.

34 Tian, M., Kumar, N., Xu, S., Wang, J., Kurtz, J.S., and Chan, M.H.W. (2005) Suppression of superconductivity in zinc nanowires by bulk superconductors. *Physical Review Letters*, **95** (7), 076802.

35 Singh, M., Wang, J., Tian, M., Zhang, Q., Pereira, A., Kumar, N., Mallouk, T.E., and Chan, M.H.W. (2009) Synthesis and superconductivity of electrochemically grown single-crystal aluminum nanowires. *Chemistry of Materials*, **21** (23), 5557–5559.

36 Michotte, S., Piraux, L., Dubois, S., Pailloux, F., Stenuit, G., and Govaerts, J. (2002) Superconducting properties of lead nanowires arrays. *Physica C Superconductivity*, **377**, 267–276.

37 Tian, M., Wang, J., Kurtz, J.S., Liu, Y., Chan, M.H.W., Mayer, T.S., and Mallouk, T.E. (2005) Dissipation in quasi-one-dimensional superconducting single-crystal Sn nanowires. *Physical Review B*, **71** (10), 104521.

38 Tian, M., Kumar, N., Wang, J., Xu, S., and Chan, M.H.W. (2006) Influence of a bulk superconducting environment on the superconductivity of one-dimensional zinc nanowires. *Physical Review B*, **74** (1), 014515.

39 Zgirski, M., Riikonen, K., Touboltsev, V., and Arutyunov, K. (2005) Size dependent breakdown of superconductivity in ultranarrow nanowires. *Nano Letters*, **5**, 1029–1033.

40 Savolainen, M., Touboltsev, V., Koppinen, P., Riikonen, K.P., and Arutyunov, K. (2004) Ion beam sputtering for progressive reduction of nanostructures dimensions. *Applied Physics A: Materials Science & Processing*, **79**, 1769–1773.

41 Lehtinen, J.S., Sajavaara, T., Arutyunov, K.Y., Presnjakov, M.Y., and Vasiliev, A.L. (2012) Evidence of quantum phase slip effect in titanium nanowires. *Physical Review B*, **85** (9), 094508.

42 Arutyunov, K.Y., Hongisto, T.T., Lehtinen, J.S., Leino, L.I., and Vasiliev, A.L. (2012) Quantum phase slip phenomenon in ultra-narrow superconducting nanorings. *Scientific Reports*, **2**, 293.

43 Chen, Y. and Goldman, A.M. (2008) A simple approach to the formation of ultranarrow metal wires. *Journal of Applied Physics*, **103** (5), 054312.

44 Haller, I., Hatzakis, M., and Srinivasan, R. (1968) High-resolution positive resists for electron-beam exposure. *IBM Journal of Research and Development*, **12**, 251.

45 Hatzakis, M. (1969) Electron resists for microcircuit and mask production. *Journal Electrochemical Society*, **116** (7), 1033–1037.

46 Cord, B., Lutkenhaus, J., and Berggren, K.K. (2007) Optimal temperature for development of poly(methylmethacrylate). *Journal of Vacuum Science Technology B: Microelectronics and Nanometer Structures*, **25**, 2013.

47 Yang, J.K.W. and Berggren, K.K. (2007) Using high-contrast salty development of hydrogen silsesquioxane for sub-10-nm half-pitch lithography. *Journal of Vacuum Science Technology B: Microelectronics and Nanometer Structures*, **25**, 2025.

48 Cord, B., Yang, J., Duan, H., Joy, D.C., Klingfus, J., and Berggren, K.K. (2009) Limiting factors in sub-10 nm scanning-electron-beam lithography. *Journal of Vacuum Science Technology B: Microelectronics and Nanometer Structures*, **27**, 2616.

49 Yang, J.K.W., Kerman, A.J., Dauler, E.A., Cord, B., Anant, V., Molnar, R.J., and Berggren, K.K. (2009) Suppressed critical current in superconducting nanowire single-photon detectors with high fill-factors. *IEEE Transactions on Applied Superconductivity*, **19**, 318–322.

50 Marsili, F., Najafi, F., Dauler, E., Bellei, F., Hu, X., Csete, M., Molnar, R.J., and Berggren, K.K. (2011) Single-photon detectors based on ultranarrow superconducting nanowires. *Nano Letters*, **11**, 2048–2053.

7
Experimental Review of Experiments on 1D Superconducting Nanowires

7.1
Introduction

In this chapter, we will try to provide an overall review of the main experimental results in the field. Many issues are being explored experimentally, of which, the most controversial to assess and pin down being the observation of quantum phase slips. Notably, *coherent* quantum phase slips have only recently (2012) been demonstrated [1–3] and we will discuss those experiments in Chapter 8. In this chapter, we plan to briefly review the last decade or so of experimental effort in the field of 1D superconducting nanowires. Relevant topics such as the possibility of observing shape resonance, variation in T_c and the general trend of superconducting correlation as the wire lateral dimensions are reduced, phase slips, particularly those of quantum-tunneling origin, as well as novel behaviors, for example, the antiproximity effect, in which superconducting leads connected to superconducting nanowires appear to quench the superconductivity in the nanowires, will be discussed. While this overview is not intended to be exhaustive, our intent is to present what we feel are the main results, emphasizing the most controversial and possibly interesting topic, that is, the observation of quantum phase slips.

Before delving into the review of experimental results, we would like to briefly discuss an often overlooked issue: appropriate filtering. The availability of high-quality filtering to remove unwanted environmental interference, for example, Johnson noise emitted into the nanowire circuit by room temperature electronics, or RF and microwave pickup from nearby radio, television stations and cellular phone towers, is essential for the observation of intrinsic behaviors of the superconducting nanowires. On the flip side, such sensitivity to noise renders the 1D superconducting nanowire system an excellent candidate for single-photon detection in the microwave range.

One-Dimensional Superconductivity in Nanowires, First Edition. F. Altomare and A.M. Chang

7.2
Filtering

While often unrecognized, proper filtering is crucial for obtaining meaningful experimental results. Unfortunately, this information is also extremely hard to come by as it is not typically commented on in great detail in published papers, that is, besides a few lines acknowledging that proper filtering has been used. This kind of information is typically found in the theses of graduate students who performed the experiments and becomes common knowledge and an asset for a particular lab until the students graduate and create their own lab in which they implement similar techniques. It is very instructive to go through those graduate students' theses and learn the tips and tricks they have developed and implemented sometimes unbeknownst (at least in the initial phase) to their advisers. While many laboratories employ similar or equivalent techniques, we will briefly mention three filtering setups that have been used in nanowire measurements.

Sahu *et al.* (see [4], Supplementary information) in Bezryadin's group extensively tested their setup while measuring the switching current distribution of $Mo_{1-x}Ge_x$ nanowires. Their setup comprises a room-temperature low-pass π-filter with an attenuation of 7 dB at the cut-off frequency of 3 MHz. The other fundamental part of the setup is a set of low temperature filters thermalized at the base temperature of the ^3He cryostat: copper powder and silver paste microwave filters. These filters were built using insulated Constantan wire, a copper/nickel alloy (55%/45%). This wire has a resistance of 18.4 Ω/ft and diameter of 0.004 in. The copper powder variety is fabricated by embedding the wire in a mixture of copper powder and epoxy, while silver paste is used for the latter variety. The overall attenuation is greater than 100 dB at a frequency above 1 GHz and below the instrumental noise floor for frequency greater than 6.25 GHz. These filters are based on a design pioneered in Clarke's group by J. Martinis [5][1)] and relies on the skin effect to damp high frequency noise: at high frequency, the (AC) current tends to be flowing within a small distance from the surface and the higher the frequency, the smaller the distance (skin depth). By using copper powder, the natural oxide of the copper powder tends to provide an extremely large surface and a high resistance, providing substantial attenuation.

In his thesis, Li [7] extensively describes the filtering apparatus and the noise performance of the various instruments used. The final filtering setup, which makes extensive use of the expertise of collaborators Gleb Finkelstein, Yuriy Bomze, and Ivan Borzenets, includes low-pass RC filters, ferrite-bead filters, room temperature copper-powder filters and Thermocoax filters. RC low-pass filters are extremely convenient for reducing unwanted noise, however when used while measuring a superconducting sample, they are not extremely effective. RC filters, in fact, rely on capacitors to shunt the high frequency as the impedance of a capacitor is $Z_C = 1/j\omega C$. At high frequency, $Z_C \to 0$. However, what happens when measuring a sample in the superconducting state and its resistance drops to zero? Now

1) A new take on microwave filters has been recently published in [6].

the device and capacitor impedance are comparable thus the filter ceases to provide any filtering effect. An additional problem lies in the discrete nature of the components used in the filters: at very high frequency, while the ideal model predicts large attenuation, they might not behave as expected. For this reason, he developed PCB-board-based RC filters which where placed at the "1 K pot" stage in the ^{3}He and the mixing chamber stage for the measurements in the dilution refrigerator. This had the added benefit of reducing Johnson noise and noise picked from adjacent cables. However, even the performance of PCB-board RC filters is limited up to 100 MHz. For attenuation of larger frequency, ferrite bead filters were used: instead of relying on the skin effect, they can be thought of as inductors which have high dissipation at the frequency they are designed to attenuate. The filtering setup is then completed by metal powder filters [5] at room temperature and Thermocoax coaxial cable [8]. The full setup is

- ^{3}He: Room temperature: Ferrite bead and metal powder filter. "1 K pot" PCB board RC filters
- DilFridge: Room temperature: Ferrite bead and metal powder filter. Thermocoax cables were used for the signal. PCB board RC filters were thermalized at the mixing chamber.

Another design, inspired by the work done by the Marcus group (then) at Harvard University [9], which only involves low-pass filtering was implemented in Rogachev's group [10]. In this case as well, π-filters were used at room temperature (cut-off frequency of 1 MHz). The low temperature measurements were performed in a ^{3}He system: the low temperature filter was provided by three 100 Ω resistors connected in a series and tightly anchored to three separate brass plates with silver paste.

Again, we would like to emphasize that filtering is as important as sample fabrication for ensuring that one is measuring the intrinsic properties of the superconducting nanowire. Improper filtering, at the minimum, produces noisy data, and in the worst case, can produce features that can be interpreted as real effects. The importance of extensive filtering has come to the forefront in recent years, as several recent works have taken substantial care in explicitly describing the filtering capability of the experimental apparatus.

7.3 Phase Slips

As indicated in the original paper by Little [11] and discussed in detail in Section 2.4.1, during a phase slip, the order parameter goes to zero in a small section of the wire, at the minimum, at one single point within the phase slip core. At the zero bias current, the condensate is localized in a minimum of the free energy potential. An applied current tilts the potential, producing what is commonly described as a tilted washboard potential: when the particle escapes from the local minimum to an adjacent minimum, the phase of the condensate changes by 2π. If the escape is due to thermal energy, the phase slips are termed thermally activated phase slips (TAPS). Following Langer and Ambegaokar [12] and McCumber and Halperin [13], the rate at which the condensate can escape from the local minimum will depend on the height of the free energy barrier (ΔF) and the thermal energy (kT) Sections 2.4.3–2.4.7

$$\Gamma(T) = \Omega(T) \exp\left(-\frac{\Delta F}{kT}\right), \tag{7.1}$$

with Ω as the attempt frequency. From this expression, and making use of the Ginzburg–Landau free energy expression: $\Delta F(T) = \Delta F_0(1-t)^{3/2}$ with $t \equiv T/T_c$, we can derive the measured resistance in the limit of low currents ($I \ll I_0 = 4ekT/h$):

$$R_{\text{LAMH}} = \frac{\pi\hbar^2\Omega}{2e^2kT} \exp\left(-\frac{\Delta F}{kT}\right) \propto \exp\left[-0.83\frac{R_Q}{R_{\xi(0)}}\frac{(1-t)^{3/2}}{t}\right]. \tag{7.2}$$

In the right-hand side of (7.2) we have expressed all the quantities in terms of the superconducting quantum resistance ($R_Q = h/(2e)^2 \approx 6.45\,\text{k}\Omega$) and the (normal) resistance of the superconductor in a length equal to the superconducting coherence length at zero temperature ($R_{\xi(0)}$) [14] and the numerical factor 0.83 is approximate.

As we can see from the above expression, for very large barriers or very small temperatures, or equivalently for very small $R_{\xi(0)}$ the exponential term will render the resistance immeasurably small.

However, it must be noted that the shape of the nanowire potential, as a function of the winding number or momentum k_n (see Chapters 2 and 3), is similar to the potential for a Josephson junction (Figure 2.12) for which quantum tunneling through the barrier has been unequivocally demonstrated [15, 16]. The potential is a tilted washboard potential when a current-bias is applied. It is therefore natural to ask if quantum tunneling through the barrier, or quantum phase slips (QPS), can also be observed in nanowires. In this case, the nanowire resistance (2.127), in the same limit of low currents ($I \ll 4e/\tau_{\text{GL}}, \tau_{\text{GL}} = (\pi/8)(\hbar/(T_c - T))$), has the form [14]

$$R_\Delta = B\frac{\pi\hbar^2\Omega_{\text{QPS}}}{2e^2\Delta} \exp\left(-\frac{a\Delta F}{\Delta}\right) \propto \exp\left(-0.83\frac{\pi}{8}a\frac{R_q}{R_{\xi(0)}}(1-t)^{1/2}\right) \tag{7.3}$$

with Δ as a suitable energy scale [17–20]; a and B are numerical factors of order unity, and the last equality assumes [18] $\Delta = \hbar/\tau_{\text{GL}}$, but similar expressions that are numerically equivalent have also been derived on a microscopic basis [19, 21]. The main difference between the two expressions is in the temperature dependence of the exponential term. While the Langer–Ambegaokar expression vanishes at low temperatures, in the same limit, (7.3) implies a resistance that tends to a constant value (the temperature dependence of the prefactor to the exponential is very weak). Note that according to the discussion in Chapter 3, this expression amounts to taking into account the contribution of the QPS fugacity only and neglecting a hydrodynamic contribution. It is valid at intermediate temperatures. At sufficiently low T, where the plasmon thermal wavelength exceeds the wire length, a power law dependence is predicted to come into play, which leads to a vanishing resistance at zero temperature.

Another useful limit to cover is the limit of high current [12, 22] which allows one to study the shape of dV/dI or VI curves at currents near the switching current. In the thermal case, the voltage measured is (2.86, 2.87)

$$V_{\mathrm{LAMH}} = I_0 R_{\mathrm{LAMH}} \sinh\left(\frac{I}{I_0}\right), \tag{7.4}$$

with $I_0 = 4ekT/h$. In the quantum case

$$V_{\mathrm{QPS}} = I_\delta R_\Delta \sinh\left(\frac{I}{I_\Delta}\right), \tag{7.5}$$

with $I_\Delta = 4e\Delta/h$ where Δ is the appropriate energy scale.

One crucial difference between superconducting nanowires and Josephson junctions (JJs) is their size. JJs used in quantum tunneling experiments are anywhere from 0.5 μm × 0.5 μm to 100 μm × 100 μm and JJs in other superconducting devices can be as small as 100 nm × 100 nm. These dimensions are readily achievable with common e-beam lithography. As pointed out in the introduction to this book, in order to display 1D effects, superconducting nanowires need to have a lateral dimension comparable to the superconducting coherence length [23]. This is typically of the order of 10–20 nm and can reach 100 nm only in aluminum nanowires [24] and a few other superconductors with a large bulk coherence length. An additional consideration that allows QPS to be discernible is the magnitude of the free energy barrier (ΔF) to the macroscopic quantum tunneling of the phase. For the materials system utilized to date, this restricts the nanowire diameter to the range quoted above. Sample fabrication is therefore nontrivial as discussed in Section 6.1, and the presence of weak links, which can complicate the interpretation of the experimental results, must be avoided.

7.4 Overview of the Experimental Results

While we will attempt to describe the experimental results in the most direct fashion, because of the many different techniques employed and the variety of results reported in the literature, this will not be straight forward. In the following, we will attempt to provide a concise historical overview and draw attention to what we believe is firmly established and what still contains an element of controversy.

The field of modern day 1D superconductivity is a relatively young one. The LAMH theory was verified in the temperature range close to T_C in the early 1970s [25, 26] and the pioneering Giordano experiments date back to the late

1980s. Despite some controversy surrounding Giordano's results, the lack (and possibly lack of interest in the development) of fabrication techniques for producing even smaller nanowires rendered the field dormant until 2000 [27]. At that point in time, the major interest was in the elucidation of the general trends of superconducting correlations in narrow superconducting wires as the diameter shrinks, and in the observation of quantum phase transitions [27] which are better observed in the regime of small current in order to avoid heating effects. The focus then shifted to the observation of quantum phase slips, basically reproducing Giordano's results in smaller and more uniform nanowires. However, using a small drive current implies measuring a small signal, possibly lowering the signal to noise ratio; it then becomes harder to extract small effects as QPS. In addition, it is easy to misinterpret artifacts (as, for example, phase slips due the unintended presence of weak-links or constriction) in the R vs. T for real physical features. For this reason, the focus has since shifted to studying, at fixed temperature, the entire voltage vs. current. Not only does this allow one to extract smaller resistance values with higher confidence, smaller than what could be measured in the small current regime but it also helps in elucidating different phase-slip mechanisms [22, 24]. However, arguably, a significant turning point in the assessment of phase slips took place with the recognition of the similarities between Josephson junctions and 1D superconducting nanowires [4, 28]. Using switching current measurements, borrowed from the physics of Josephson junctions [15, 16], it is now well-established that in aluminum and $Mo_{1-x}Ge_x$ nanowires, quantum phase slips can be responsible for switching the nanowire from the superconducting to the normal state [28, 29]. Other issues, for example, the antiproximity effect [30] and superconductivity reentrance [31, 32], are still open issues that are the subjects of current studies. The following is a very coarse timeline of what we believe are the major works:

1988 Giordano [17] suggested the existence of QPS in superconducting nanowires.

2000 Bezryadin *et al.* [27] demonstrated the molecular templating technique and published results suggesting the existence of a critical resistance above which $Mo_{1-x}Ge_x$ nanowires are not superconducting.

2001 Lau *et al.* [14] suggested the presence of QPS in $Mo_{1-x}Ge_x$ nanowires.

2003 Melosh *et al.* [33] demonstrated the SNAP fabrication.

2005
- Rogachev *et al.* [22] demonstrated that $Mo_{1-x}Ge_x$ nanowires follow the LAMH model, even in the presence of a magnetic field and suggested the use of measurement in nonlinear regime for nanowires connected to large superconducting leads.
- Zgisrki *et al.* [34] demonstrated the ion milling technique.
- Altomare *et al.* [35] demonstrated their semiconducting stencil technique.
- Tian *et al.* [30] demonstrated the antiproximity effect in Zn nanowires and QPS behavior in Sn nanowires [36].

2006 Altomare *et al.* [37] demonstrated superconducting behavior in a long nanowire with resistance larger than the superconducting quantum resistance, and a nonlinear V–I relationship consistent with a power law.

2008 Bollinger *et al.* [38] suggested the existence of a superconductor/insulator phase diagram.

2008
- Chen *et al.* [39] developed a novel technique for the fabrication of small nanowires.
- Xu *et al.* [40] applied the SNAP technique to the study of superconducting arrays of niobium.

2009
- Sahu *et al.* [4] used the switching current technique to ascertain the presence of QPS in superconducting nanowires.
- Chen *et al.* [31] performed initial studies on Zn nanowires.

2011 Li *et al.* [28] demonstrated that there is a regime in which the switching current distribution has a constant width (i.e., the existence of a constant escape temperature).

2012 Aref *et al.* [29] confirmed the results of Li *et al.* in $Mo_{1-x}Ge_x$ nanowires.

2012 Kim *et al.* [41] demonstrated that in long $Mo_{1-x}Ge_x$ nanowires 1–25 μm in length with a width as narrow as 9 nm, a superconductor to insulator transition occurs at a critical value of normal-state wire resistance per unit length as the wire cross section is reduced.

7.4.1 Giordano's Experiments

The first experimental verification of the LAMH theory[2] was reported by Lukens *et al.* [25] and Newbower *et al.* [26]. They performed experiments on tin whiskers at a temperature very close to the critical temperature. At that time, QPS was not taken into consideration. Interest in quantum tunneling phenomena commenced in the mid 80s following the work on Josephson junctions, and the first work claiming observation of quantum phase slips was Giordano's pioneering experiments in late 1980s [17]. Using the step-edge lithographic technique, he was able to fabricate both narrow superconducting In, Pb and InPb wires; the normal-superconducting (N-S) transition of the thinnest wires showed an unexpected behavior. His results are exemplified by Figure 7.1, which shows transition for different nanowires fabricated in indium with wire uniformity comparable to AuPd nanowires [42] fabricated with the same process: in particular, sample roughness was of the order of 10 nm. The evaporation of indium was performed on cooled substrates and, during warming up to room temperature after metal deposition, O_2 was admitted to the chamber to reduce agglomeration. While the transition in thicker wires would follow LAMH theory, thinner wires would follow LAMH theory only near T_c: at lower temperatures, the slope would change drastically, becoming considerably less steep. In order to explain these data, Giordano proposed a model (Section 2.4.7) analogous to the quantum tunneling already observed in Josephson

2) See also the brief history in Section 6.1.

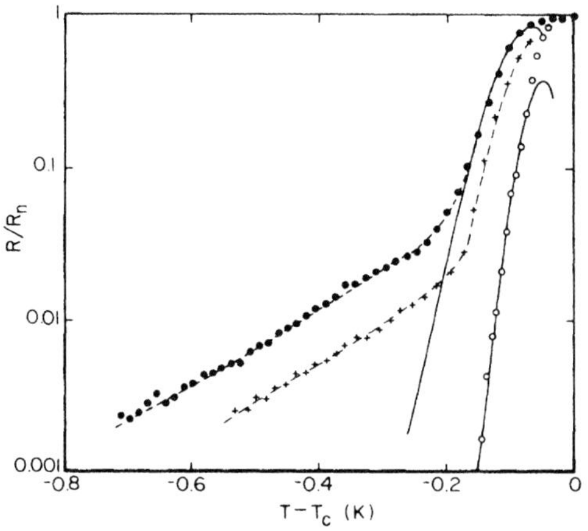

Figure 7.1 Resistance, normalized by the normal state value, as a function of temperature for three In wires; the sample diameters were 410 Å (•), 505 Å (+), and 720 Å (○). The solid curves are fits to the thermal activation theory (1), while the dashed curves are fits to (7). Where the solid and dashed curves overlap, only the former is shown for clarity. The samples had lengths of 80, 150, and 150 μm, and normal-state resistances of 5.7, 7.1, and 1.2 kΩ, where we list the values in order of sample size, with the value for the smallest sample first. From [17].

junctions [18, and references therein]. In this model, he replaced the temperature scale in the LAMH with an energy scale related to the Ginzburg–Landau relaxation time (see (2.127)); this provided a resistance value that saturates at lower temperatures. The main criticism to these experimental results came from Duan [43]. Looking at the expression for Giordano's model (see Sections 2.4.7 and 7.3), the slope of the R vs. T curve should depend on the cross section of the wire. However, from the graph in Figure 7.1, it is possible to observe that the slope is constant for wires of different cross section; this suggests the existence of another length scale, possibly the grain size dominating the experimental results.

Despite this objection, the work of Giordano is extremely important because for the first time, it suggested the possibility of quantum tunneling effects in nanowires and because it proposed a model that is still widely used today in the analysis of quantum phase slips.

7.4.2
Recent Experiments on QPS

After a hiatus of about 20 years, the field was revived by the pioneering work of Bezryadin and Tinkham in 2000 [27]. Using carbon nanotubes as a stencil, they were able to fabricate nanowires of the order of 10 nm in width, and length in the 100 nm–1 μm range. Typical values for the superconducting coherence length ξ is $\sim$ 10 nm.

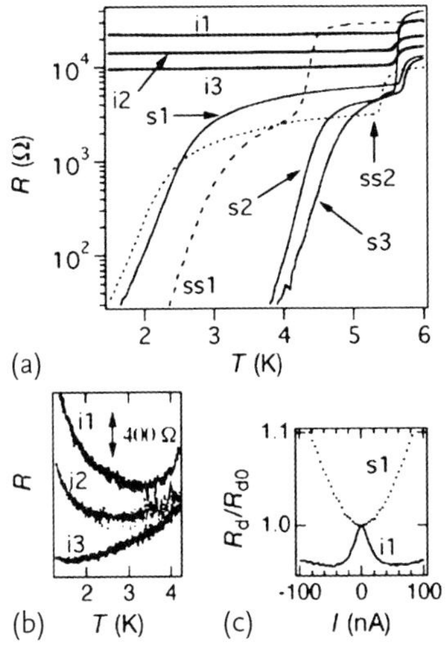

Figure 7.2 Transport properties of superconducting and insulating $Mo_{1-x}Ge_x$ nanowires. (a) Resistance vs. temperature curves for eight different samples. The superconducting transition of the leads takes place at $T_c < 5.5$ K. Sample ss1 has a different T_c (< 4.3 K), presumably due to a different substrate treatment. Samples ss1 (dashed curve) and ss2 (short dashed curve) contain two parallel wires. All other samples only have a single wire. The samples i1, i2, i3, s1, s2, s3, ss1 and ss2 have the following parameters. Apparent widths (nm; measured under SEM) are $W = 11, 11.4, 13.2, 18, 21, 16.2, 13.3$ and 15, respectively. The wires in each pair ss1 and ss2 have about the same width. The normal-state resistances are $R_N = 22.6, 14.79, 10.29, 6.42, 4.53, 5.66, 3.09$ and 3.2 kΩ, respectively. The effective lengths are $L = 185, 135, 130, 168, 165, 146, 57$ and 72 nm , respectively. For the pairs of parallel wires, the quoted length is calculated as $1/L = 1/L11/L2$. The lengths of the ss1 wires are $L_1 = 96$ nm and $L_2 = 139$ nm. For the ss2 pair, the lengths are $L1 \approx L2 \approx 143$ nm. (b) Magnification of the $R(T)$ curves i1, i2 and i3. The curves are vertically displaced for clarity. All three samples show $dR/dT < 0$ at low enough temperatures. The upturn takes place at $T = 3.2, 2.8$ and 1.6 K for the samples i1, i2 and i3, respectively. (c) A plot of normalized differential resistance ($R_d \equiv dV/dI$, $R_{d0} \equiv dV/dI|_{I=0}$) vs. the bias current ($I$) for two samples: i1 ($T = 1.2$ K) and s1 ($T = 4.2$ K). From [27].

The nanowires are formed by sputter deposition of $Mo_{1-x}Ge_x$ on a carbon nanotube suspended across a slit (Figure 6.2) and a section of the film is measured in series with the nanowires (see Section 6.4). The main result of this work, shown in Figure 7.2, is an apparent dichotomy in behavior between nanowires with normal resistance larger or smaller than R_Q. The resistance of nanowires i1, i2 and i3, all with $R_N > R_Q$, does not go to zero as the temperature is lowered and instead has an upturn in the differential resistance at $T = 3.2, 2.8$ and 1.6 K; this upturn is characteristic of a superconducting to insulator transition (SIT). The other samples, all with a resistance (after the transition of the series film resistance toward zero) smaller than R_Q, are superconducting at the lowest temperature. While these results were interpreted as a dissipative phase transition governed by the nanowire

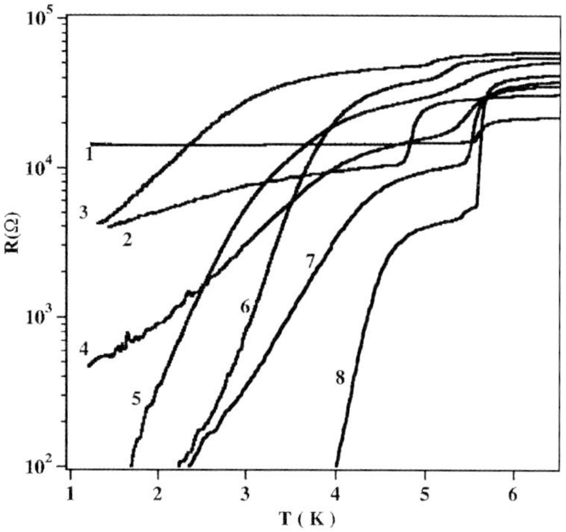

Figure 7.3 Resistances as a function of temperature for eight different samples. The $Mo_{1-x}Ge_x$ nanowire samples' normal state resistances and lengths are 1: 14.8 kΩ, 135 nm; 2: 10.7 kΩ, 135 nm; 3: 47 kΩ, 745 nm; 4: 17.3 kΩ, 310 nm; 5: 32 kΩ, 730 nm; 6: 40 kΩ, 1050 nm; 7: 10 kΩ, 310 nm; 8: 4.5 kΩ, 165 nm. From [14].

normal resistance, a later work from the same group [14], in which a larger number of samples were examined, produced a different result.

In Figure 7.3, we can see how even nanowires with $R_N > R_Q$ show a superconducting behavior down to the lowest measured temperature. Using the expression put forth by Giordano [17, 18] for the resistance due to QPS, and including the contribution of both the thermal phase-slip channel and of the normal channel (this is important if it is comparable to $R_{TPS} + R_{QPS}$), they express the total resistance as

$$\frac{1}{R_{\text{measured}}} = \frac{1}{R_N} + \frac{1}{R_{TPS} + R_{QPS}}. \tag{7.6}$$

Using the two free parameters a and B (7.3) for all the measured traces, they obtain the best fits with $a = 1.3$ and $B = 7.2$. It should be noted that these parameters, of the order of unity, are specified by the Giordano model or by microscopic models [19, 21].

After switching to a fluorinated carbon nanotube as the template, which is more advantageous with respect to unfluoridated nanotubes as they are always insulating, Bezryadin's group demonstrated the suitability of the molecular templating technique for the fabrication of sub-10 nm Nb nanowires [44].

However, neither Nb (Figure 7.4) nor $Mo_{1-x}Ge_x$ (Figure 7.5) nanowires, fabricated on the new stencil, displayed any deviation from the LAMH model below T_c [22, 44]. In the most dramatic set of data, the resistance drops by over 10 orders of magnitude without showing any hint of a deviation.

One important result emerging from [22] is the analysis of the nonlinear $dv/dI = R(T) \times \cosh(I/I_0)$ traces (equivalent to (7.4)). The authors apply a small sinusoidal current superimposed to a DC current that increases until the nanowire switches to the normal state: $R(T)$ together with a small correction to

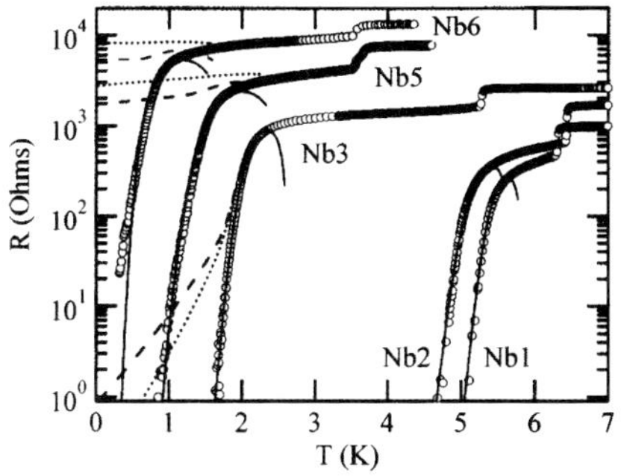

Figure 7.4 Temperature dependence of the resistance of superconducting Nb nanowires. Solid lines show the fits to the LAMH theory. The samples Nb1, Nb2, Nb3, Nb5, and Nb6 have the following fitting parameters. Transition temperatures are $T_C = 5.8, 5.6, 2.7, 2.5$, and 1.9 K, respectively. Coherence lengths are $\xi(0) = 8.5, 8.1, 18, 16$, and 16.5 nm, respectively. The dashed lines are theoretical curves that include the contribution of quantum phase slips into the wire resistance [14], with generic factors $a = 1$ and $B = 1$. The dotted lines are computed with $a = 1.3$ and $B = 7.2$. From [44].

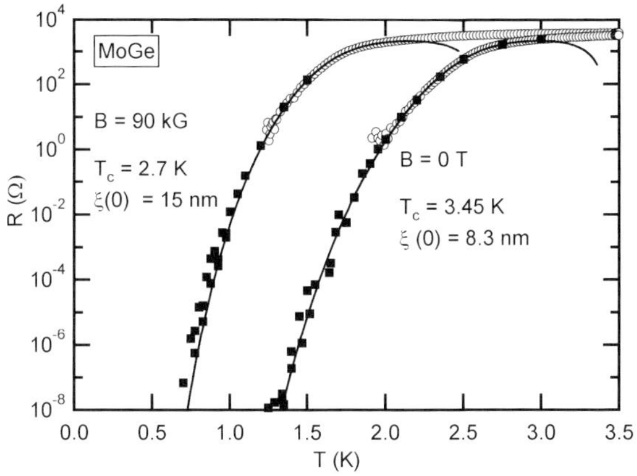

Figure 7.5 Resistance vs. temperature for the MoGe nanowire (same as in Figure 3 in [22]) in magnetic fields 0 and 9 T. Open circles represent zero bias current measurements and black squares indicate the resistance values obtained from the fit of the nonlinear portion of dV/dI curves (Figure 3 in [22]). Solid lines are the fits to the LAMH theory (Eq. (1) in [22]). Extracted fitting parameters T_c and $\xi(0)$ are indicated. The inset shows the experimental dependence of the parameter I_0 on temperature (solid and open symbols) and the theoretical value $I_0 = 4ekT/h$ (solid line). Adapted from [22].

I_0 is the only fitting parameter. The value thus extracted is consistent with the resistance vs. temperature measured in the limit of small currents (solid squares in Figure 7.5). It is also important to notice that the application of a magnetic field up to 9 T only introduces a decrease in the critical temperature consistent with the theory of pair-breaking perturbations.

As discussed at the beginning of this paragraph, early results from [27] indicated the existence of a phase transition for wires with normal resistance greater than R_q. However, the sampling size of only six samples did not allow for sufficient statistics; after collecting larger samples of about 100 nanowires, the group was finally able to confirm the existence of a phase diagram with a clear separation between insulating and superconducting samples [38]. The summary of the cumulative results from this group on homogeneous nanowires is presented in [45]. Short wires, shorter than an empirical length of 200 nm, exhibit a clear dichotomy:

a) If $R_N < R_Q = h/4e^2$, the wires are superconducting and their $R(T)$ characteristics are described in some cases by TAPS and in other cases by some form of QPS-based theory.
b) If $R_N > R_Q = h/4e^2$, the wires are insulating and their behavior is described by the Coulomb blockade.
c) Longer wires have mixed behavior and, as their diameter is reduced, their behavior changes from superconducting to insulating.

Notice that results on wires shorter than 200 nm, which were intentionally rendered inhomogeneous, do not show a clear dichotomy. One recent development in this direction comes from Kim *et al.* [41] in A. Rogachev's group at the University of Utah. In particular, they studied long $Mo_{1-x}Ge_x$ nanowires. These nanowires have been fabricated using negative HSQ resist (see Section 6.10.3); here, length is not a limiting factor and the thinnest width they have been able to achieve is 9 nm. The main result is that in long nanowires (from 1 to 25 μm long), the superconductor to insulator transition is governed by the $R/L = \varrho_l = 40\,\Omega/\text{nm}$; wires with resistance per unit length larger than this value would be insulating at low temperature and superconducting otherwise (smaller than this value). By measuring the differential resistance (dV/dI), the superconducting nanowires would show a dip at zero voltage while an insulating nanowire would show a peak (zero bias anomaly), as shown in Figure 7.6a.

While the origin of the ZBA is not well-understood, a qualitatively similar behavior is displayed by applying a magnetic field to a superconducting sample

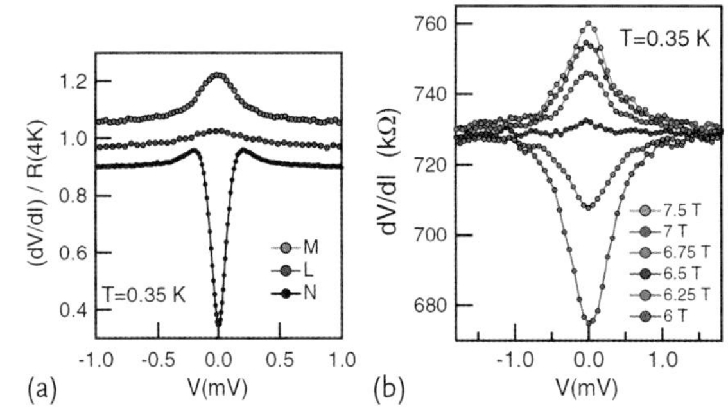

Figure 7.6 (a) Normalized differential resistance at $T = 0.35$ K for indicated MoGe wires (M, L insulating, N superconducting). Data for wires L and N are downshifted by 0.05 and 0.1, respectively. (b) Differential resistance as a function of bias voltage at $T = 0.35$ K in the transitional regime of the SIT for superconducting nanowires F. Adapted from [41]. For a color version of this figure, please see the color plates at the beginning of the book.

both in the R vs. T and in the differential resistance (Figure 7.6b). As the authors mentioned, these results are not compatible with existing SIT (superconductor-insulator transition) theories and while they are not consistent with the entirety of experimental results from Bollinger *et al.* [38], they are consistent with the subset obtained on nanowires longer than 100 nm. In both sets of data, smaller T_c correspond to the decreasing cross section and a similar value of R/L governs the SIT transition. As a side note, only the thickest wire fabricated (cross section 175 and 152 nm^2) had a transition consistent with the LAMH model: the fit of thinner wires, while not showing a positive curvature in the log-linear plot, produced unreasonable fit parameters. Surprisingly, they did not find any indications of a crossover behavior on the superconducting side, where an LAMH behavior evolves into a gentler QPS behavior as the temperature decreases when the cross section of the nanowires is systematically reduced.

In 2005, using the ion beam milling technique (Section 6.10.1), Zgirski *et al.* [34] attempted to trace the crossover between the TAPS and QPS mechanism in the same sample. The starting point is a nanowire with dimensions 60 nm × 100 nm × 10 μm evaporated on a silicon substrate with conventional e-beam lithography and e-gun evaporation of 99.995% pure aluminum. The nanowire's lateral size is progressively reduced down to a cross section of $\sqrt{\sigma} \approx 8$ nm (Figure 6.7a). After each sputtering session, the nanowire transition is measured. The results are shown in Figure 7.7.

From the graphs in Figure 7.7, we can observe that, as expected, the normal resistance increases with the reduction of the nanowire cross section and at the

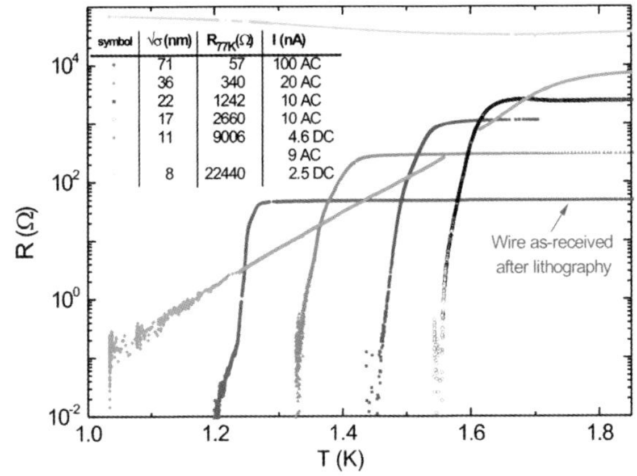

Figure 7.7 Resistance vs. temperature for the same aluminum wire of length $L = 10$ μm after several sputtering sessions. The sample and the measurement parameters are listed in the table. For low-Ohmic samples, lock-in AC measurements with the front-end preamplifier with input impedance 100 kΩ were used; for resistance above ∼ 500 Ω, a DC nanovolt preamplifier with input impedance ∼ 1 GΩ was used. The absence of data for the 11 nm sample at $T \sim 1.6$ K is due to switching from a DC to AC setup. Note the qualitative difference of $R(T)$ dependencies for the two thinnest wires from the thicker ones. From [34]. For a color version of this figure, please see the color plates at the beginning of the book.

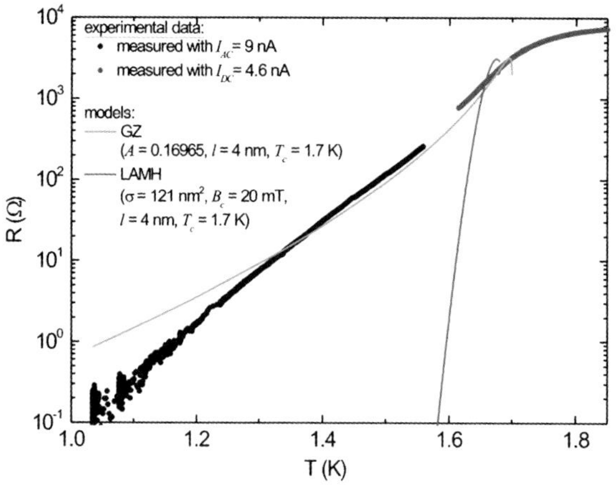

Figure 7.8 $R(T)$ dependence for the $\sqrt{\sigma} \sim 11$ nm aluminum sample. The green line shows the result of fitting to a renormalization theory (Reference 18 in [34]) with A, l, and T_C being the fitting parameters. The same set of parameters together with the critical magnetic field B_C measured experimentally is used to show the corresponding effect of thermally activated phase slips on the wire's $R(T)$ transition (red line) (References 1,2 in [34]). The parameter σ is obtained from the normal state resistance value and the known sample geometry (Reference 12 in [34]). The estimation for σ is in reasonable agreement with SPM analysis as well as with evolution of σ over all sputtering sessions (see Figure 7.7). From [34]. For a color version of this figure, please see the color plates at the beginning of the book.

same time, the critical temperature of the nanowire increases as commonly observed in aluminum. More importantly, the transition drastically changes character as the wire cross section is progressively reduced. When the nanowire cross section is about 11 nm (Figure 7.8), fits based on the LAMH theory fail to reproduce the experimental data while somewhat better agreement is observed using a simplified version of a model based on the renormalization theory by Golubev and Zaikin for the QPS (Section 3.2). Another interesting effect observed by the authors in the thinnest nanowire is that both the $R(T)$ and $V(I)$ depend on the current used for the measurement in a nonmonotonic manner; however, for such a thin nanowire, the authors do not exclude the possibility of inhomogeneities. In a later paper [46], the authors examine another set of nanowires produced with the same technique.[3] The results are substantially similar: transition curves of nanowires with $\sqrt{\varrho} \ll 15$ nm can only be reproduced by the renormalization model by Golubev and Zaikin. The transition is much wider than what was predicted by TAPS and wider than what a simple calculation, assuming experimentally measured roughness, could suggest [47]. Agreement with the GZ theory is also observed in both the $R(T)$ and $V(I)$ curves. The novelty of this work consists of the new data reporting a negative magnetoresistance at a temperature well below

3) While the authors reported to have performed four-terminal measurements, the generally accepted configuration for four-terminal measurements involves voltage leads smaller than or comparable to the device under study and separated from the current leads.

T_c and is only present in the thinnest wires for a field smaller than 25 mT. Since aluminum is well-known to be immune to the creation of localized magnetic moments, and because T_c is not enhanced, this effect is qualitatively different from what was observed in $Mo_{1-x}Ge_x$ and Nb nanowires [48]. The author's hypotheses are that this effect is due to (i) a competition between a reduction in ΔF and a suppression in the superconducting gap Δ or (ii) due to suppression of a dissipation channel caused by a region of charge imbalance [49]. A similar negative magnetoresistance was recently observed in titanium nanowires fabricated using the same technique [50], but no quantitative argument is provided; thus, this effect remains unexplained. One interesting observation from [50] is that even relatively wide nanowires, fabricated with the ion beam milling technique displayed deviations from TAPS theory, however, a more ideal experiment, easily implementable in Ti rather than Al, would be a direct comparison between as-deposited and ion beam polished nanowires.

Around the same time frame, Altomare *et al.* [35] developed a new technique for the fabrication of narrow and long nanowires (see Section 6.8). In particular, the authors were able to fabricate two aluminum nanowires, sample s1 of length 10 μm and cross-sectional area 130 nm^2 and sample s2, 100 μm long with a cross section of 58 nm^2. They focused on the behavior in the presence of a finite magnetic field, applied perpendicular to the nanowire length. The magnetic field has two main effects: (i) It drives the large, superconducting electrical leads in to the normal state, and thus, the configuration is an N-SCNW-N configuration, where N stands for normal metal. This is in contrast to the system in Bezryadin's group, for which all nanowires are connected to superconducting leads. In view of the importance of the leads, as discussed in Chapters 3 and 4, one may expect qualitatively different behaviors. (ii) Increasing the magnetic strength enables one to tune the gap of the superconducting nanowire, allowing one to systematically investigate the behavior as the gap is reduced to zero. The analysis followed the one performed by Rogachev *et al.* [22], ensuring consistency between measurements in low and high-current regimes.

The consistency between measurements in linear and nonlinear regimes, shown in Figure 7.9, allows the authors to establish the importance of quantum phase slips in the transition curve of this device. A similar analysis on the wider device s1 did not show presence of QPS.

An alternative interpretation of the data from this work is shown in Figure 7.10. As suggested independently by Zaikin and Golubev [51], by Khlebnikov and Pryadko [21], and Refael, Demler, *et al.* [52], the V–I characteristic can be modeled by (Sections 3.2.6 and 3.5.2)

$$V = V_S + \left(\frac{I}{I_k}\right)^{\nu} . \tag{7.7}$$

Here, ν is a fitting exponent (not to be confused by that defined in (3.152)). The fit of the V–I curves to (7.7) are indistinguishable from the best fit to the Giordano expression (GIO) in Figure 7.9. It should also be pointed out that the heating effect cannot explain those results of nonlinear behavior, as was extensively discussed in [37]. At the same time, this nonlinearity cannot be interpreted as a result of the proximity effect from the normal pads into the superconducting nanowire[4)], which produces a quasi-particle to Cooper-pair conversion resistance at the N–S interface: recent analysis from Klapwijk's group [53] suggest that in the I–V trace, during an up-sweep of the applied current, the perturbation of the normal leads into the superconducting nanowire is minimal until the nanowire switches to the normal state. This was experimentally verified using a somewhat wider aluminum nanowire with normal contacts and a tunnel probe located at a distance of only 2.4ξ

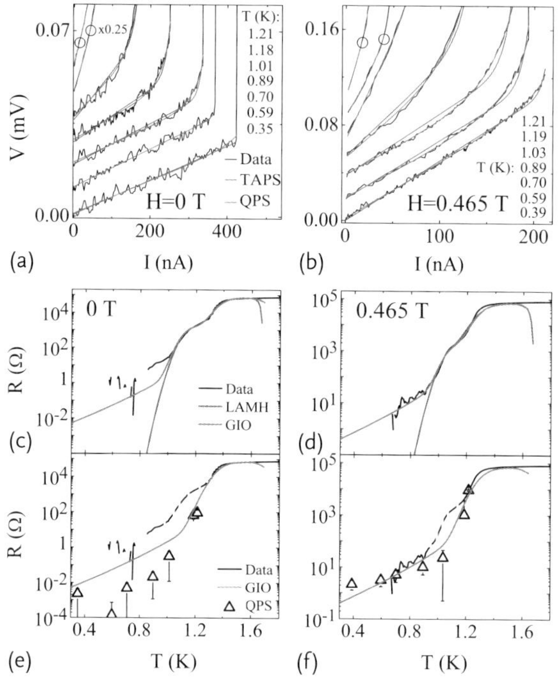

Figure 7.9 Nonlinear I–V curves and linear resistance for Al nanowire sample at different magnetic fields (H): Black curves – data; red curves – fits to the GIO (≡ GIO + LAMH; GIO denotes Giordano) expressions for QPS (≡ TAPS + QTPS; TAPS denotes thermal-activated phase-slips and QTPS quantum-tunneling of phase-slips), and blue curves – fits to the LAMH expressions for TAPS alone. (a) and (b) show I–V curves offset for clarity. The fits to QPS are of higher quality compared to TAPS; each fit includes a series resistance term V_S. These I–V curves can be fitted equally well by a power law form $V = V_S + (I/I_k)^{\nu}$ where $12 < \nu < 3.2$ for $0 \leq H \leq 1.05$ T [21, 37]. (c) and (d) show linear resistance after background subtraction (see Figure 1a in [24] and [37]). The LAMH fits are poor at low T. (e) and (f) show the resistance contribution due to phase slips (R_{QPS}) extracted in (a) and (b) from fits to the I–V curves using the GIO expressions (discrete points, Δ). RQPS and the linear resistance from (c) and (d) are refitted using the GIO expressions (red) with the same a_{GIO} (= 1.2) while disregarding the irrelevant shoulder feature (dashed line). Adapted from [24]. For a color version of this figure, please see the color plates at the beginning of the book.

4) The nanowire was contacted by a trilayer Ti/AuPd/Al which became normal at a field of ∼ 0.11 T

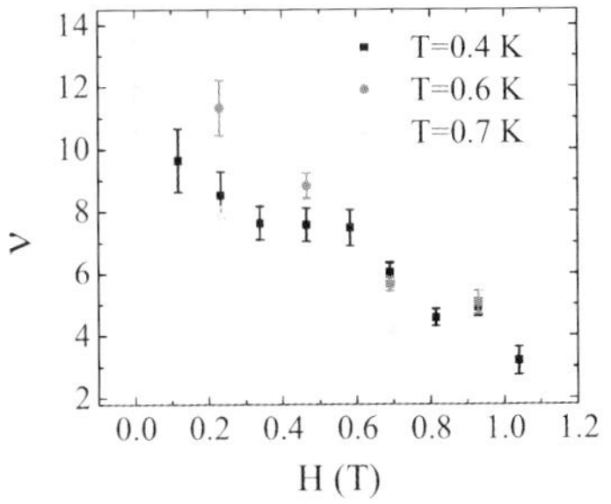

Figure 7.10 The exponent ν, extracted from the fit of the V–I curves to the power law expression, plotted as a function of the magnetic field. Different symbols correspond to V–I curves at different temperatures [37].

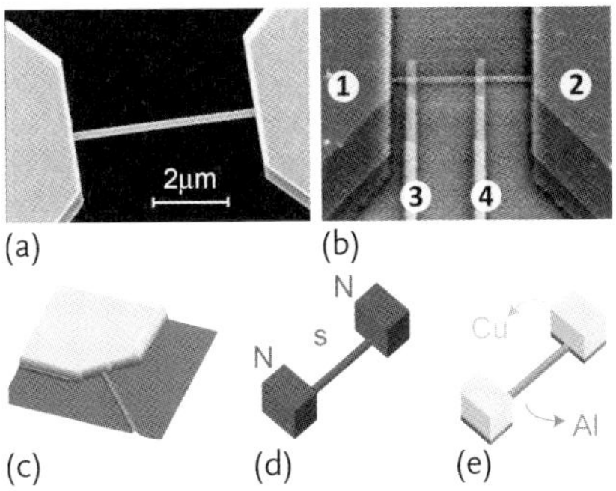

Figure 7.11 A superconducting Al nanowire connected to two massive normal reservoirs, consisting of the same Al, covered by a normal metal Cu layer: (a) SEM-picture, (c) AFM-picture, and (d) and (e) show a schematic representation. The thin Al of the pads is driven normal by the inverse proximity effect of the thick normal Cu. Normal tunneling probes are attached for local measurements (b). From [53]. For a color version of this figure, please see the color plates at the beginning of the book.

from the contacts. The sample configuration and experimental results from [53] are reported in Figures 7.11 and 7.12. Using the findings from [53] as a guide, where the interface resistance reflected in the dI/dV was found to hardly change for different bias currents, the nonlinearity observed here in the 100 μm long nanowire cannot be attributed to the interface resistance at the N-SCNW interface as the nanowire length is $L \gtrsim 1100\xi(T = 0)$, but most likely reflects the intrinsic resistance of the nanowire itself due to QPS processes. More recent measurements on a shorter Al nanowire used extensively for switching current studies [7] to reveal the existence of QPSs produced similar nonlinear power law dependences with an exponent roughly half of the values of $3.2 < \nu < 12$ for $0\,\text{T} \leq H \leq 1.05\,\text{T}$ found here.

Another experiment aimed at the observation of QPS in niobium nanowires was performed in 2008 by Heath's group. The SNAP technique they developed (Section 6.7) naturally lends itself to the study of arrays, although it could be as easily

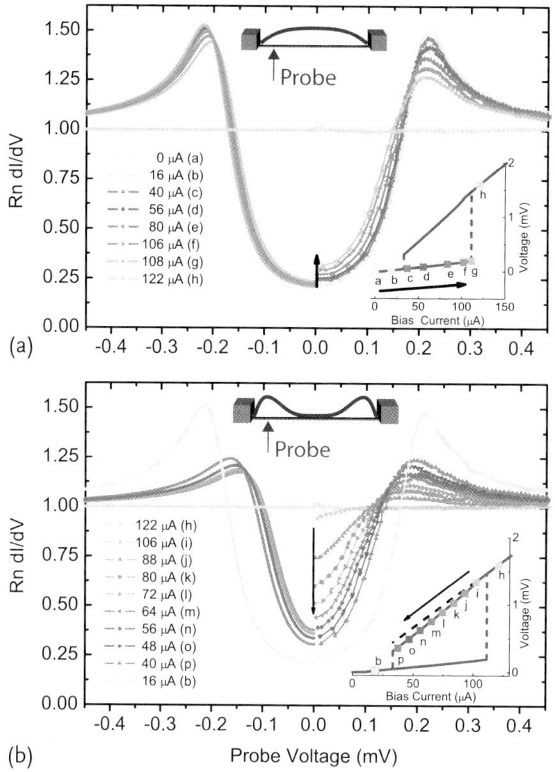

Figure 7.12 The local density of states (DOS $\propto dI/dV$) (a) the global superconducting state and (b) the bimodal state for different bias currents I_{12} of a 100 nm wider Al nanowire, measured at 200 mK. For the global superconducting state, the gap is only weakly dependent on the bias current, while for the bimodal state, for which only regions adjacent to the normal contacts are superconducting while the middle of the nanowire is in the normal state, one observes a DOS gradually changing from a normal into a superconducting state. From [53]. For a color version of this figure, please see the color plates at the beginning of the book.

adapted to study single nanowires. The group's main results are presented in [40] in which the authors perform an exhaustive analysis of arrays of niobium nanowires with varying length, cross section, and electrode composition. The first observation is that arrays of cross section $30 \times 16\,\text{nm}^2$ (thickness × width) contacted with normal electrodes (Pt) have a critical temperature $T_c \approx 3.5$ K, while their critical temperature is $T_c \approx 5$ K when niobium contacts are used (Figure 7.13b,c). After the transition, a residual resistance of about 18 Ω is measured for the Pt-contacted arrays, which might suggest a superconducting-normal mixed state localized at the contacts. Given the strong influence of the contacts' composition, the author focused their attention on an array with superconducting electrodes. The bulk of their results is obtained in an array of 100 NW with a cross section of $11 \times 16\,\text{nm}^2$ and length varying from 1.6 to 100 μm (Figure 7.13a). As can be seen in the inset of Figure 7.13a, the normal resistance of the NW arrays vs. their length is linear. This represents a healthy check that the fabrication does not introduce any obvious artifacts. In Figure 7.13c, it can be seen that in the case of an array with a cross section of $30 \times 16\,\text{nm}^2$ (1.3 μm long), the TAPS (thermally-activated phase-slips) and TAPS + QPS theories are indistinguishable. Looking at Figure 7.13e, it can be observed as predictions from TAPS and TAPS + QPS start to diverge that the data for a 2.4 μm long array follows the prediction of the TAPS + QPS model more closely. The agreement between the experimental results for the 20 μm long array (Figure 7.13d) and the TAPS + QPS theory is even clearer. As an additional check, the slope for the $11 \times 16\,\text{nm}^2$ array is about one-third of the slope for the $30 \times 16\,\text{nm}^2$ as it follows from the free energy expression [14]. The suppression of QPS in the shorter arrays is probably due to local enhancements of superconductivity caused by the superconducting contacts.

The two longest nanowire arrays, 50 and 100 μm long, display a shoulder feature similar to the one observed by Altomare *et al.* [24] in both 100 and 10 μm long nanowires.[5] The explanation of this feature requires looking at the I–V curves in Figure 7.14. While arrays with wide cross section ($30 \times 16\,\text{nm}^2$) exhibit a I–V curve with hysteresis, thinner arrays with a length of 2.4 μm do not display any hysteresis, but only a single jump at I_c; as all the nanowires switch at the same critical current, this is interpreted as an additional demonstration of the uniformity of the nanowires in the array.

However, longer arrays with the same cross section exhibit a completely different behavior: the NW enter an intermediate state in which multiple small jumps are performed until the array become normal; this is characteristic of a transition localized at a phase-slip center (PSC). The number of PSC that a nanowire can sustain is $\approx L/\Lambda_Q$, where L is the wire length and Λ_Q is the quasiparticle diffusion length. In the 100 μm long array, 4 PSC are found, which suggests a quasiparticle diffusion length of $\approx$ 10 μm with a PSC occupying a length $2\Lambda_Q$. This is consistent with the results for the 50 μm long array of nanowires where only 2 PSC are found and none are found for smaller wires. However, in 20 and 10 μm long arrays, small branching features are observed which are consistent with the length scale for PSCs in longer arrays. In both the 50 and 100 μm long arrays, the shoulder feature coincides with the onset of a stable PSC line.

It is important to point out that these results seem to go in a different direction than the results obtained with molecular templating [22]. While there are obvious differences between the two works, for example, the study of single nanowires vs. the study of arrays and different experimental setups, these results seems to indicate the possibility of observing QPS even in the low current, regime.

5) Xu *et al.* report that in Altomare *et al.* [24], this feature is present only in the 100 μm long nanowire. The shoulder feature is not obvious in linear-linear scale, but only in log scale: see the plot in [54].

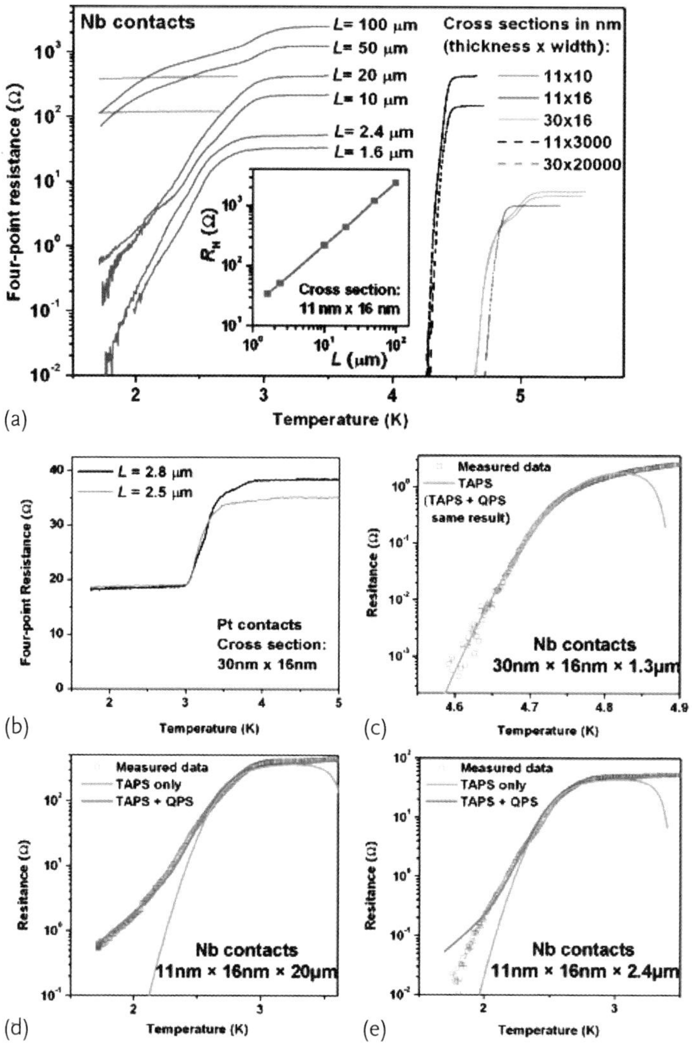

Figure 7.13 Temperature dependence for the four-point resistance of Nb NW arrays and films. (a) Superconducting Nb contacted NW arrays and films. Red lines: arrays of 12 NWs of cross section 11 nm × 10 nm and length L (from top to bottom) = 3 and 0.9 μm. Blue lines: arrays of 100 NWs of cross section 11 nm × 16 nm and L = 100, 50, 20, 10, 2.4, and 1.6 μm. Green lines: arrays of 250 NWs of cross section 30 nm × 16 nm and L = 1.5 and 1.3 μm. Black dashed lines: 11 nm thick films with width of 3 μm, L = 60 and 20 μm. Purple dashed line: a 30 nm thick film with a width of 20 μm, L = 2.5 μm. Inset: Length dependence of the normal-state resistance for arrays of 100 NWs of cross section 11 nm × 16 nm. (b) Normal-state Pt contacted NW arrays of cross section 30 nm × 16 nm. (c) 30 nm × 16 nm × 1.3 μm data in (a) fitted to TAPS theory (TAPS + QPS gives an indistinguishable result). (d) 11 nm × 16 nm × 20 μm data fitted to the theories. (e) 11 nm × 16 nm × 2.4 μm data fitted to the theories. From [40]. For a color version of this figure, please see the color plates at the beginning of the book.

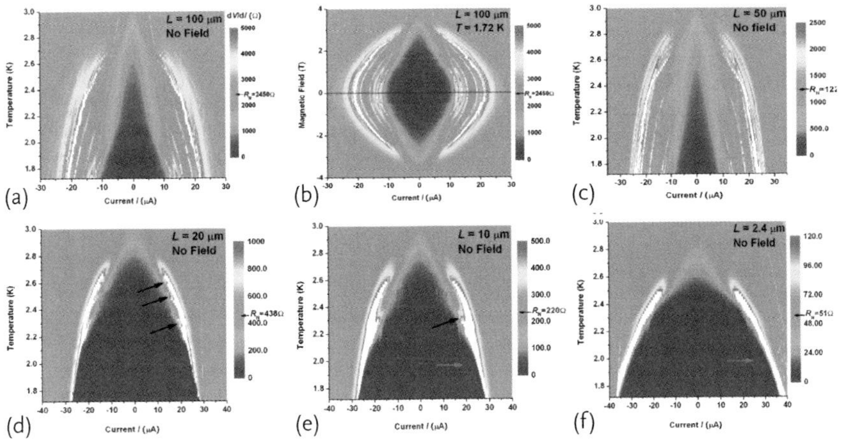

Figure 7.14 Differential resistance dV/dI of arrays of 100 Nb NWs of cross section 11 nm × 16 nm, as a function of current, temperature and applied magnetic field. (a) and (c–f) temperature is varied at zero field. (b) An applied magnetic field is varied at constant temperature T) 1.72 K. The lengths of the arrays are labeled on each plot. Black arrows point to the small branching features observed close to I_c at higher temperatures for arrays of intermediate lengths, and red arrows point to the very faint (light blue) peak lines found only on the two shortest arrays. From [40].

7.4.3
QPS Probed via Switching Current Measurements

Up to approximately 2008, the nanowire characterization was performed either by examining the $R(T)$ curve using low excitation current or by fitting the voltage–current curves at a current lower than the switching events. A complementary and perhaps more definitive approach would be one that led to the demonstration of quantum tunneling in a Josephson junction; it is therefore natural to follow in the footsteps of Clarke's group [15, 16, 55] and look at the switching current distribution. The experiments by Devoret, Martinis and Clarke [15, 16, 55] represented the first convincing experimental evidence of macroscopic quantum tunneling; macroscopic, not in the sense that they are a macroscopic manifestation of a series of microscopic effects (as superconductivity for example), but rather that a macroscopic object has a single degree of freedom that behaves quantum mechanically (from [55]). Today, we take these results for granted as JJs are ubiquitous in the superconducting devices realm and quantum computing is a reality. In the following, wherever possible, we will draw a parallel between recent experiments on nanowires and these pioneering experiments on JJs. The goal is to prepare the reader for the experimental results described in Chapter 8, in which a recent proposal for a qubit, using coherent quantum phase slips, is demonstrated.

The first experiment we will discuss was reported in [4]. In this paper, Bezryadin's groups studied the switching current distribution of five $Mo_{1-x}Ge_x$ nanowires fabricated using molecular templating techniques. The wires range in length between

100 and 200 nm and in resistance between 1.4 and 4.1 kΩ. Those nanowires are in the regime were the nanowires are superconducting as described above and their low temperature characterization is entirely consistent with LAMH behavior. Before we discuss the experimental results on the switching current distribution, we would like to provide an intuitive understanding of the mechanism involved. As we have already mentioned, the potential landscapes of a nanowire and a Josephson junction (JJ) with an applied current, are analogous, that is, a sinusoidal corrugation is superimposed to a linear slope provided by the applied current – the so-called tilted washboard potential (Figure 2.12). Increasing the magnitude of the current is equivalent to applying a larger tilt to the washboard, with the phase trapped in one of the local minima.

Once the bias current is large enough to remove any minima, the phase will run downhill along the tilted-washboard potential without interruption: this corresponds to the nanowire/JJ being in the normal state, as the critical current I_c is exceeded. However, the interesting regime to explore is that of a bias current close to, but smaller than I_c. In this case, the minima will be very shallow and small fluctuations, as those caused by a thermal Johnson noise, for example, can cause the phase particle to overcome the local minimum and allow the nanowire to undergo a phase slip (PS). As a PS event will depend on a purely stochastic process, it is necessary to accumulate a large statistic of switching events in order to measure their distribution. In fact, it can be shown that if temperature is the main energy scale governing the switching process, the width of the distribution is proportional to $kT^{2/3}$ (see also the discussion in [56]). This will be true even in the presence of quantum effects as long as the dominant energy scale is the temperature. However, when the temperature becomes sufficiently low, the switching will be independent of the temperature and other effects, like quantum tunneling, may become relevant. This method was pioneered by Fulton and Dunkleberger while studying the switching current distribution of JJs [57] and later applied in Clarke's group to the study of quantum tunneling in JJs. The result of this experiment showed how the width of the switching distribution decreases with decreasing temperature down to a critical temperature; below this value, the width of the switching distribution does not change Figure 7.15.

Therefore, it is natural to perform a similar experiment on nanowires. Bezryadin's group has pioneered the use of this technique in nanowires [4] and their results, indeed, show a clear difference with respect to a simple thermal activation model. However, instead of observing a monotonic decrease of the width of the distribution, they observe a monotonic increase (Figure 7.16). While this is quite surprising and unexpected from the simple picture painted above, the authors are able to exclude experimental artifacts and to provide an explanation of this effect considering a mechanism driven by phase-slip fluctuations, which depend on the temperature.

In particular, they observe that at $T > 1$ K, the rate of TAPS is expected to be much larger than the measured switching rate. This means that more than a single phase slip is required to drive the wire in the normal state, at $T > 2.7$ K, these

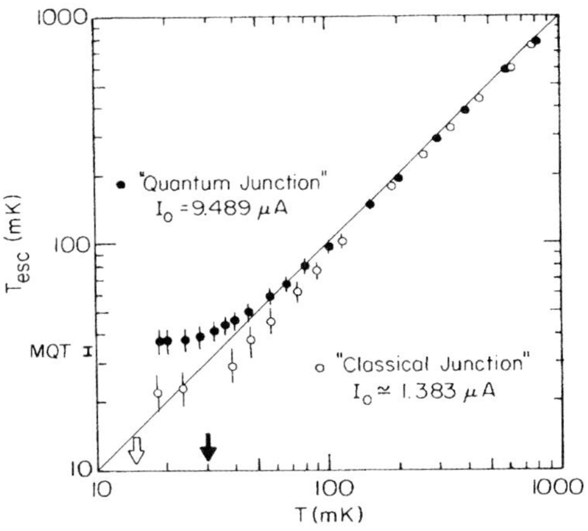

Figure 7.15 T_{esc} vs. T for two values of critical current in an under-damped JJ for $\ln(\omega_p/2\pi\Gamma) = 11$. The solid and open arrows indicate the predicted crossover temperatures for the higher and lower critical currents, respectively. The prediction of Eq. (5) in [16] for the higher critical current is indicated on the left. Adapted from [16].

phase-slip events can be observed in the V–I switching curves. The model [4, 58], which has to account for the above observations, has two main components:

a) Stochastic phase slip that turns a segment of length of the order of ξ into a normal wire and occurs at random time and location, but depends on the temperature, and
b) the heat produced by the phase slips which is carried away by the leads. After a phase slip has occurred, the local temperature of the wire is higher, which in turn increases the rate of phase slip, until the wire has time to cool down. If another phase slip happens while the wire temperature is still high, it will increase the temperature even further and the process will repeat in a thermal runaway process until the entire wire has moved in the voltage (resistive) state.

At high temperatures, I_c (and therefore I_s, which always falls slightly below I_c) is quite small, and the heat is better transferred to the leads: therefore, more PSs are required to drive the nanowire out of the superconducting state. The critical current I_c is, in fact, determined by the depairing mechanism, whereby the superfluid velocity reaches a critical value, enabling Cooper pairs to be broken up into individual quasiparticles. When the temperature is lowered, the critical current tends to increase and the ability of the nanowire to dissipate heat becomes lower as few PSs are required to drive the wire in the voltage state. Numerical calculation suggests that in the competition between the lower phase-slip rate (at low temperatures) and the larger heat deposited in the wire, the latter wins, and therefore the nanowire exhibits a regime in which a single PS can trigger the switch. From the fit to the data, the single regime PS with TAPS rate matches the data well, down to 1.3 K (Figure 7.17). However, for even lower temperatures, the agreement with data is

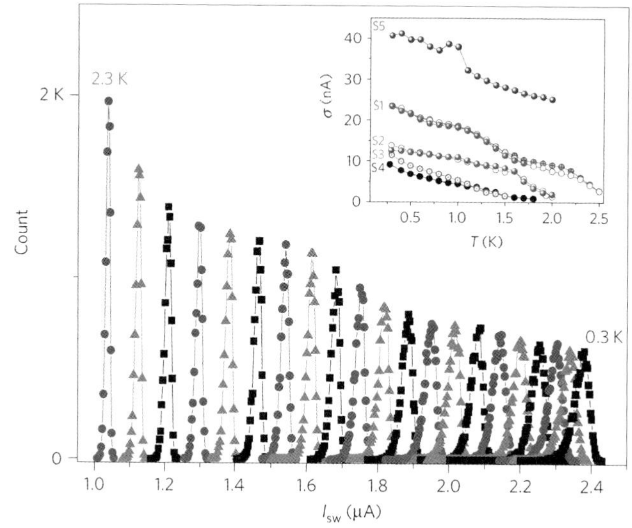

Figure 7.16 Switching-current distributions at different temperatures. Switching-current distributions $P(I_{SW})$ for temperatures between 0.3 K (right-most) and 2.3 K (left-most) with $\Delta T = 0.1$ K for sample S1. For each distribution, 10 000 switching events were recorded and the bin size of the histograms was 3 nA. Inset: Standard deviation $\sigma\{= \sqrt{[\sum_{i=1}^{n}(I_{SW,i} - \bar{I}_{SW})^2/(n-1)]}\}$ of $P(I_{SW})$ vs. T for five different $Mo_{1-x}Ge_x$ nanowires including sample S1. For samples S1 and S2, the measurements were repeated a few times to verify the reproducibility of the temperature dependence of σ. For all wires, the width of the distributions increases as the temperature is decreased. From [4].

marginal. At this temperature, fitting using the expression for QPS by Giordano instead (7.3) and an effective temperature [15, 16, 55] T^* at which QPS dominate over TAPS, good agreement with the data is obtained. The main difference between these results and the ones obtained in JJs is that T^* is not constant, but linearly increasing with temperature until a maximum temperature is reached. For one of the samples measured (S5), the slope is close to zero. Additionally, the authors observe that samples with larger critical current have larger T^*, which allows them to exclude any artifact due to the presence of weak links; a weak link would, in fact, have a lower critical current. Moreover, if the fluctuations were due to externally uncontrolled causes, thicker wires with larger critical current would have a smaller T^*.

The same experiment was performed by Li *et al.* in Chang's group on five aluminum nanowires of varying width (between 5 and 10 nm) and length (between 1.5 and 10 μm). Measurement in both a ^{3}He system and a dilution refrigerator showed consistent results. The main results from this work are presented in Figure 7.18.

In particular, Figure 7.18c shows the width of the switching current distribution as a function of the temperature. Three distinct regions are observed. At high temperature ($T > 0.6\,T_C$), the width of the distribution decreases quite rapidly; this points to the necessity of multiple phase slips to trigger the switching. For intermediate temperatures ($0.3\,T_c < T < 0.6\,T_c$), for all the devices, the switching is dominated by single thermally activated phase slips, as can be see by the agreement with the dashed lines calculated assuming that the distribution is entirely

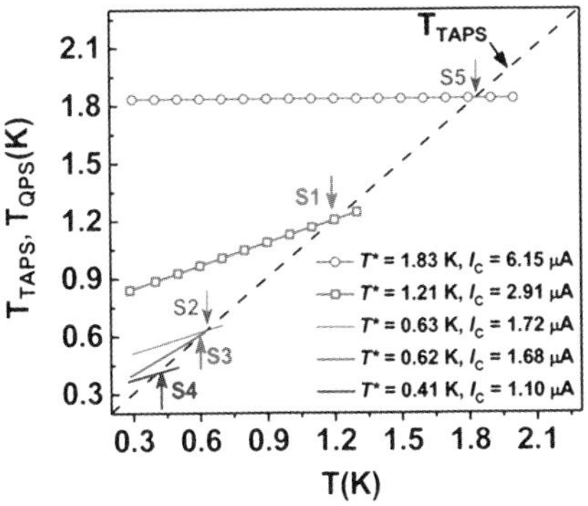

Figure 7.17 The best-fit effective temperature for fluctuations at different bath temperatures for five different $Mo_{1-x}Ge_x$ nanowire samples (S1–S5). For all TAPS rate calculations, the effective temperature is chosen as the bath temperature (shown by the black dashed line). For the QPS rates, the effective temperature TQPS, used in the corresponding QPS fits, similar to the blue-line fits of b [4], is shown by the solid lines. For each sample, below the crossover temperature T^* (indicated by arrows), QPS dominates the TAPS. They find that the T^* decreases with decreasing critical depairing current of the nanowires, which is the strongest proof of QPS. The trend indicates that the observed behavior of TQPS below T^* is not due to extraneous noise in the setup or granularity of wires, but, indeed, is due to QPS. Adapted from [4]. For a color version of this figure, please see the color plates at the beginning of the book.

due to thermal effects $\delta I_s \sim (k_B T/\phi_0)^{2/3} I_c(T)^{1/3}$ ($\phi_0 = \hbar/e$ in SI units). At lower temperatures ($T < 0.3T_c$), for three devices, the authors observe a flattening of the switching current I_s distribution. This is quite a large departure from what is expected for TAPS and consistent with the presence of QPS. These results are also more in line with the intuitive expectation that, below a certain temperature, the thermal fluctuations are not expected to play a significant role.

Here, no magnetic field is present, and the device configuration is S-SCNW-S, where S denotes the superconducting leads. These aluminum nanowires are remarkably uniform in their characteristics, as evidenced by the fact that the product of the resistance per unit length ($\propto A^{-1}$, where A is the cross-sectional area of the wire) and the switching current I_s at 300 mK ($\propto$ A) is nearly identical, that is independent of A, to within $\sim$ 20% for the entire set of nanowires. The resistivity of the nanowires is comparable to coevaporated 2D films with $\rho \sim 3.5\ \mu\Omega$ cm. This scaling of the resistance per unit length and the I_s enables one to rule out the presence of weak links or constrictions and attests to the uniformity of the nanowires.

It is extremely interesting to note that two almost identical experiments, under pretty much the same assumption that the heat generated in the nanowire by the phase slip is not efficiently dissipated through the substrate but only through the leads, produce two qualitatively different behaviors as far as the switching distribution is concerned. The main differences between the two experiments are wire dimensions and material. While the cross section of the two nanowires are comparable, the material is quite different. The former experiment uses $Mo_{1-x}Ge_x$ nanowires for which ξ is of the order of 10 nm and with mean free path of

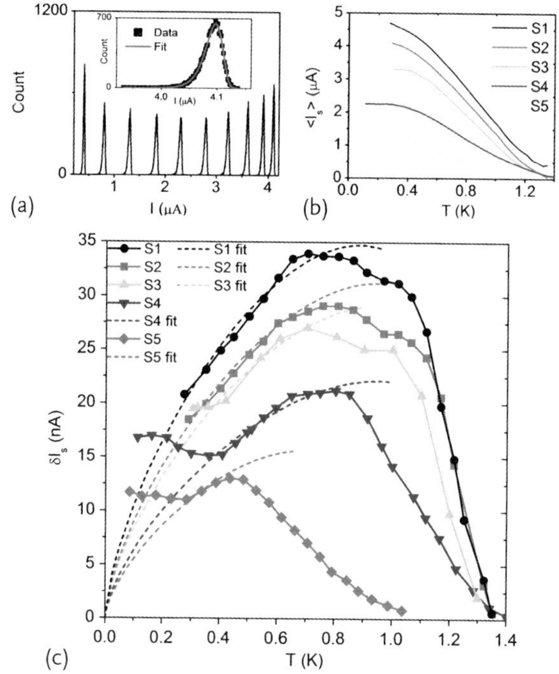

Figure 7.18 (a) I_s distribution for Al nanowire S2 at different temperatures: right to left: 0.3–1.2 K in 0.1 K increments. The inset shows the 0.3 K distribution, fitted by the Gumbel distribution (Reference 20 in [28]). (b) $\langle I_s \rangle$ vs. temperature – top to bottom S1–S5. (c) Symbols: δI_s vs. temperature. Dashed lines: fittings in the single TAPS regime using (Eq. 2 in [28]). An additional scale factor of 1.25, 1.11, 1.14, 0.98, and 1.0, for S1–S5, respectively (average 1.1 ± 0.1), is multiplied to match the data. Alternatively, a $\approx$ 6% adjustment in the exponent fits the data without the scale factor. Adapted from [28]. For a color version of this figure, please see the color plates at the beginning of the book.

0.3 nm, the latter uses aluminum nanowires with ξ of the order of 100 nm and a mean free path of 10 nm. The product $k_F l_{mfp}$ is about 1 for $Mo_{1-x}Ge_x$ and > 60 for the Al. The lengths are also quite different with L/ξ of the order of 10 in the $Mo_{1-x}Ge_x$ nanowires and 100 in the aluminum nanowires. Additionally, while in both cases the lateral dimensions are smaller than about 4.4 times the superconducting coherence length, and thus can be considered 1D [23], the ratio $w/\xi \sim 10$ in aluminum nanowires and is of the order ~ 1 in $Mo_{1-x}Ge_x$ nanowires. Another possibly important difference lies in the structural composition of the nanowires: while $Mo_{1-x}Ge_x$ nanowires used in this experiment are amorphous, the TEM cross section of aluminum nanowires shows quite a regular structure suggesting a polycrystalline structure (see Figure 1 in [28]). Are these differences sufficient to explain the different results? The situation is not clear: recent experiments by Bezryadin's group [29] confirm the results of Li *et al.* [28], both in terms of the 2/3 power law dependency and the saturation of the width of the switching current distribution. These results were performed on a

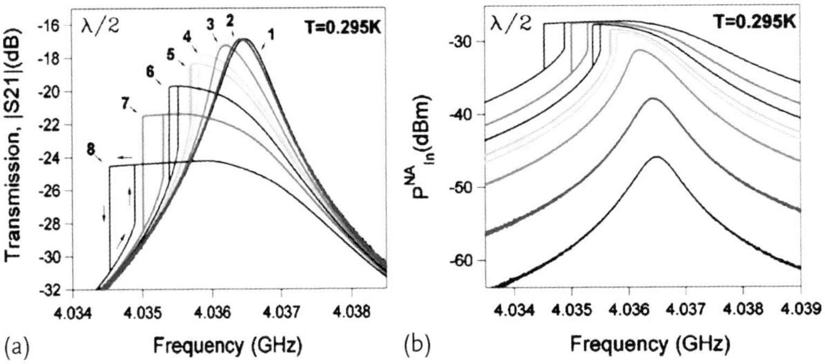

Figure 7.19 (a) ($Mo_{1-x}Ge_x$ sample S2) Transmission amplitude S_{21} (dB) in forward and backward frequency sweep for various driving powers. The graph shows Duffing bifurcation occurring at higher driving powers. The curves correspond to different driving powers: 1: $P_{out}^{NA} = -29$ dBm (black); 2: −21 dBm (blue); 3: −14 dBm (red); 4: −11 dBm (orange); 5: −10 dBm (green); 6: −8 dBm (black); 7: −6 dBm (violet); and 8: −3 dBm (black). (b) (sample S2) Replotting of the data from (a) as the transmitted power P_{in}^{NA} measured at the network analyzer input vs. frequency. From [60]. For a color version of this figure, please see the color plates at the beginning of the book.

nanowire fabricated by a molecular templating technique. It should be pointed out that two samples, treated with a newly developed pulsed current melting technique [59] which transforms an amorphous $Mo_{1-x}Ge_x$ nanowire into a single crystal Mo_3Ge, showed a different power law dependence 5/4 following recrystallization. This power law dependence derives from the slightly different shape of the tilted-washboard potential in a nanowire when compared with the usual JJ [56]. More experimental and theoretical work is definitely needed in this field.

Two recent experiments we want to touch upon from Bezryadin's group, although not directly related to QPS, are described below. In the first one, Ku *et al.* [60] create a nonlinear resonator embedding a nanowire in a superconducting resonator. In the experiment, a $Mo_{1-x}Ge_x$ nanowire is placed at the center of a coplanar waveguide (CPW) which is capacitively coupled to an input and output electrode. While the first excited mode, with wavelength λ, has current nodes in correspondence of the input and output capacitance and at the center of the CPW, the fundamental mode, with wavelength $\lambda/2$, has current nodes at the input and output capacitor and an antinode at the center of the CPW and is thus sensitive to the presence of non-linear elements at the center of the CPW.[6] At the frequency corresponding to the $\lambda/2$ mode, varying the power through the CPW will vary the current through the nanowire: as the inductance of the nanowire depends on the current flowing through it [23], the transmission peak will be a function of the applied power as shown in Figure 7.19.

6) From an experimental point of view, this difference in behavior is extremely useful as a sanity check.

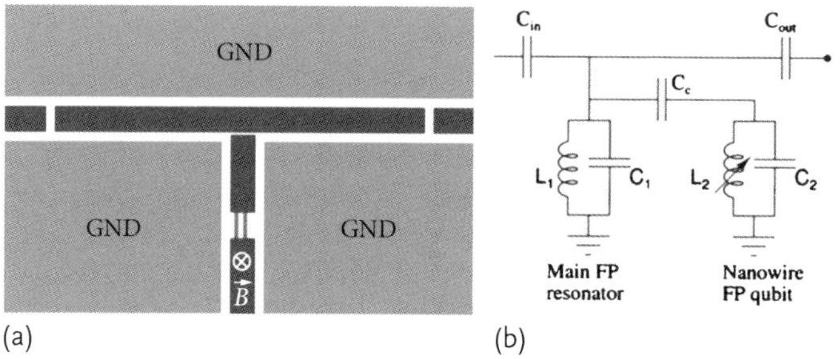

Figure 7.20 (a) Schematic of a nanowire-FP qubit (vertical) coupled to the main resonator (horizontal). The qubit is a FP-type coplanar waveguide resonator having two parallel nanowires (red) in the center. The nanowires make the resonator anharmonic. Thus, two levels can be addressed by a proper choice of frequencies. The critical current of the pair of nanowires can be controlled by the perpendicular magnetic field (Reference 16 in [20]). Thus, the qubit resonator can be tuned in resonance with the main resonator, if desired. The ground planes (gray) are indicated by "GND." Although the bottom end of the qubit resonator is shown ungrounded, it can be linked to the ground plane if desired. Note that both the gray color and the blue color represent the MoGe film, but the regions shown in gray are grounded, while blue regions are not. (b) Simplified equivalent circuit of the sample. The inductor in the qubit can be tuned with the magnetic field, and it is current-dependent, and thus termed "nonlinear." The arrow crossing the inductor symbolizes the current dependence of the inductor. Adapted from [60].

In particular, we can see how for sufficiently large powers, the curve becomes hysteric when the frequency is swept in two different directions. This is a well-known effect in nonlinear oscillators and is usually modeled as a Duffing oscillator [61]. An important suggestion derived by this work is the possibility of a new type of qubit in which the active element consists of a nanowire loop [62] capacitively coupled to a CPW. In the design show in Figure 7.20, closely resembling the *transmon* qubit originally developed in Schoelpkopf's group at Yale [63, see, for example, Figure 2a], the flux threading a nanowire loop changes the critical current of the nanowires and thus the inductance of the loop.

A final experiment we want to touch upon is the study of current phase relations (CPRs) in the presence of phase slips from Berzryadin's group [64]. What is extremely interesting is that for the first time, two different materials, $Mo_{1-x}Ge_x$ and aluminum, are used for the nanowire fabrication using the same fabrication technique and measured by the same group using the same experimental apparatus. Since aluminum does not adhere properly to carbon (fluoro)nanotubes, a scaffold created by peeling off adhesive tap from a substrate with a trench was used. When the tape is peeled off, polymer strings remain attached to the substrate across the trenches. After isolating a single scaffold across a trench, $Mo_{1-x}Ge_x$ or aluminum can be deposited on the scaffold. The only drawback of this technique is that the width of the scaffold and consequently of the nanowire is about 20 nm. The group prepared 2 $Mo_{1-x}Ge_x$ wires and a single aluminum wire for this study and measured them both in linear regime and at high current in an applied microwave field.

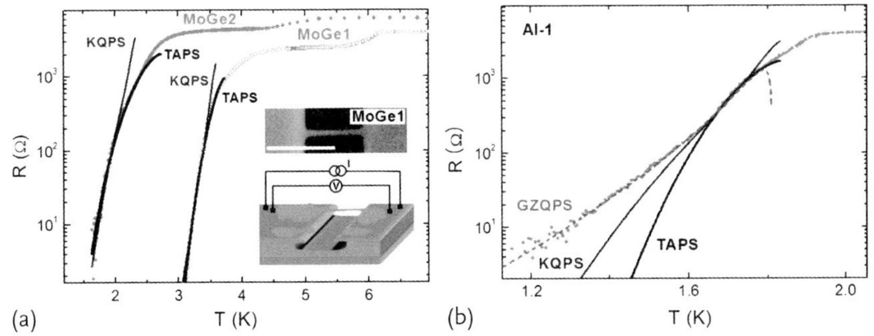

Figure 7.21 (a) $R(T)$ curves of two $Mo_{1-x}Ge_x$ nanowires (MoGe1, open circles, and MoGe2, closed circles). The solid black line on each $R(T)$ curve is the best fit produced by the Arrhenius–Little (AL) formula $R_{AL}(T)$ (Eq. (1) in [64]), which is due to TAPS, with the three fitting parameters listed in the text. The solid blue lines are obtained from the Khlebnikov QPS theory (KQPS), $R_{KQPS}(T)$ (References 3,13 in [64]) (upper inset): SEM image of wire MoGe1. The scale bar is 200 nm. (lower inset): Schematic for the transport measurement setup. The lower nanowire, drawn in Figure 1d, is removed. A focused ion beam is used to remove unwanted nanowires. (b) $R(T)$ curve of an Al nanowire (Al-1, scattered points). The dashed green curve is the prediction of the Golubev–Zaikin QPS theory ($R_{GZQPS}(T)$) (Reference 28 in [64] with five fitting parameters (listed in the text), and the solid black curve corresponds to the phenomenological AL fit ($R_{AL(T)}$) (Reference 23 in [64]). The solid blue line is the best fit predicted by KQPS theory. Adapted from [64].

Looking at the data in the linear regime (Figure 7.21), we can see how the two $Mo_{1-x}Ge_x$ nanowires fit with both the Khlebnikov quantum theory of phase slips (KQPS) for wires shorter than the thermal plasmon wavelength [65] (Section 3.5.3) and the Arrhenius–Little formula [11]: $R_{AL} = R_N \exp \Delta F/(k_B T)$ [7] with R_N the nanowire's normal resistance. When attempting to fit the aluminum nanowire, the Arrhenius–Little formula deviates significantly for temperatures smaller than ≈ 1.6 K and so does the KQPS model. The Golubev–Zaikin theory [19] instead is able to reproduce the transition using $a = 3$ and $B = 5.1$ (Section 3.2). The authors conclude that weak links, whose presence can deeply affect the rate of TAPS and therefore QPS, may be responsible for such a different behavior. This is once more a demonstration of the importance of sample fabrication which can complicate the interpretation of the data. However, the authors use other tools to further investigate the experimental results reproduced in Figure 7.21. In particular, they tried using microwave excitation to observe Shapiro steps [67] whose amplitude depends on the current–phase relation (CPR) of the nanowire. For the $Mo_{1-x}Ge_x$ nanowires as $L/\xi(T) > 3.6$ (long wire limit), the appropriate CPR is

$$I_S(\phi) = I_C \left[\frac{\phi\xi(T)}{L_S}\right] \left\{1 - \left[\frac{\phi\xi(T)}{L_S}\right]^2\right\} ,$$

7) The authors use this expression as, in the mean time, it was pointed out that the LAMH expression is not applicable for $T < 0.9T_c$ [66]. However, there is little difference with the LAMH as the major temperature dependence resides in the exponential part.

(References 11–13 in [64])[8] with ϕ as the phase and L_S as the nanowire length. While the data qualitatively agree with the theory, some details, such as the presence of fractional Shapiro steps, do not match. Instead, better agreement is obtained with a new $Mo_{1-x}Ge_x$ nanowire fabricated on a fluorocarbon-nanotube. It is noteworthy to notice that KQPS theory suggests that the CPR should be single-valued if QPS are present. As such, $Mo_{1-x}Ge_x$ nanowires in this regime do not show any evidence of QPS!

The appropriate CPR in the QPS limit based on KQPS is [64]

$$I_S = \left(\frac{I_C \phi}{\pi}\right) - \left(\frac{2\alpha^2}{4\pi^3 I_C}\right)\left\{\frac{\phi}{[1-(\phi/\pi)^2]^2}\right\},$$

with $\alpha \ll 1$ denoting the bare QPS fugacity ($\alpha \equiv f_{QPS}$) for the Al nanowire. Except for the Shapiro step corresponding to $n = 0$, the simulation agrees with the experimental data. However, as the length of the nanowire is comparable to the $\xi(T)$, its CPR could be multivalued and the data could be equally explainable with the multivalued CPR described above. The conclusions of this paper are therefore mixed: 1) the $Mo_{1-x}Ge_x$ nanowire which in linear regime follows both the TAPS and KQPS model, when studied under microwave irradiation, displays effects compatible with a multivalued CPR; QPS are therefore not present in the regime observed. 2) Aluminum nanowires, which in linear regime are not consistent with either the TAPS or KQPS model, but are consistent with the Golubev–Zaikin model, when studied under microwave irradiation, are compatible with both KQPS and multivalued (TAPS) CPR. As suggested by the authors, further studies are needed before drawing any firm conclusions. A later study by the same group focused on the $Mo_{1-x}Ge_x$ nanowire; not only the multivalued CPR has been confirmed, but fractional Shapiro steps have also been observed [68].

7.5 Other Effects in 1D Superconducting Nanowires

The template synthesis method has also provided new and interesting physics to study. The two main results achieved in experiments performed on samples fabricated with this technique are the observation of a peculiar *S-shape* in *I–V* measurement and the "antiproximity effect" (APE) in which superconductivity in a superconducting nanowire is weakened if a different superconductor is used as the electrical contact.

7.5.1 S-Shaped Current–Voltage Characteristic

Michotte *et al.* [69] first used the template synthesis method to study transport in lead nanowires of diameter $\approx$ 40 nm and length 22 μm, and later also examined

8) A similar expression in [68] has a numerical prefactor.

tin nanowires [70] (55 nm diameter and 50 μm long). The nanowire characteristics are examined both at low and high current. In the low excitation regime, the authors observe a clear superconducting transition at a temperature of about 7.2 K with a shape consistent with the appearance of TAPS. Upon decreasing the temperature, the measured resistance flattens out to a value which is almost temperature independent and is attributed either to the contribution of the normal contacts or the metallic side of a S-M-I (superconductor-metal-insulator) transition. Quite interestingly, in the high excitation regime, the authors uncovered a truly surprising effect.

The main results are shown in Figure 7.22. The curves in a current-biased regime exhibit finite jumps that the author interpreted as the emergence of phase-slip centers in the nanowires. However, when measuring in a voltage-biased regime, while the overall shape of the curves resembles the ones measured in a current-bias regime, the straight vertical jumps are replaced by a curved transition, almost an "S" shape (Figure 7.22a,b). While PSCs (phase-slip centers) had already been observed in microbridges [71], this peculiar "S" shaped behavior had not been observed. The authors interpret this "S" shape as a manifestation of two critical currents: the critical current necessary for the emergence of phase slips and the critical current for the destruction of the superconductivity. The interpretation of the I–V in the current-biased regime does not present any difficulty: when the current is increased, no voltage is measured, as the superconductor resistance is zero. However, every time the current exceeds the critical current for the emergence of a phase slip, a voltage is measured until the nanowire is eventually brought to the normal state.

In the voltage-biased regime, the situation is slightly different. For any given applied voltage, there will be a current carried by the Cooper pairs; this requires the Cooper pairs to move with a certain velocity. When the velocity is larger than the maximum local velocity sustainable by the condensate (the depairing current), possibly due to fluctuations, the pair will break (label 1 in Figure 7.22c) and the current will be carried by normal electrons (producing a phase slip) which will lower the voltage applied to the nanowire. A reduction in the voltage will also reduce the current (label 2 in Figure 7.22c); when the current is lower than the depairing current, the condensate can carry it and the cycle will repeat. If we continue to increase the voltage, the process will repeat (label 3 in Figure 7.22c), but the time between successive pair-breaking will be lower until we reach a point in which after the phase-slip event the current will be exactly equal to the depairing current (label 4 in Figure 7.22c). A further increase in the voltage, since the Cooper pairs cannot sustain the high current without breaking, will produce a more stable phase-slip center (label 5 in Figure 7.22c). Increasing the voltage even further will bring the system back to the point where the PSC is completely stable; the I–V in voltage mode will be similar to the one in current mode, until the next phase-slip center is formed. Figure 7.22d shows the dependence of the order parameter ψ in the center of the wire for voltage corresponding to 1 and 2 Figure 7.22c. A similar behavior has also been observed in aluminum nanowires [72]. The full theoretical description is provided in [70].

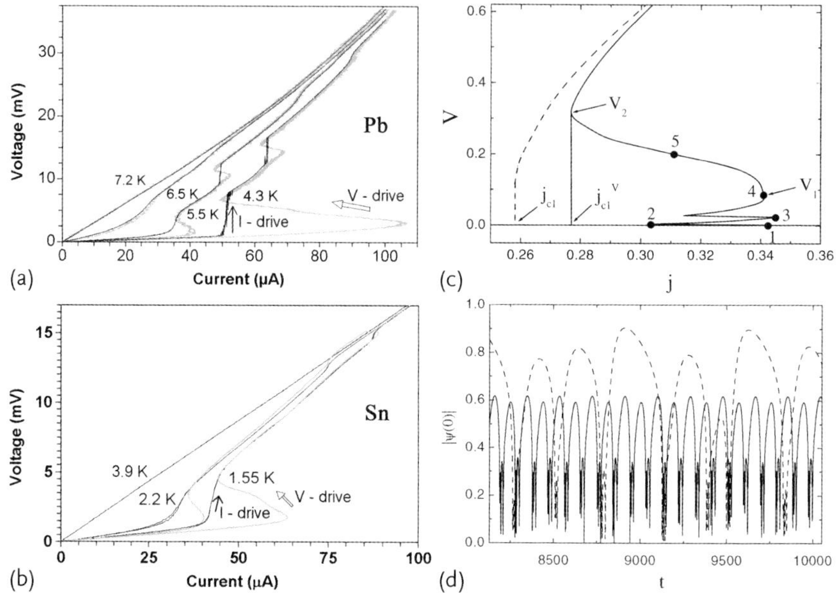

Figure 7.22 Current–voltage characteristics at different temperatures of (a) a Pb nanowire (diameter 40 nm, length 22 μm) and (b) a Sn nanowire (diameter 55 nm, length 50 μm). Results are shown for the current driven mode (black curves) and the voltage driven mode (gray curves). (c) The current–voltage characteristics of a superconducting wire of length $L = 40\xi$ and $\gamma = 10$ ($\lambda_Q \simeq 2.3\xi$). These parameters and associated with the time-dependent Ginzurg–Landau equations used in the analysis. Dotted (solid) curve for the I–V = const regime. (d) Dependence of the order parameter in the center of the wire (in the minimal point of $j\,|\psi(s)|$) as a function of time, where ψ is the superconducting order parameter. The solid (dotted) curve corresponds to the voltage in point 1 (2) [$V = 0.0024$ ($V = 0.0028$)] of Figure 7.3. From [70].

7.6
Antiproximity Effect

QPS, in the regime of low currents, have also been observed in Sn nanowire (w $\leq$ 40 nm) prepared with a similar technique by Moses Chan's group at Penn State University [36]. However a more intriguing result termed "anti-proximity" effect (APE) has also been reported. The proximity effect is normally observed by placing a normal metal in contact with a superconductor; in this case, a region of the normal metal in close proximity to the superconductor will retain some superconducting characteristics. Unexpectedly, when measuring 40 nm diameter superconducting zinc nanowires (ZnNW) contacted with superconducting bulk electrodes (Sn), the authors could not observe the transition between the normal and superconducting state until they employed a magnetic field to suppress the superconductivity of the electrodes (Figure 7.23). This effect was absent in ZnNWs of diameter $\approx$ 70 nm, pointing to the existence of a critical size below which this effect could be observed. A qualitatively similar effect was observed in ZnNWs with In (Figure 7.24) and Pb electrodes. When using In electrodes, the APE was switched off exactly when the critical field for the electrodes was reached. When using Pb, the effect was weaker and observed only in the value of the critical current of the nanowires; I_c would increase with increasing field and then decrease as soon as the field was larger than the critical field of the electrodes.

To rule out any explanation that tied this effect to the particular superconductor used in these experiments, Moses Chan's group also measured single crystal aluminum nanowires fabricated with the template synthesis method. In addition, to further exclude any effect due to the nanowires being embedded in the template, they confirmed this result on nanowires released from the membrane.

The authors start by studying single crystal aluminum nanowires with a diameter of 200 and 80 nm [73]. For the array with larger nanowires, applying a magnetic field along the axis of the NWs (perpendicular to the contact), the I_c decreases monotonically as one would expect by considering the effect of the magnetic field on the superconductor. When the same experiment was performed on the smaller diameter nanowires, the critical current increased by a small amount (1 μA out of 20 μA), but it was significantly larger than the experimental error. This result is consistent with observation on Zn in which the effect was only observed in smaller NW: while the diameter of the aluminum nanowire is different from the diameter of Zn nanowire, it is natural that the difference in superconductor and thus in coherence length might play a role.

In the second experiment, a single nanowire from each of the 80 nm and the 200 nm arrays was released. After transferring it on a Si substrate covered by SiN, normal (Pt) or superconducting electrodes (W with $T_c = 4.8$ K and $H_c = 8$ T) were fabricated on the NW, using a focused ion beam deposition and etching system. Examining the nanowire critical current, the authors noticed that the 80 nm showed

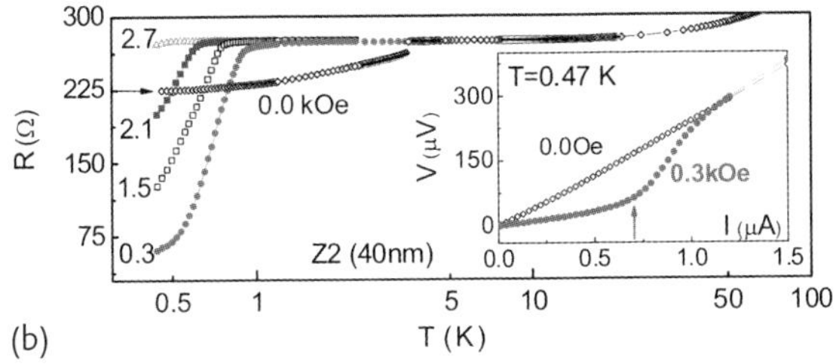

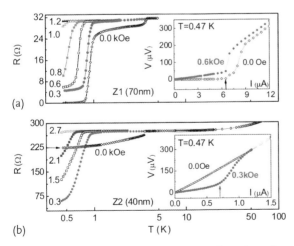

Figure 7.23 *R–T* curves of (a) Z1 and (b) Z2 with Sn electrodes under different magnetic fields ($H \perp$ ZNWs). The length of the ZNWs (zinc nanowires) is 6 μm. The insets show the *V–I* curves at $H = 0.0$ and 0.3 kOe, respectively. From [30].

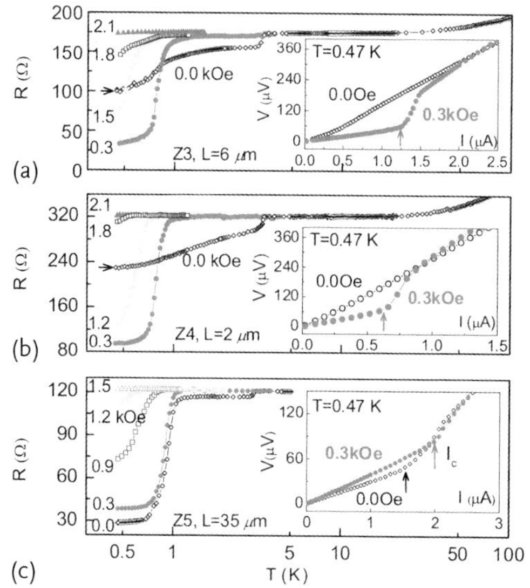

Figure 7.24 *R–T* curves of (a) Z3, (b) Z4, and (c) Z5 zinc nanowire with indium electrodes under different fields ($H \perp$ ZNWs). The diameter of the ZNWs is 40 nm. The insets show the *V–I* curves of the specific sample measured at 0.0 and 0.3 kOe, respectively. From [30].

a much larger critical current with normal electrodes than with superconducting electrodes (1.5 vs. 0.4 μA at 0.1 K). The 200 nm nanowire, instead, showed an I_c of 2 μA when measured with normal electrodes and $I_c \sim 13$ μA when measured with superconducting electrodes.

In order to shed further light on these results, another experiment was designed in which both superconducting and normal electrodes were used to contact the crystalline nanowire. Again, it was observed how the critical current of the nanowire segment between superconducting electrodes was smaller than in segments between the normal electrodes. One flaw of this experiment was that the overall critical current densities were smaller than in the case when all superconducting or all normal electrodes were connected to the nanowire array; this effect is attributed to the additional FIB processing used to fabricate the electrodes.

It is important to point out that in these latest experiments, the APE was observed without any magnetic field. A few theoretical models have been suggested to explain this effect; unfortunately, these models are not able to distinguish between different types of superconductors [74], or they do not take into account that APE can be observed without magnetic field [75]. This is definitely an area where theoretical effort is needed in order to fully understand this mechanism.

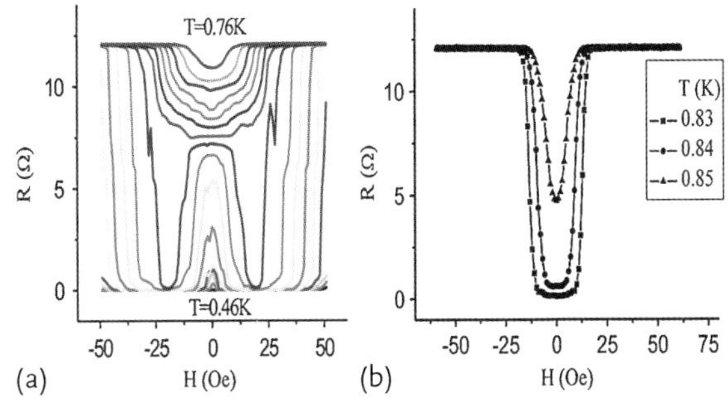

Figure 7.25 (a) Magnetic-field dependence of Zn wire resistance, at a high current $I \approx 4.4\,\mu A$, with temperatures ranging from 0.46 to 0.76 K, every 0.02 K. (b) Magnetic-field dependence of wire resistance, at a low current $I = 0.4\,\mu A$, with temperatures ranging from 0.83 to 0.85 K, every 0.01 K. From [31]. For a color version of this figure, please see the color plates at the beginning of the book.

7.6.1
Stabilization of Superconductivity by a Magnetic Field

In an attempt to reproduce the results from Moses Chan's group in wires fabricated with a different method and in a different geometry, Goldman's group at the University of Minnesota used an e-beam lithography and thermal evaporation to fabricate Zn nanowires. While this group has developed an ingenious technique to beat the limit of conventional e-beam lithography, we believe that the main reason they did not use their technique is because they were set on fabricating Zn nanowires. Zn tends to grow in grains and even on cooled substrate, the grain size would probably be too large to fabricate continuous sub-10 nm NW. Another difficulty in working with Zn is that it readily oxidizes and the oxide penetrates inside the metal. This is in contrast to material like titanium and aluminum in which the oxidation is mostly confined to the surface.

The group examined various lithographically defined Zn NW [31, 32] and the only "trick" employed was the use of a bilayer PMMA resist and quenched deposition; the substrate temperature was held at 77 K and the deposition rate was about 5 Å/s. The chosen geometry was such that both NW and electrical contacts would be deposited at the same time and made of the same material. Curiously enough, instead of confirming the previous results on ZnNW, they ended up finding new intriguing results.

Their main observation is reported in Figure 7.25: at high current (about 4 μA), in zero magnetic field the transition between the normal and the superconducting state is quite broad. Additionally, defining two characteristic currents, a current at which the $R/R_N > 0.99$ and a current at which $R/R_N < 0.01$, the authors observed that at low magnetic field, the first characteristic current would decrease (reentrance of superconductivity) and the second one would increase. This effect would decrease with increasing temperature. An important point to make is that

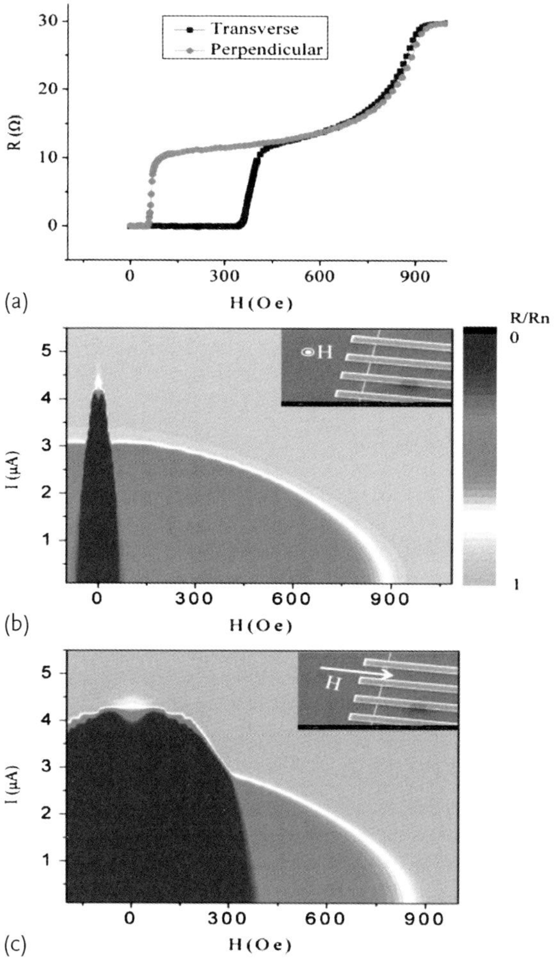

Figure 7.26 (a) Magnetoresistance of a 1.5-μm long wire (sample B) with different field orientations, at temperature 460 mK and with a current of 0.4 μA. (b) RCCM for this sample at 460 mK in a perpendicular field as indicated in the inset. (c) RCCM (resistance color contour map) for this sample at 460 mK in a parallel field transverse to the axis of the wire as indicated in the inset. From [32].

this effect is qualitatively different from the APE, as in that case, the wire would abruptly become superconducting.

By varying the orientation of the magnetic field in a direction perpendicular and parallel to the contact (but always perpendicular to the nanowire), the authors established that the response of the nanowire is dependent on effects taking place in the electrodes rather than being intrinsic to the nanowire. The nanowire cross-section shape does not change appreciably in the two-field orientation, as the nanowire is symmetric in its width and height directions, while the response is quite different in the two cases, as can be seen in Figure 7.26. While the current at which

$R/R_N > 0.01$ is different for different lengths of wire, the application of a small field makes them similar; this should be expected for coevaporated wires with similar cross section. Another useful observation is that this effect is weak in both short (1.5 μm) and long (10 μm) wires and is stronger for wires of intermediate length which is clearly indicative of the presence of two length scales.

Again, in this case, a theory that fully explains the mechanism is missing. However the authors suggest a qualitative scenario for their observation. Increasing the current produces an increase in the rate of phase slip which increases the transition width. This increase in transition width decreases the temperature at which the measured resistance is below the experimental resolution. By increasing the magnetic field, quasiparticles are produced in the electrodes; the quasiparticles dampen the phase slips by providing an additional mechanism for dissipation, and thus the wire recovers the zero resistance state. If the wire is short (on the order of ξ), the superconductivity is sustained by the electrodes, and phase slips are rare as Cooper pairs propagate coherently through the length of the wire without scattering (see also Chapters 3 and 4 on short superconducting nanowires connected to large superconducting leads). If the wire is longer than the quasiparticle relaxation length, the quasiparticles will decay while traveling through the length of the nanowire and will be converted in Cooper pairs; therefore, they cannot provide a dissipation mechanism that can stabilize superconductivity. For this reason, the effect is stronger in wires of intermediate length. It is noteworthy that the authors have observed a similar effect in aluminum nanowires. However, they have chosen to not report on these nanowires in detail because the aluminum films were deposited using an evaporator which had previously been used to deposit zinc.

7.6.2
Shape Resonance Effects

One interesting detail of nanowires fabricated in aluminum is the dependence of the critical temperature as a function of the wire dimension. It is, in fact, a well-known effect that in aluminum, the critical temperature increases as the cross section decreases. Using the Bogoliubov–de Gennes equations (Section 2.2.2, (2.31)) Peeters and collaborator showed [76] how this increase is due to "shape resonances": quantization of the electron motion in the direction transverse to the nanowires produce resonances which have the direct effect of increasing the critical temperature of the nanowires. Before discussing this work in detail, we want to focus our attention on a closely related work by Chen *et al.* [77] from Peeters' group. In this work, the authors examine the superconducting transition temperature in Pb nanofilms reported in the literature using, as a starting point, in this case as well, the BdG equations.

Lead films, grown on silicon Si(111) substrates, have shown superconducting properties down to a single monolayer of thickness [78]; additionally, oscillations in the superconducting properties due to the quantization in the direction perpendicular to the film were been predicted since 1963 by Blatt and Thompson [79]. The electron–phonon enhancement factor is related to the density of states at the

Fermi level and to the electron–phonon coupling **g**. In a 2D film with a thickness of a few monolayers, quantization in the vertical direction changes the density of states ($N(0)$) as the band diagram is parabolic in the x–y coordinate, but discrete in the z coordinate; increasing the thickness will change the discrete energy levels and for some particular thickness, the density of states will increase, enhancing superconductivity ($\Delta \approx 2\hbar\omega_c \exp(-1/N(0)VV)$). At the same time, electron–phonon coupling is also dependent on the thickness of the material as we cannot expect a few monolayer films to have the same coupling as a bulk superconductor. Using a simple step function approximation for **g** so that near the interface it assumes a different value with respect to the bulk $\mathbf{g}_{\text{if}} \neq \mathbf{g}_{\text{bulk}}$, and assuming that $\mathbf{g}_{\text{if}}$ oscillate with a period equal to half the Fermi wavelength (2.34), the authors are able to reproduce the experimental data as shown in Figure 7.27. It should be noted that this argument is tightly related to the one presented in Section 2.2.2 except that in this case, the solution of the BdG equation is far more simplified.

In 1D nanowires, quantum confinement is much stronger. In addition to this, many subbands are degenerate: this means that upon varying the lateral dimension

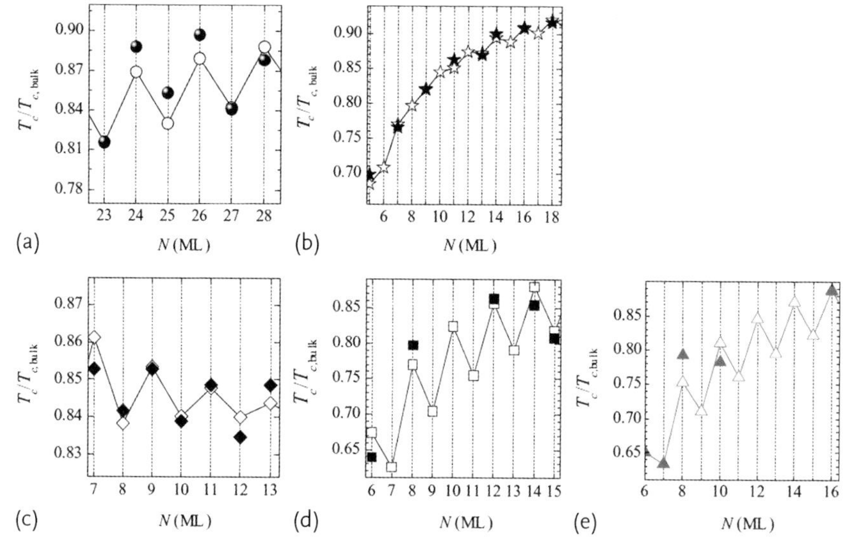

Figure 7.27 Theoretical results for the even-odd oscillations of T_c (open symbols) vs. the experimental data (solid symbols): (a) Theoretical data were calculated for $g_1(\pi) = 1.67g_0$ (odd numbers) and $g_1(2\pi) = 2.13g_0$, and the experimental set is from Reference 9 in [77]; (b) theoretical data were calculated for $g_1(\pi) = 0.64g_0$ (odd) and $g_1(2\pi) = 1.46g_0$ (even), and experimental results are from Reference 4 in [77]; (c) the same as in the previous panels, but for $g_1(\pi) = -0.12g_{\text{eff}}$ and $g_1(2\pi) = 0.84g_{\text{eff}}$ (where $g_{\text{eff}}N_{\text{bulk}}(0) = 0.36$), and the experimental set is from Reference 5 in [77]; (d) the same as in previous panels, but for $g_1(\pi) = 1.08g_0$ and $g_1(2\pi) = 1.54g_0$, and the experimental set is from Reference 9 in [77] (for the interface Si(111)(7 × 7) Pb); (e) the theoretical data were calculated for $g_1(\pi) = 1.05g_0$ and $g_1(2\pi) = 1.59g_0$, and the experimental set is from Reference 9 in [77] (for the interface Si(111)($\sqrt{3}\times\sqrt{3}$) Pb). Adapted from [77].

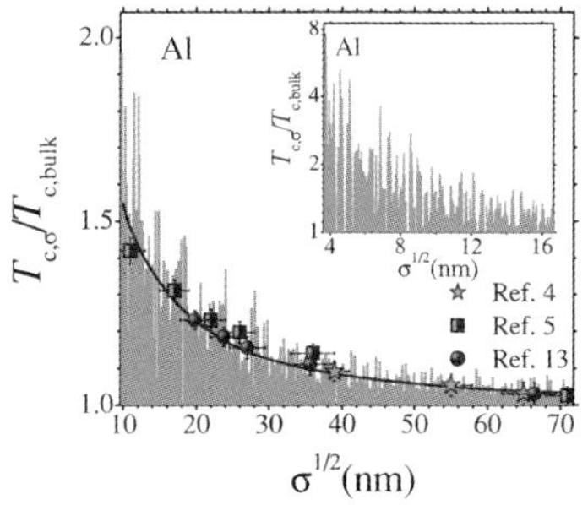

Figure 7.28 Critical temperature $T_{c,\sigma}/$, $T_{c,bulk}$ vs. the square root of the cross section of a cylindrical Al nanowire. Symbols are the experimental results from References 4, 5, and 13 in [76], the solid curve is a guide to the eye and indicates the general trend of the experimental results. Inset: the same dependence, but in the small-σ region. From [76].

on the nanowire, more bands can be occupied and thus $N(0)$ increases even more. Moreover subbands are not equidistant, as in thin films, so that resonance can enter into resonance before the previous one has completely decayed. Using the method described in Section 2.2.4, Shanenko *et al.* [76] calculated the variation in critical temperature as a function of the cross-sectional area for a nanowire with a mean free path comparable to the lateral cross section. The theoretical data are shown together with experimental data on aluminum nanowires obtained via the ion beam sputtering method in Figure 7.28.

From Figure 7.28, it is possible to see how both the theory and the experimental results point to an increase of T_c with decreasing cross-sectional area. However, the oscillation are smeared out by fluctuation in the cross section. In this respect, we can look back at the results of Li *et al.* [28]: the authors measures five nanowires of decreasing cross section and the thinnest device, s5, has a critical temperature *smaller* than the others. This indicates, possibly, that the nanowire is fully in the ballistic regime or that the fluctuations in the cross-sectional area are not sufficient to wash out the resonances.

References

1 Hongisto, T.T. and Zorin, A.B. (2012) Single-charge transistor based on the charge-phase duality of a superconducting nanowire circuit. *Physical Review Letters*, **108** (9), 097001.

2 Arutyunov, K.Y., Hongisto, T.T., Lehtinen, J.S., Leino, L.I., and Vasiliev, A.L. (2012) Quantum phase slip phenomenon in ultra-narrow superconducting nanorings. *Science Report*, **2**, 293.

3 Astafiev, O.V., Ioffe, L.B., Kafanov, S., Pashkin, Y.A., Arutyunov, K.Y., Shahar, D., Cohen, O., and Tsai, J.S. (2012) Coherent quantum phase slip. *Nature*, **484**, 355–358.

4 Sahu, M., Bae, M.H., Rogachev, A., Pekker, D., Wei, T.C., Shah, N., Goldbart, P.M., and Bezryadin, A. (2009) Individual topological tunnelling events of a quantum field probed through their macroscopic consequences. *Nature Physics*, **5**, 503–508.

5 Martinis, J.M. (1985) Macroscopic quantum tunneling and energy-level quantization in the zero voltage state of the current-biased Josephson junctions, PhD Thesis, University of California at Berkeley.

6 Lukashenko, A. and Ustinov, A.V. (2008) Improved powder filters for qubit measurements. *Review of Scientific Instruments*, **79** (1), 014701.

7 Li, P. (2010) Fluctuation effects in one-dimensional superconducting nanowires, PhD Thesis, Duke University.

8 Bladh, K., Gunnarsson, D., Hürfeld, E., Devi, S., Kristoffersson, C., Smålander, B., Pehrson, S., Claeson, T., Delsing, P., and Taslakov, M. (2003) Comparison of cryogenic filters for use in single electronics experiments. *Review of Scientific Instruments*, **74**, 1323–1327.

9 Johnson, A.C. (2005) Charge sensing and spin dynamics in GaAs quantum dots, PhD Thesis, Harvard University.

10 Kim, H., Jamali, S., and Rogachev, A. (2012) Superconductor–insulator transition in long MoGe nanowires. *Physical Review Letters*, **109** (2), 027002, supplementary information.

11 Little, W.A. (1967) Decay of persistent currents in small superconductors. *Physical Review*, **156**, 396–403.

12 Langer, J.S. and Ambegaokar, V. (1967) Intrinsic resistive transition in narrow superconducting channels. *Physical Review*, **164**, 498–510.

13 McCumber, D.E. and Halperin, B.I. (1970) Time scale of intrinsic resistive fluctuations in thin superconducting wires. *Physical Review B*, **1**, 1054–1070.

14 Lau, C.N., Markovic, N., Bockrath, M., Bezryadin, A., and Tinkham, M. (2001) Quantum phase slips in superconducting nanowires. *Physical Review Letters*, **87** (21), 217003.

15 Martinis, J.M., Devoret, M.H., and Clarke, J. (1985) Energy-level quantization in the zero-voltage state of a current-biased Josephson junction. *Physical Review Letters*, **55**, 1543–1546.

16 Devoret, M.H., Martinis, J.M., and Clarke, J. (1985) Measurements of macroscopic quantum tunneling out of the zero-voltage state of a current-biased Josephson junction. *Physical Review Letters*, **55**, 1908–1911.

17 Giordano, N. (1988) Evidence for macroscopic quantum tunneling in one-dimensional superconductors. *Physical Review Letters*, **61**, 2137–2140.

18 Giordano, N. (1990) Dissipation in a one-dimensional superconductor: Evidence for macroscopic quantum tunneling. *Physical Review B*, **41**, 6350–6365.

19 Golubev, D.S. and Zaikin, A.D. (2001) Quantum tunneling of the order parameter in superconducting nanowires. *Physical Review B*, **64** (1), 014504.

20 van Otterlo, A., Golubev, D.S., Zaikin, A.D., and Blatter, G. (1999) Dynamics and effective actions of BCS superconductors. *European Physical Journal B*, **10**, 131–143.

21 Khlebnikov, S. and Pryadko, L.P. (2005) Quantum phase slips in the presence of finite-range disorder. *Physical Review Letters*, **95** (10), 107007.

22 Rogachev, A., Bollinger, A.T., and Bezryadin, A. (2005) Influence of high magnetic fields on the superconducting transition of one-dimensional Nb and MoGe nanowires. *Physical Review Letters*, **94** (1), 017004.

23 Likharev, K.K. (1979) Superconducting weak links. *Reviews of Modern Physics*, **51**, 101–161.

24 Altomare, F., Chang, A.M., Melloch, M.R., Hong, Y., and Tu, C.W. (2006) Evidence for macroscopic quantum tunneling of phase slips in long one-dimensional superconducting Al wires. *Physical Review Letters*, **97** (1), 017001.

25 Lukens, J.E., Warburton, R.J., and Webb, W.W. (1970) Onset of quantized thermal fluctuations in "One-Dimensional" superconductors. *Physical Review Letters*, **25**, 1180–1184.

26 Newbower, R.S., Beasley, M.R., and Tinkham, M. (1972) Fluctuation effects on the superconducting transition of tin whisker crystals. *Physical Review B*, **5**, 864–868.

27 Bezryadin, A., Lau, C.N., and Tinkham, M. (2000) Quantum suppression of superconductivity in ultrathin nanowires. *Nature*, **404** (6781), 971–974.

28 Li, P., Wu, P.M., Bomze, Y., Borzenets, I.V., Finkelstein, G., and Chang, A.M. (2011) Switching currents limited by single phase slips in one-dimensional superconducting Al nanowires. *Physical Review Letters*, **107** (13), 137004.

29 Aref, T., Levchenko, A., Vakaryuk, V., and Bezryadin, A. (2012) Quantitative analysis of quantum phase slips in superconducting $Mo_{76}Ge_{24}$ nanowires revealed by switching-current statistics. *Physical Review B*, **86**, 024507.

30 Tian, M., Kumar, N., Xu, S., Wang, J., Kurtz, J.S., and Chan, M.H.W. (2005) Suppression of superconductivity in zinc nanowires by bulk superconductors. *Physical Review Letters*, **95** (7), 076802.

31 Chen, Y., Snyder, S.D., and Goldman, A.M. (2009) Magnetic-field-induced superconducting state in Zn nanowires driven in the normal state by an electric current. *Physical Review Letters*, **103** (12), 127002.

32 Chen, Y., Lin, Y.H., Snyder, S.D., and Goldman, A.M. (2011) Stabilization of superconductivity by magnetic field in out-of-equilibrium nanowires. *Physical Review B*, **83** (5), 054505.

33 Melosh, N.A., Boukai, A., Diana, F., Gerardot, B., Badolato, A., Petroff, P.M., and Heath, J.R. (2003) Ultrahigh-density nanowire lattices and circuits. *Science*, **300**, 112–115.

34 Zgirski, M., Riikonen, K., Touboltsev, V., and Arutyunov, K. (2005) Size dependent breakdown of superconductivity in ultranarrow nanowires. *Nano Letters*, **5**, 1029–1033.

35 Altomare, F., Chang, A.M., Melloch, M.R., Hong, Y., and Tu, C.W. (2005) Ultranarrow AuPd and Al wires. *Applied Physics Letters*, **86** (17), 172501.

36 Tian, M., Wang, J., Kurtz, J.S., Liu, Y., Chan, M.H.W., Mayer, T.S., and Mallouk, T.E. (2005) Dissipation in quasi-one-dimensional superconducting single-crystal Sn nanowires. *Physical Review B*, **71** (10), 104521.

37 Altomare, F., Chang, A.M., Melloch, M.R., Hong, Y., and Tu, C.W. (2006) Evidence for macroscopic quantum tunneling of phase slips in long onedimensional superconducting Al wires. *Physical Review Letters*, **97** (1), 017001, supplementary information.

38 Bollinger, A.T., Dinsmore, III, R.C., Rogachev, A., and Bezryadin, A. (2008) Determination of the superconductor–insulator phase diagram for one-dimensional wires. *Physical Review Letters*, **101** (22), 227003.

39 Chen, Y. and Goldman, A.M. (2008) A simple approach to the formation of ultranarrow metal wires. *Journal of Applied Physics*, **103** (5), 054312.

40 Xu, K. and Heath, J.R. (2008) Controlled fabrication and electrical properties of long quasi-one-dimensional superconducting nanowire arrays. *Nano Letters*, **8**, 136–141.

41 Kim, H., Jamali, S., and Rogachev, A. (2012) Superconductor–insulator transition in long MoGe nanowires. *Physical Review Letters*, **109** (2), 027002.

42 Lin, J.J. and Giordano, N. (1987) Localization and electron–electron interaction effects in thin Au–Pd films and wires. *Physical Review B*, **35**, 545–556.

43 Duan, J.M. (1995) Quantum decay of one-dimensional supercurrent: Role of electromagnetic field. *Physical Review Letters*, **74**, 5128–5131.

44 Rogachev, A. and Bezryadin, A. (2003) Superconducting properties of polycrystalline Nb nanowires templated by carbon nanotubes. *Applied Physics Letters*, **83**, 512.

45 Bezryadin, A. (2008) Topical Review: Quantum suppression of superconductivity in nanowires. *Journal of Physics Condensed Matter*, **20** (4), 043 202.

46 Zgirski, M., Riikonen, K.P., Touboltsev, V., and Arutyunov, K.Y. (2008) Quantum fluctuations in ultranarrow superconducting aluminum nanowires. *Physical Review B*, **77** (5), 054508.

47 Zgirski, M. and Arutyunov, K.Y. (2007) Experimental limits of the observation of thermally activated phase-slip mechanism in superconducting nanowires. *Physical Review B*, **75** (17), 172509.

48 Rogachev, A., Wei, T.C., Pekker, D., Bollinger, A.T., Goldbart, P.M., and Bezryadin, A. (2006) Magnetic-field enhancement of superconductivity in ultranarrow wires. *Physical Review Letters*, **97** (13), 137001.

49 Arutyunov, K.Y. (2007) Negative magnetoresistance of ultra-narrow superconducting nanowires in the resistive state. *Physica C*, **468**, 272–275.

50 Lehtinen, J.S., Sajavaara, T., Arutyunov, K.Y., Presnjakov, M.Y., and Vasiliev, A.L. (2012) Evidence of quantum phase slip effect in titanium nanowires. *Physical Review B*, **85** (9), 094508.

51 Arutyunov, K.Y., Golubev, D.S., and Zaikin, A.D. (2008) Superconductivity in one dimension. *Physics Reports*, **464**, 1–70.

52 Refael, G., Demler, E., Oreg, Y., and Fisher, D.S. (2007) Superconductor-to-normal transitions in dissipative chains of mesoscopic grains and nanowires. *Phys. Rev. B*, **75** (1), 014522.

53 Vercruyssen, N., Verhagen, T.G.A., Flokstra, M.G., Pekola, J.P., and Klapwijk, T.M. (2012) Evanescent states and nonequilibrium in driven superconducting nanowires. *Physical Review B*, **85** (22), 224503.

54 Altomare, F., Chang, A.M., Melloch, M.R., Hong, Y., and Tu, C.W. (2005) Evidence for macroscopic quantum tunneling of phase slips in long one-dimensional superconducting al wires. *arXiv:cond-mat/0505772v1*.

55 Martinis, J.M., Devoret, M.H., and Clarke, J. (1987) Experimental tests for the quantum behavior of a macroscopic degree of freedom: The phase difference across a Josephson junction. *Physical Review B*, **35**, 4682–4698.

56 Tinkham, M., Free, J.U., Lau, C.N., and Markovic, N. (2003) Hysteretic *I–V* curves of superconducting nanowires. *Physical Review B*, **68** (13), 134515.

57 Fulton, T.A. and Dunkleberger, L.N. (1974) Lifetime of the zero-voltage state in Josephson tunnel junctions. *Physical Review B*, **9**, 4760–4768.

58 Shah, N., Pekker, D. and Goldbart P.M. (2008) Inherent stochasticity of superconductor–resistor switching behavior in nanowires. *Physical Review Letters*, **101**, 207001.

59 Aref, T. and Bezryadin, A. (2011) Precise in situ tuning of the critical current of a superconducting nanowire using high bias voltage pulses. *Nanotechnology*, **22**, 5302.

60 Ku, J., Manucharyan, V., and Bezryadin, A. (2010) Superconducting nanowires as nonlinear inductive elements for qubits. *Physical Review B*, **82** (13), 134518.

61 Nayfeh, A.H. and Mook, D.T. (1995) *Nonlinear Oscillations (Wiley Classics Library)*, Wiley-VCH.

62 Hopkins, D.S., Pekker, D., Goldbart, P.M., and Bezryadin, A. (2005) Quantum interference device made by DNA templating of superconducting nanowires. *Science*, **308**, 1762–1765.

63 Wallraff, A., Schuster, D.I., Blais, A., Frunzio, L., Huang, R.S., Majer, J., Kumar, S., Girvin, S.M., and Schoelkopf, R.J. (2004) Strong coupling of a single photon to a superconducting qubit using circuit quantum electrodynamics. *Nature*, **431**, 162–167.

64 Bae, M.H., Dinsmore, R.C., Aref, T., Brenner, M., and Bezryadin, A. (2009) Current-phase relationship, thermal and quantum phase slips in superconducting nanowires made on a Scaffold created using adhesive tape. *Nano Letters*, **9**, 1889–1896.

65 Khlebnikov, S. (2008) Quantum mechanics of superconducting nanowires. *Physical Review B*, **78** (1), 014512.

66 Meidan, D., Oreg, Y., and Refael, G. (2007) Sharp superconductor–insulator transition in short wires. *Physical Review Letters*, **98** (18), 187001.

67 Shapiro, S. (1963) Josephson currents in superconducting tunneling: The effect of microwaves and other observations. *Physical Review Letters*, **11**, 80–82.

68 Dinsmore, R.C., Bae, M.H., and Bezryadin, A. (2008) Fractional order Shapiro steps in superconducting nanowires. *Applied Physics Letters*, **93** (19), 192505.

69 Michotte, S., Piraux, L., Dubois, S., Pailloux, F., Stenuit, G., and Govaerts, J. (2002) Superconducting properties of lead nanowires arrays. *Physica C Superconductivity*, **377**, 267–276.

70 Vodolazov, D.Y., Peeters, F.M., Piraux, L., Mátéfi-Tempfli, S., and Michotte, S. (2003) Current–voltage characteristics of quasi-one-dimensional superconductors: An S-shaped curve in the constant voltage regime. *Physical Review Letters*, **91** (15), 157001.

71 Skocpol, W.J., Beasley, M.R., and Tinkham, M. (1974) Phase-slip centers and nonequilibrium processes in superconducting tin microbridges. *Journal of Low Temperature Physics*, **16**, 145–167.

72 Altomare, F. (2004) Superconducting ultra-narrow aluminum nanowires, PhD Thesis, Purdue University.

73 Singh, M., Wang, J., Tian, M., Zhang, Q., Pereira, A., Kumar, N., Mallouk, T.E., and Chan, M.H.W. (2009) Synthesis and superconductivity of electrochemically grown single-crystal aluminum nanowires. *Chemistry of Materials*, **21** (23), 5557–5559.

74 Fu, H.C., Seidel, A., Clarke, J., and Lee, D.H. (2006) Stabilizing superconductivity in nanowires by coupling to dissipative environments. *Physical Review Letters*, **96** (15), 157005.

75 Vodolazov, D.Y. (2007) Negative magnetoresistance and phase slip process in superconducting nanowires. *Physical Review B*, **75** (18), 184517.

76 Shanenko, A.A., Croitoru, M.D., Zgirski, M., Peeters, F.M., and Arutyunov, K. (2006) Size-dependent enhancement of superconductivity in Al and Sn nanowires: Shape-resonance effect. *Physical Review B*, **74** (5), 052502.

77 Chen, Y., Shanenko, A.A., and Peeters, F.M. (2012) Superconducting transition temperature of Pb nanofilms: Impact of thickness-dependent oscillations of the phonon-mediated electron–electron coupling. *Physical Review B*, **85** (22), 224517.

78 Zhang, T., Cheng, P., Li, W.J., Sun, Y.J., Wang, G., Zhu, X.G., He, K., Wang, L., Ma, X., Chen, X., Wang, Y., Liu, Y., Lin, H.Q., Jia, J.F., and Xue, Q.K. (2010) Superconductivity in one-atomic-layer metal films grown on Si(111). *Nature Physics*, **6**, 104–108.

79 Blatt, J. and Thompson, C. (1963) Shape resonances in superconducting thin films. *Physical Review Letters*, **10** (8), 332–334.

8
Coherent Quantum Phase Slips

8.1
Introduction

In the last few years, experiments aimed at demonstrating the existence of coherent quantum phase slips (cQPS) have flourished. In particular, in 2012, three experiments published within a few months of each other provided convincing evidence that the effects of cQPS are readily observable and the duality proposed by Mooij and Nazarov [1], and later by Khlebnikov [2, 3], has experimentally observable consequences. Those experiments are:

- A single-charge transistor based on the charge-phase duality of a superconducting nanowire circuit
- Quantum phase slip (QPS) phenomenon in ultranarrow superconducting nanorings
- Coherent quantum phase slip.

8.2
A Single-Charge Transistor Based on the Charge-Phase Duality of a Superconducting Nanowire Circuit

In 2006, Mooij and Nazarov suggested that "if coherent quantum phase slips exist, they are the exact dual to Josephson tunneling" [1]. In their paper, they also suggested a device to be used as a standard of current – dual to the voltage standard based on the Josephson junction (JJ). From this suggestion and the later proposal by Hriscu and Nazarov [4], Hongisto and Zorin [5] demonstrated a QPS transistor, dual to a DC SQUID.

From the circuit presented in Figure 8.1a, we can see that the transistor is composed of two QPS elements embedded in a high-dissipative environment ($R_1 \sim R_2 \sim 0.4\,\mathrm{M\Omega}$): the modulation is provided by a voltage gate symmetric with respect to the two QPS elements. Looking at Figure 8.2, we can see that the two halves of the circuit (left and right of the central capacitor) are each equivalent to a current biased JJ (Josephson junction). Remembering that the dual of a capacitor

One-Dimensional Superconductivity in Nanowires, First Edition. F. Altomare and A.M. Chang

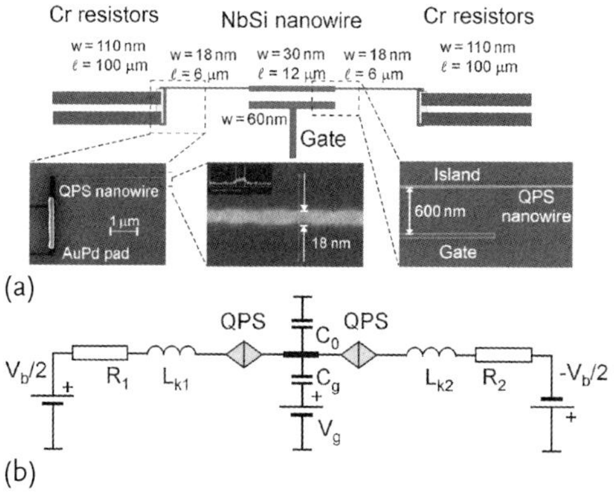

Figure 8.1 (a) The layout with three fragments of the SEM image and (b) the simplified electric circuit diagram of the QPS transistor embedded, in fact, in the four-terminal network of on-chip resistors. The device includes two QPS elements denoted by diamond symbols, kinetic inductances of the nanowire segments, and a capacitive gate. Thicknesses of Nb_xSi_{1-x} and Cr films are 10 and 30 nm, respectively. From [5].

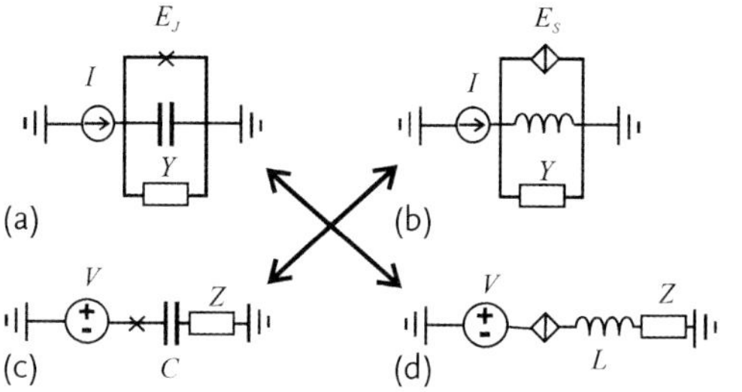

Figure 8.2 (a) Current-biased Josephson junction; (b) voltage biased Josephson junction; (c) current-biased QPS junction; (d) voltage-biased QPS junction. Circuit (a) is the exact dual of circuit (d). Circuit (b) is the exact dual of circuit (c). From [1].

is an inductance and the dual of the gate voltage will be a current flowing in the inductance – and hence a flux – the equivalence of this circuit with a DC SQUID is clear.

As the DC SQUID is sensitive to the phase across the junctions, it depends on the critical current and capacitance of the JJs, it depends on the shunting resistor across the JJs and it operates in a low impedance environment; the QPS transistor is sensitive to the charge through the QPS elements, it depends on the voltages drop and inductances of the QPS, it has series resistances and it operates in a high-impedance environment.

To make sure that the nanowires were in the 1D regime, the authors fabricated an 18 nm wide wire, smaller than the superconducting coherence length ($\sim$ 20 nm). If the circuit is asymmetric, as the fabrication of an 18 nm wide wire is nontrivial, that is, one of the two QPS junctions has an energy higher than the other ($E_{QPS1} \gg$

E_{QPS2}), the equation of motion describing this circuit becomes ((4.16) with $Q = -2eq$)

$$L_k \ddot{Q} + R\dot{Q} + V_m(Q_g) \sin\left(\frac{\pi Q}{e}\right) = V_b\,, \tag{8.1}$$

with $L_k = L_{k1} + L_{k2}$, $R = R_1 + R_2$, V_m the modulation of the blockade voltage, $Q_g = q_1 - q_2$, $Q = q_{1,2} \pm Q_g/2$, L_{ki} the kinetic inductance of each of the wire segments, R_i the resistors, q_i the charge of each QPS element, and V_b the driving voltage. Equation (8.1) describes the dynamics of a nonlinear oscillator with finite damping and is fully analogous to the equation describing the DC SQUID (see [7], for example).

Two devices (differing from each other for the length of the island and nanowire) were reported by Hongisto and Zorin [5]. Device fabrication consisted of the initial deposition of chromium resistors together with gold micropads that were used to make contact to the superconductor Nb_xSi_{1-x}. This is deposited in a subsequent step in a sputtering chamber in which the relative composition of the two targets had been calibrated. Finally, the nanowire elements and the island are defined with e-beam lithography using negative tone resist hydrogen silsesquioxane (HSQ). Once exposed with the electron beam during the lithography step, HSQ would harden and the unexposed region would be washed away by the developer. Dry etching is then used to transfer the pattern in the film, etching away the excess superconductor. The main experimental results are shown in Figure 8.3.

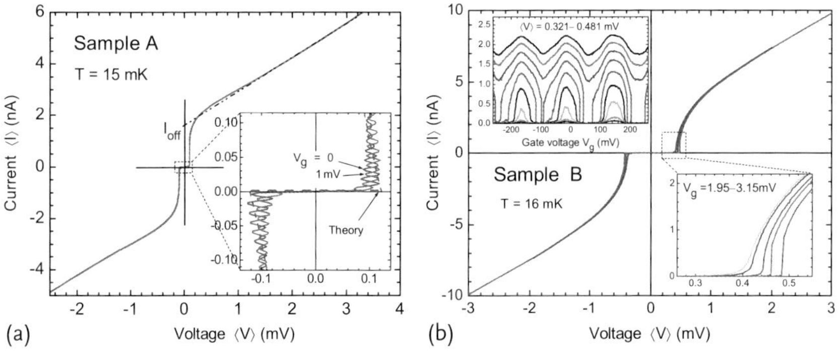

Figure 8.3 (a) The charge-modulated I–V curves of sample A recorded in a current-bias regime for two gate voltages shifted by a half-period. In the region of small currents (enlarged in the inset), one can see the modulation with period $\Delta I = 13.5$ pA, which is due to the asymmetry of off-chip biasing circuitry, resulting in the current dependence of the electric potential of the transistor island, $\delta V = (R_{bias1} - R_{bias2})I$ and, therefore, of the effective gate charge, $\delta Q_g = (C_g + C_0)\delta V$. The green dashed line shows the shape of the bare I–V curve given by the RSJ model (Equation 10 in [5]) with fixed Q_g. (b) The I–V curves of sample B measured in the voltage bias regime at different values of gate voltage V_g. The bottom right inset shows details of the Coulomb blockade corner. Upper left inset: the gate voltage dependence of the transistor current measured at different bias voltages V_b, providing a steady increase of $\langle V \rangle$ from 0.321 up to 0.481 mV in 20 μV steps (from bottom to top). Adapted from [5]. For a color version of this figure, please see the color plates at the beginning of the book.

As can be seen from Figure 8.3a,b, both devices qualitatively exhibit the same behavior. The I–V curves of both devices show the Coulomb blockade ($\langle I \rangle = 0$) until a certain critical voltage is reached. For larger voltage, the nanowires become normal. This is precisely the reverse of what takes place in a JJ, for which the voltage is zero until a critical current is reached. As seen in many superconducting nanowires, the intercept between the linear extrapolation of the I–V curve at high voltage with the current axis is positive, that is, a fraction of the electrons are still superconducting; this is a clear indication that the gating effect has its origin in the superconducting nanowire portion of the circuit (e.g., [8]). The insets in Figure 8.3b clearly show the effect due to the gating voltage. However, while the period of modulation for sample A is about 2 mV, somewhat consistent with the estimate of the capacitance, for sample B, it is 150 mV with a capacitance two orders of magnitude smaller than expected. While the authors cannot explain this discrepancy, this experiment definitely demonstrates the behavior expected from a QPS-based transistor and its sensitivity to the gate voltage.

8.3
Quantum Phase-Slip Phenomenon in Ultranarrow Superconducting Nanorings

As it is well-known, a superconducting ring can sustain a persistent current. Changing the flux through the ring will change the value of the persistent current until the critical current of the ring is reached. If one were to further increase the flux above this value, the total flux through the ring will increase by one or more flux quanta and its persistent current will drop. This simple picture is heavily modified by the presence of quantum phase slips as suggested by Matveev *et al.* [9]: "The superconductivity in very thin rings is suppressed by quantum phase slips. As a result, the amplitude of the persistent current oscillations with flux becomes exponentially small, and their shape changes from sawtooth to a sinusoidal one."

As shown in Figure 8.4, the effect of quantum phase slips is to open a gap where the energy levels cross so that the persistent current, for a system evolving in flux while following the ground state, becomes sinusoidal. This effect was derived for nanorings composed of a chain of weak links: however, for high transparency between the grains, this is equivalent to a uniform nanowire in a ring geometry [10]. Starting from this observation, Arutyunov *et al.* [11] fabricated nanorings of both aluminum and titanium. The nanoring is embedded between two highly resistive contacts in order to perform tunnel spectroscopy of the system as a function of the magnetic field. Starting from a nanoring composed of a large diameter nanowire (not to be confused with the diameter of the nanoring) and using the ion milling technique [12], the tunneling current through the nanoring is measured as a function of the magnetic field. The results for aluminum and titanium nanorings are shown in Figures 8.5 and 8.6.

In the case of an aluminum nanoring for a wide nanowire (Figure 8.5), the tunneling current exhibits a behavior similar to the one in Figure 8.4: the current increases until a current close to the critical current is reached at which point

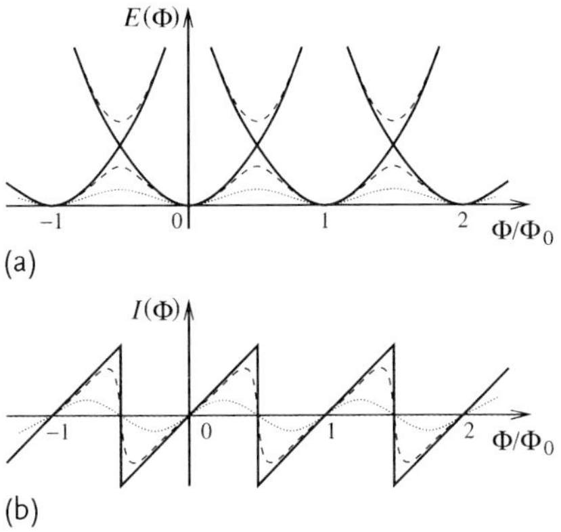

Figure 8.4 (a) The energy of a nanoring as a function of the flux Φ through it. In the limit of the relatively large nanowire cross section A, the energy shows the classical behavior shown by a solid line. Quantum phase slips result in level splitting, shown by a dashed line, and then lead to sinusoidal dependence of the ground state energy on the flux (dotted line). (b) The persistent current in the nanoring, $I(\Phi) = c dE/d\Phi$ (in cgs units). In the classical limit, the current shows sawtooth behavior, shown by a solid line. As the wire becomes thinner, the sawtooth is rounded and eventually transforms to a sinusoidal oscillation. From [9].

the current drops. Further increasing the flux through the nanoring, produces a similar trace. The increase in the baseline is attributed to a gap reduction due to the applied magnetic field. Reducing the cross-sectional area from $\sigma_{\text{fit}}^{1/2} = 91\,\text{nm}$ to $\sigma_{\text{fit}}^{1/2} = 43\,\text{nm}$ produces a reduction of the overall amplitude of the tunneling current. A further reduction in area completely changes the characteristics of the tunneling current and the shape of the persistent current. Even more impressive are the results for a titanium nanoring in Figure 8.6: even the $\sigma_{\text{fit}}^{1/2} = 67\,\text{nm}$, comparable to the superconducting coherence length in Ti [6] (see Table 6.1), show a behavior close to the QPS regime. A simple reduction of 10% in cross-sectional area, a device that can still easily be fabricated with conventional lithography, shows a changes in the amplitude of the oscillations by a factor of ≈ 6. The appearance of cQPS at such relatively large cross sectional area, is probably to be expected given the results on titanium nanowires reported by the same group (see [6] and Section 7.4.2).

8.4 Coherent Quantum Phase Slip

In 2005, Mooij and Harmans [13] suggested a (relatively) simple circuit, a superconducting loop interrupted by a nanowire, in which the effect of coherent quantum phase slips would change the number of quantized flux in a loop. Recently, Astafiev *et al.* [14] not only fabricated this circuit, but showed that its operation is consistent

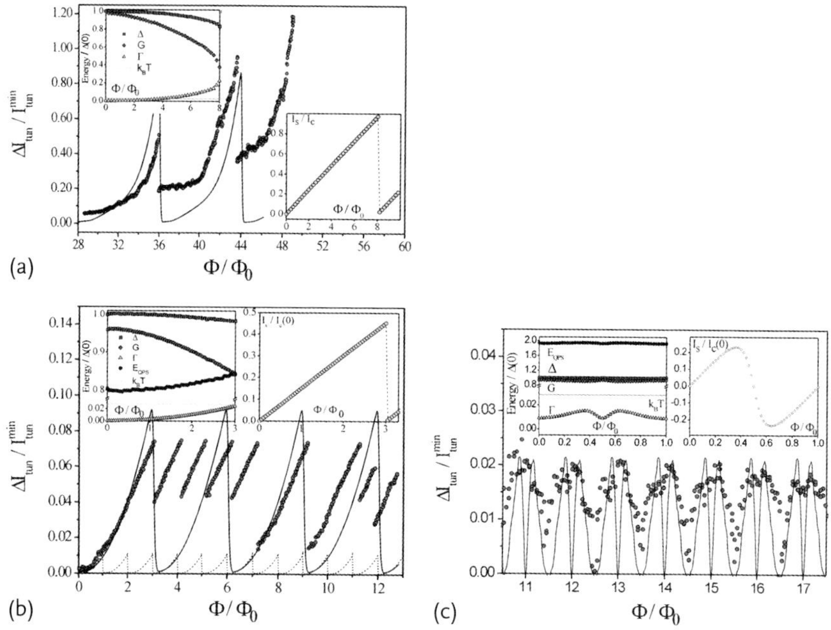

Figure 8.5 Oscillations of the normalized tunnel current $|I_{\text{tun}}(W) - I_{\text{tun}}^{\text{min}}|/I_{\text{tun}}^{\text{min}}$ in the external magnetic flux Φ/Φ_0 ($\Phi_0 = \hbar/e$ in SI units) Al-AlO$_x$-Al-AlO$_x$-Al structures with the same area of the aluminum loop $S = 19.6\,\mu\text{m}^2$. Experimental data are shown by circles (o), calculations – by lines. (a) Large period ($\Phi/\Phi_0 = 8$) and magnitude ($\Delta I_{\text{tun}}/I_{\text{tun}}^{\text{min}} \approx 0.8$) oscillations in the structure with loops formed by 110 nm × 75 nm wire, $V_{\text{bias}} = 780\,\mu\text{V}$, $T_{\text{bath}} = 65 \pm 5\,\text{mK}$, $T_e = 70\,\text{mK}$, $\sigma_{\text{fit}}^{1/2} = 90.8\,\text{nm}$. The monotonous increase of the base line is due to the gap reduction by the magnetic field noticeable at biases close to the gap edge $eV_{\text{bias}} \approx 2(\Delta_1/\Delta_2)$. (b) Oscillations with the variable period in the narrower (ion-milled) sample, $T_{\text{bath}} = 52 \pm 5\,\text{mK}$, $V_{\text{bias}} = 608\,\mu\text{V}$, $\sigma_{\text{fit}}^{1/2} = 42 \pm 30\,\text{nm}$. The solid line represents calculations at the intermediate limit with $T_e = 70\,\text{mK}$ and $\sigma_{\text{fit}}^{1/2} = 12.49\,\text{nm}$, resulting in the $\Delta\Phi/\Phi_0 = 3$ period, dashed line – calculated $\Delta\Phi/\Phi_0 = 1$ oscillations in a slightly narrower loop $\sigma_{\text{fit}}^{1/2} = 12.37\,\text{nm}$. (c) The same sample as in (b), but further gently ion-milled at $T_{\text{bath}} = 54 \pm 5\,\text{mK}$ and $V_{\text{bias}} = 666\,\mu\text{V}$. The solid line corresponds to calculations in the QPS limit with $\sigma_{\text{fit}}^{1/2} = 12.15\,\text{nm}$ and $T_e = 70\,\text{mK}$ with the same parameters used to fit $R(T)$ dependencies of Al nanowires. Left insets – flux dependencies of the characteristic energies: superconducting pairing potential Δ (□), spectral gap G (◇), depairing energy Γ (△), rate of quantum fluctuations E_{QPS} (·) and the corresponding thermal energy $k_B T$ (—). Right insets – calculated flux dependence of the persistent current $I_s(\Phi)$ normalized by the critical current $I_c(0)$. Note the change of the shape, reduction of the magnitude and the period of both the tunnel and the persistent current oscillations. In the classic (a) and intermediate (b) regimes, the periodicity of oscillations is defined by Equation 1 in [11], while in the essentially QPS regime (c) – by Equation 4 in [11]. For further details, see the text. Adapted from [11].

with the theoretical model. The circuit together with the theoretical dependence of the energy levels on the flux is shown in Figure 8.7.

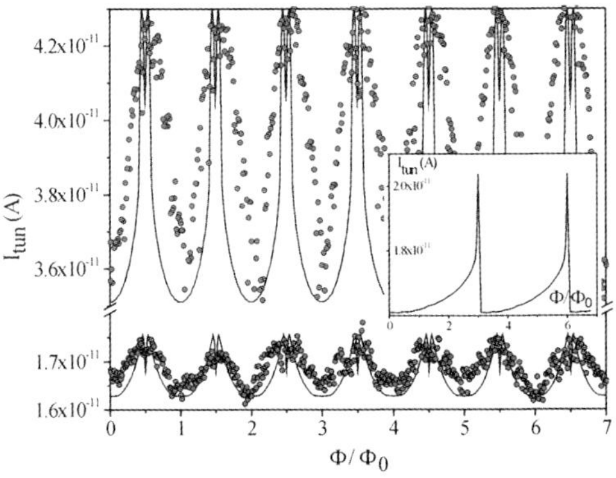

Figure 8.6 Tunnel current I_{tun} oscillations in external magnetic flux Φ/Φ_0 of Al-AlO_x-Ti-AlO_x-Al structure with the area of the titanium loop $S = 18.9\,\mu\mathrm{m}^2$. Top data – for non ion-milled sample $\sigma_{\mathrm{fit}}^{1/2} = 66 \pm 8$ nm, bottom data – for sputtered sample $\sigma_{\mathrm{fit}}^{1/2} = 62 \pm 8$ nm, $T_{\mathrm{bath}} = 65 \pm 5$ mK at the same bias $V_{\mathrm{bias}} = 105\,\mu$V. Solid lines represent QPS limit calculations, assuming $T_{\mathrm{e}} = 70$ mK, $\sigma_{\mathrm{fit}}^{1/2} = 63$ and 56 nm, respectively, and the same parameters used to fit $R(T)$ dependencies of Ti nanowires [6]. Note the drop of the oscillation amplitude by a factor of ≈ 6 when the diameter $\sigma_{\mathrm{fit}}^{1/2}$ is reduced by just $\approx 10\%$. The inset shows simulation: for the same fitting parameters of the ion-milled sample, which would be the tunnel current oscillations in the intermediate limit (similar to Figure 8.5b). For further details see the text and [11]. Adapted from [11].

The distance between the energy bands is designed to be in the GHz range; by tuning the external flux, it is possible to map the difference between the ground and the first excited state. Notice that the minimum difference in energy is equal to the amplitude of the quantum phase-slip process E_{QPS}; this provides an important consistency check with the value of the QPS provided by the theory. The Hamiltonian of the circuit is

$$H = -\frac{1}{2} E_{\mathrm{QPS}} \left(|N+1\rangle\langle N| + |N\rangle\langle N+1| \right) + E_n\, |N\rangle\langle N| \,, \tag{8.2}$$

where $|N\rangle$ is the flux state with $\Phi = N\Phi_0$ in the loop, $E_n = (\Phi_{\mathrm{ext}} - N\Phi_0)^2/2L_{\mathrm{k}}$, and L_{k} is the kinetic inductance of the loop. By exchanging E_{QPS} with E_{j} (the Josephson energy), considering N the number of Cooper pairs on a superconducting island $E_n = (V_{\mathrm{ext}} - NV_{\mathrm{c}})/eC_{\mathrm{island}}$, this equation becomes the Hamiltonian of a Cooper pair box which is the dual of this circuit. This naturally suggests that the same technique used to measure a Cooper box can (and was) used to measure the QPS qubit. One important observation is that coupling of the $|N\rangle$ and $|N+1\rangle$ states is only possible if $E_{\mathrm{QPS}} \neq 0$; this is therefore a direct proof of the existence and a direct measurement of the amplitude of the quantum phase-slips process. The main results of the paper are reported in Figure 8.8.

A small superconducting loop (of area $\mathrm{A} = 32\,\mu\mathrm{m}^2$), in which a nanowire of dimension $\mathrm{w} = 40$ nm, $\mathrm{h} = 35$ nm and length $L = 400$ nm long is galvanically connected to a transmission line with characteristic impedance $Z_1 = 1600\,\Omega$. This is galvanically coupled to a transmission line of impedance $Z_0 = 50\,\Omega$. Given

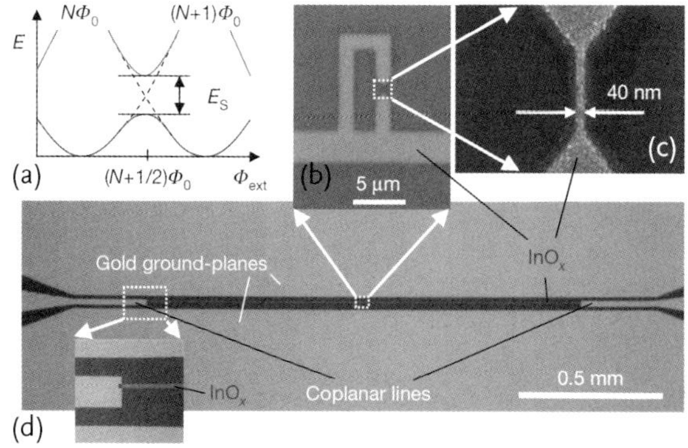

Figure 8.7 The device. (a) Energies of the loop versus external flux, Φ_{ext}. Blue and red lines: ground and excited energy levels, respectively. The degeneracy between states with N and $N + 1$ flux quanta (Φ_0), seen at $\Phi_{ext} = (N + 1/2)\Phi_0$, is lifted by the phase-slip energy, E_S. (b) InO_x loop with a narrow wire segment on the right side is attached to the resonator (horizontal line) at the bottom. (c) False-color scanning-electron micrograph of the narrow InO_x segment. (d) Step-impedance resonator comprising a 3-mm-wide InO_x strip with wave impedance $Z_1 \approx 1600\,\Omega$ galvanically coupled to a gold coplanar line with impedance $Z_0 = 50\,\Omega$. The boundaries of the resonator are defined by the strong impedance mismatch ($Z1 \gg Z_0$). From [14].

the large difference in impedance, a standing wave, with maximum current at the boundaries, is formed. As the loop is placed in the center of the step-impedance resonator, only even modes of the standing wave with frequency $f_m = m \times 2.4\,\text{GHz}$ will couple with the loop. The mode with $m = 4$ and $f_4 = 9.08\,\text{GHz}$ is used to measure the transmission coefficient ($|t|$) as a function of the applied magnetic field in Figure 8.8b; the dips in $|t|$ correspond to absorption of the microwave by the QPS qubit. The observed flux period is consistent, within the error, with the area of the loop with a 40 nm nanowire. In fact, in each resonator, several loops, with varying loop areas and nanowire widths are embedded in the resonator, but no signal has been observed from a loop with a wider nanowire.

Using two-tone spectroscopy, with a tone at $f = f_4$ to monitor the state of the step-impedance resonator and a second tone to monitor the qubit state, Astafiev *et al.* were able to map the energy spectrum as a function of the applied field in Figure 8.8c. Here, they also report the theoretical results assuming an energy splitting $\Delta E/\hbar = 4.9\,\text{GHz}$. The agreement between theory ($\delta E = \sqrt{(2 I_p \delta \Phi)^2 + E_{QPS}^2}$ with $I_p = \Phi_0/2L_k$) and experiment is quite good. The single trace at $\Phi/\Phi_0 = 0.52$ is reported in Figure 8.8d, together with a Gaussian fit, suggesting low-frequency Gaussian noise as the mechanism of decoherence.

One aspect on which we have not focused our attention is the particular choice of superconductor. While the experiment reported in [14] only discussed results obtained in devices fabricated in InO_x ($\xi = 20$–$30\,\text{nm}$), the NEC group has fabricated and measured other devices with similar qubit geometry and different materials (O.V. Astafiev, personal communication). In particular, while devices fabricated in NbTi did manifest a qubit behavior, a device with titanium nanowire of size

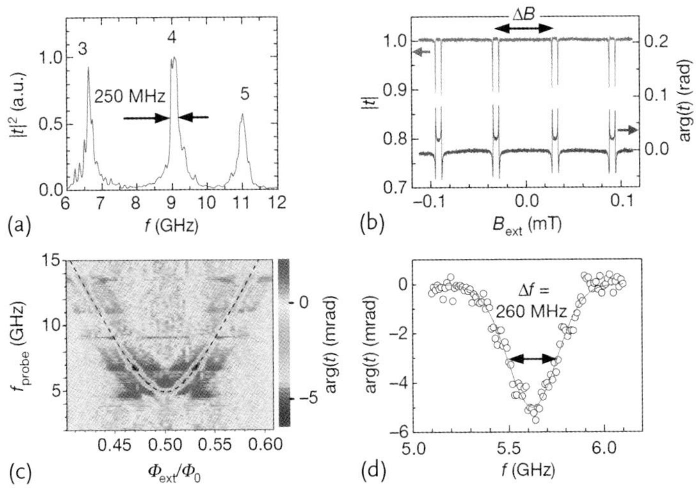

Figure 8.8 Experimental data. (a) Power transmission through the resonator measured within the bandwidth of our experimental setup. Peaks in transmission power coefficient, $|t|^2$, correspond to resonator modes, with mode number m indicated for each peak (a.u., namely, arbitrary units). (b) Transmission through the resonator as function a of external magnetic field B_{ext} at $m = 4$ ($f_4 = 9.08$ GHz). The periodic structure in amplitude ($|t|$) and phase ($\arg(t)$) corresponds to the points where the lowest-level energy gap $\Delta E/h$ matches f_4. The period $\Delta B = 0.061$ mT ($= \Phi_0/S$) indicates that the response comes from the loop (shown in Figure 1b) with the effective loop area $S = 32\ \mu m^2$. (c) The two-level spectroscopy line obtained in two-tone measurements. The phase of transmission, $\arg(t)$, through the resonator at f_4 is monitored, while another tone with frequency f_{probe} from an additional microwave generator, and B_{ext}, are independently swept. The plot is filtered to eliminate the contribution of other resonances ($2 < m < 6$), visible as horizontal red features. The dashed line is the fit to the energy splitting, with $\Delta E/h = 4.9$ GHz, $I_p = 24$ nA. (d) The resonant dip is measured at $\Phi/\Phi_0 = 0.52$. The red curve is the Gaussian fit. From [14]. For a color version of this figure, please see the color plates at the beginning of the book.

$t = 20$–30 nm, $w = 30$–60 nm, $l = 2$–$5\ \mu m$ (thickness t, width w and length l) and T_C varying between 0.4 and 0.7 K depending on the evaporation conditions, did not exhibit qubit behavior. A similar fate was suffered by devices fabricated in Arutyunov's group (coauthor of the study) in aluminum with the ion-beam milling technique and size $t = 10$ nm, $w \lesssim 10$–20 nm, $l = 2$–$5\ \mu m$, although, in this case, wire reliability might have been an issue. The superconductor used in the study is a crucial issue as emphasized by Astafiev *et al.* [14] and Bezriadyn [15]. InO_x is a very disordered superconductor and is not properly described by the BCS theory. In this material, which is very close to the superconductor-insulator transition, electrons form localized pairs even in the absence of superconductivity. Once the superconducting transition takes places, "lakes" of small condensate form, in stark contrast to the BCS superconductor in which we have only one "lake" encompassing the entire superconductor. This influences the QPS rate which for a moderately disordered ($k_F l_{mfp} > 1$) BCS conductor is ((3.37) and (3.46)) [16, 17]

$$\gamma_{QPS} \approx \frac{\Delta}{\hbar} \frac{R_Q}{R_\xi} \frac{L}{\xi} \exp\left(-\alpha \frac{R_Q}{R_\xi}\right), \tag{8.3}$$

with $R_Q = h/e^2$ and R_ξ, the resistance of a wire with length equal to the coherence length [18]. While moderate agreement can be reached between this expression and the experiment, a better agreement is obtained using the expression for disordered superconductors in the supplementary information of [14]. However, the exponential factor in (8.3) possibly provides a clue as to why neither aluminum or titanium in such a short nanowire manifested a qubit behavior. For these materials, this ratio is extremely large, thus suppressing the rate below what is measurable. In some sense, we are not properly designing the device and we are letting the material used in the device dictate the energy scale for QPS.

8.5
Conclusion

The final question that can be asked [15] is whether it is possible to create qubits out of more common superconductors such as Al, Ti, $Mo_{1-x}Ge_x$ or Nb, instead of a somewhat exotic superconductor such as InO_x or Nb_xSi_{1-x}, both close to the superconductor–insulator transition. This would be important for practical applications. While it can be argued that the presence of cQPS is necessary for the results of Arutyunov *et al.* in nanorings, the fact that QPS flux-qubits fabricated in Ti and Al have not, *so far*, shown a qubit behavior (O.V. Astafiev, personal communication) remains a mystery. As we have mentioned, the gap between the ground state and the first excited state is a direct measurement of E_s: however, if E_s in Ti were too small, the nanoring experiment would not have succeeded.

We would like to argue that this is a definite possibility, or at least we hope so. The key, in our opinion, lies in a more profound exploitation of the duality between Josephson junction devices and QPS devices. Several researchers have published diagrams of equivalent circuits between Josephson devices and QPS devices (see e.g., [4, 19]), however, in the works presented above, the scale of E_S is determined by the material properties. However, what we have presented here is just the beginning, akin to the initial development of superconducting qubits in early 2000 when the initial proof of the concepts were demonstrated. An entire class of devices dual to the more common Josephson-based devices is waiting to be exploited and the papers briefly presented in this chapter, together with the seminal work of Mooij and Nazarov, and Khlebnikov, will be regarded in a few years from now as the foundation for a new type of electronics based on quantum phase-slip devices.

References

1 Mooij, J.E. and Nazarov, Y.V. (2006) Superconducting nanowires as quantum phase-slip junctions. *Nature Physics*, **2**, 169–172.

2 Khlebnikov, S. (2008) Quantum mechanics of superconducting nanowires. *Physical Review B*, **78** (1), 014512.

3 Khlebnikov, S. (2008) Quantum phase slips in a confined geometry. *Physical Review B*, **77** (1), 014505.

4 Hriscu, A.M. and Nazarov, Y.V. (2011) Coulomb blockade due to quantum phase slips illustrated with devices. *Physical Review B*, **83** (17), 174511.

5 Hongisto, T.T. and Zorin, A.B. (2012) Single-charge transistor based on the charge-phase duality of a superconducting nanowire circuit. *Physical Review Letters*, **108** (9), 097001.

6 Lehtinen, J.S., Sajavaara, T., Arutyunov, K.Y., Presnjakov, M.Y., and Vasiliev, A.L. (2012) Evidence of quantum phase slip effect in titanium nanowires. *Physical Review B*, **85** (9), 094508.

7 Tinkham, M. (2004) *Introduction to Superconductivity*, 2nd edn, Dover Publications.

8 Rogachev, A. and Bezryadin, A. (2003) Superconducting properties of polycrystalline Nb nanowires templated by carbon nanotubes. *Applied Physics Letters*, **83**, 512, doi:10.1063/1.1592313.

9 Matveev, K.A., Larkin, A.I., and Glazman, L.I. (2002) Persistent current in superconducting nanorings. *Physical Review Letters*, **89** (9), 096802.

10 Arutyunov, K.Y., Golubev, D.S., and Zaikin, A.D. (2008) Superconductivity in one dimension. *Physics Reports*, **464**, 1–70.

11 Arutyunov, K.Y., Hongisto, T.T., Lehtinen, J.S., Leino, L.I., and Vasiliev, A.L. (2012) Quantum phase slip phenomenon in ultra-narrow superconducting nanorings. *Science Reports*, **2**, 293.

12 Zgirski, M., Riikonen, K., Touboltsev, V., and Arutyunov, K. (2005) Size dependent breakdown of superconductivity in ultranarrow nanowires. *Nano Letters*, **5**, 1029–1033.

13 Mooij, J.E. and Harmans, C.J.P.M. (2005) Phase-slip flux qubits. *New Journal of Physics*, **7**, 219.

14 Astafiev, O.V., Ioffe, L.B., Kafanov, S., Pashkin, Y.A., Arutyunov, K.Y., Shahar, D., Cohen, O., and Tsai, J.S. (2012) Coherent quantum phase slip. *Nature*, **484**, 355–358.

15 Bezryadin, A. (2012) Quantum physics: Tunnelling across a nanowire. *Nature*, **484**, 324–325.

16 Zaikin, A.D., Golubev, D.S., van Otterlo, A., and Zimányi, G.T. (1997) Quantum phase slips and transport in ultrathin superconducting wires. *Physical Review Letters*, **78**, 1552–1555.

17 Golubev, D.S. and Zaikin, A.D. (2001) Quantum tunneling of the order parameter in superconducting nanowires. *Physical Review B*, **64** (1), 014504.

18 Lau, C.N., Markovic, N., Bockrath, M., Bezryadin, A., and Tinkham, M. (2001) Quantum phase slips in superconducting nanowires. *Physical Review Letters*, **87** (21), 217003.

19 Kerman, A.J. (2012) Flux-charge duality and quantum phase fluctuations in one-dimensional superconductors. *arXiv:1201.1859v4*.

9
1D Superconductivity in a Related System

9.1
Introduction

In this chapter, we will provide a brief overview of S-SmNW-S (superconductor–semiconductor nanowire–superconductor) devices (see Chapter 5) and superconducting (nanowire) single-photon detectors (SNSPD). Superconductor–nanotube–superconductor devices (Section 5.3) belong to the first class, the understanding of which requires the use of nonequilibrium Usadel equations (Section 5.2.1); besides the intrinsic importance of nanotubes from a technological perspective, this system has shown indication of a Majorana mode (Section 5.4), which represents a finding of substantial fundamental importance. SNSPDs are all superconducting devices constituted, in their simplest form, by a single nanowire wrapped to cover a large area. The linewidth of the nanowire is approaching the limit in which they can be considered 1D (10–30 nm) and as such, relevant to the topics treated in this book. In arrays of carbon nanotubes, coupled with each other through an insulating medium, the individual nanotubes can be single-walled and as small as 0.4 nm in diameter.

9.2
Carbon Nanotubes

Carbon nanotubes were originally discovered in 1991 by S. Iijima [1] and since then, the literature on the subject has flourished. Nanotubes are created by wrapping a single sheet of graphene on itself (as can be done with a single sheet of paper). As such, their band structure combines the band characteristics of graphene together with the quantization condition imposed on the electron momentum in the direction perpendicular to the nanotube. This is also dependent on the angle between the graphene unit cell and the angle of rotation of the wrapping which determines the chirality of the nanotube. All this variability directly translates into extremely varying behavior regarding their electrical properties, metallic or semiconducting. Not only can they be used as switches or transistors (see [2] for recent results), but they can also emit light and be used as optical and electrical interconnects; we de-

One-Dimensional Superconductivity in Nanowires, First Edition. F. Altomare and A.M. Chang

fer to a paper by P. Avouris [3] for a recent review of the possibilities offered by nanotubes.

In this section, we will review the main results pertaining to 1D superconductivity in nanotube systems. Either intrinsic or due to the proximity effect, superconductivity has been observed in:

- single nanotubes: a single rolled graphene sheet
- a rope of nanotubes: several concentric nanotubes
- nanotube bundles: nanotubes embedded in a polymer matrix.

As in the case of superconducting nanowires, a certain fragmentation exists in the field: different groups are working on different types of nanotubes which makes it difficult to have a comprehensive overview of the field. Additionally, nanotube fabrication is nontrivial. Typically, nanotube growth starts from a seed particle which acts as a catalyst in a flow of gas containing carbon and other elements to which both the nanotube growth and quality are extremely sensitive. And, as if these details were not enough, results are highly dependent on the type of nanotube under study – single wall, rope, embedded in a matrix –, on its type – metallic, semiconducting –, if it is suspended or not. In addition, some effects are only seen if the nanotube is gated. We will only present the main results, providing references to the original work for the interested reader.

9.2.1
Proximity Effects in SWNT

The first observation of proximity induced superconductivity in nanotube dates back to 1999 [4, 5]. Kasumov *et al.* [4] measured metallic single-wall nanotubes (SWNT) and crystalline ropes connected to superconducting contacts: rhenium/gold for the ropes and tantalum/gold for the SWNT. They found out that in a rope ($L = 1.7\,\mu m$, $d = 23\,nm$, 200 tubes) with normal resistance smaller than $6.5\,k\Omega$ and with a thermal length greater than the rope length[1], the resistance of the rope becomes negligible below 1 K (Figure 9.1a). For the SWNT ($L = 0.3\,\mu m$, $d = 1\,nm$), the zero resistance state is attained for temperatures smaller than 0.4 K. Increasing the magnetic field would lead to a reduction of T_C up to a field of 2 T, at which point there would be a small upturn in the measured resistance for a temperature smaller than 0.2 K (Figure 9.1b). While the behavior of I_c as a function of temperature was consistent with the Ambegaokar–Baratoff expression [6, 7], for the SWNT, the extracted superconducting gap was 40 times larger than in the contact. Whereas the overall behavior of the devices was somewhat consistent with the large critical field necessary to suppress superconductivity in the SWNT, the large gap and critical field are not consistent with the idea that the superconductivity in the nanotube is merely due to the proximity effect of the superconducting contacts. One suggestion

1) Superconductivity via the proximity effect penetrates in the nanotube for a distance of the order of the thermal length.

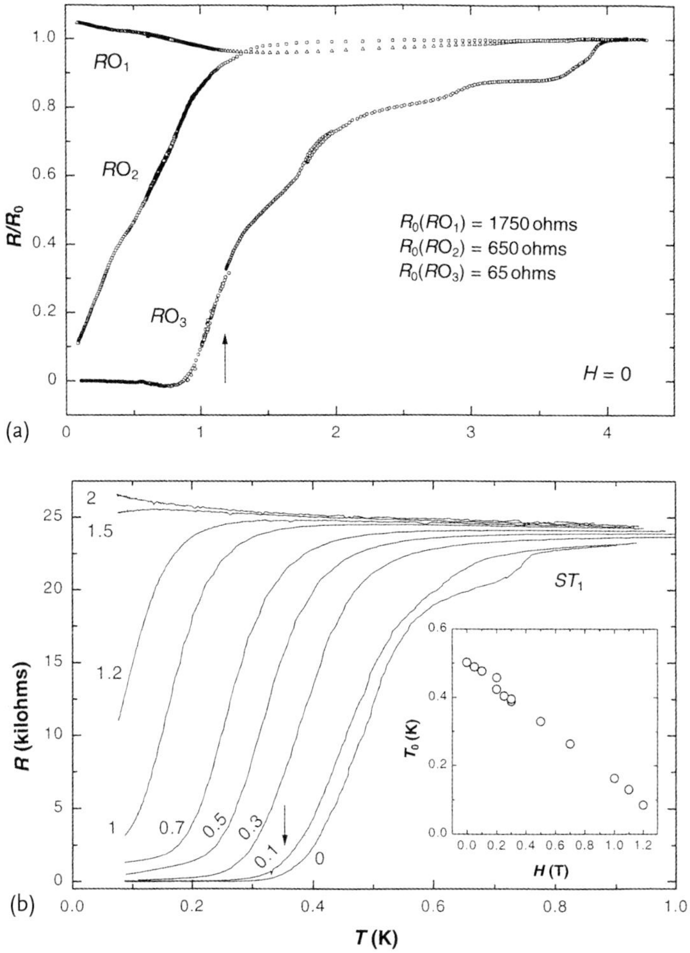

Figure 9.1 (a) Temperature dependence of the resistance of ropes RO_1, RO_2, and RO_3 mounted on Au/Re contacts measured for zero magnetic field. The arrow indicates the transition of the Au/Re bilayer. The sample RO_3 becomes superconducting below 1 K, but the existence of resistance steps above the transition indicates that superconductivity may not be homogeneous. (b) Temperature dependence of the resistance of the single tube ST_1 mounted on Au/Ta measured for different values of the magnetic field perpendicular to the tube axis. (The labels on the curves correspond to the value of magnetic field in Tesla.) The arrow indicates the transition temperature of the contact in zero magnetic field. (Inset) Field dependence of the transition temperature (defined as the inflexion point of $R(T)$). From [4].

proposed by the authors is that intrinsic superconductivity might be stabilized by the superconducting contacts. Another suggestion is that repulsive interaction in a Luttinger liquid[2)] can become attractive and thus lead to a superconducting transition (see also Section 2.3). The induced proximity effect can be exploited also for

2) Metallic carbon nanotubes behave as Luttinger liquid rather then Fermi liquid.

the fabrication of a sensitive nanoelectromechanical resonator. Reulet *et al.* [8] irradiated a suspended carbon nanotube rope with a radio frequency comparable to the mechanical vibration frequency of the nanotube and observed the acoustoelectric effect when measuring the DC resistance and the nanotube critical current (for the sample that becomes superconducting). The quality factor of the nanotube with superconducting contacts is larger than that achieved in nanotubes with normal contacts.

In the same year, Morpurgo *et al.* [5] studied gated metallic SWNT ($L \approx 0.3$ μm, $D < 1.8$ nm). The nanotubes are grown on a 0.5 μm silicon oxide layer separating the nanotubes from a doped silicon substrate which acts as a gate. Two important results are observed in this work. The first is the observation of a dip in the differential resistance which depends both on the contacts and the applied gate. This dip is indicative of a proximity effect mediated by Andreev reflection. The other interesting effect is the observation, at even lower temperature, of a peak superimposed to the Andreev features. In Figure 9.2a,b, we display the differential resistance of a sample with Nb contacts at two different gate voltages and the resistance of a different sample with normal contact (Figure 9.2c). The effect of the gate voltage between panel (a) and (b) is dramatic. In Figure 9.3, we display the effect of the temperature for temperatures close to the critical temperature of the contacts. Above T_c of the contacts, the shape of the dV/dI is quite similar to the differential resistance of the sample with normal contacts. The effect of the gate voltage is therefore to change the transparency of the contacts as the applied voltage would not be sufficient to appreciably change the population of the electronic subbands in the carbon nanotube[3]. High-transparency contact leads to efficient Andreev conversion, which therefore produces a dip in dV/dI. Low transparency, leading to a small probability of Andreev reflection, produces a peak in the differential resistance.

However, this picture, valid in the absence of electron–electron interaction, fails to explain the emergence of small peaks in the Andreev dip (Figure 9.4). Hence, the role of interaction in the nanotubes or possibly intrinsic superconductivity are to be examined in detail. In the case of moderate contact transparency, interplay between proximity induced superconductivity and the Kondo effect has been observed in SWNT [9].

9.2.2 Intrinsic Superconductivity in SWNT

This effect is quite intriguing and although reported in 2006 [10], it has, so far, not been reproduced by other groups. Quite startling, a zero bias anomaly (ZBA), commonly associated with Andreev reflection and, therefore, a signature of superconductivity [5], has been reported in single-wall nanotubes (SWNT). Let's look at the experiment in detail.

3) The resistance of nanotubes fabricated with this process is independent of the tube length up to ≈ 10 μm. The electrons move balistically along the nanotube and the resistance is dominated by the contacts.

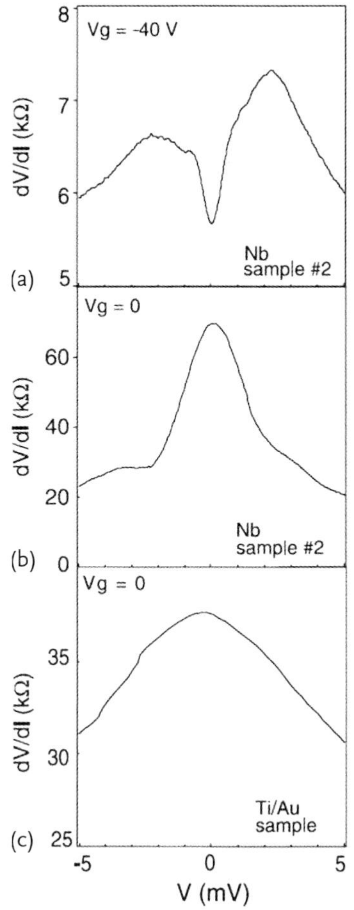

Figure 9.2 (a) and (b) show the differential resistance dV/dI vs. bias voltage V for a carbon nanotube sample of diameter $D < 1.8$ nm with Nb electrodes at (a) $V_g = -40$ V and (b) $V_g = 0$ V. (c) For comparison, the differential resistance of a sample with normal metal electrodes (Ti-Au) over the same range of bias voltage ($T = 4.2$ K). Note that (a) and (b), respectively, show a dip and a peak at bias voltages below 2–3 mV, not seen in (c). The expected range of bias voltages for these features is ≈ 2.9 mV, twice the gap of Nb (the factor of 2 accounting for two Nb-SWNT interfaces in series). Over this range of bias voltages, the relative size of the peak in dV/dI in the normal sample (c), one-tenth of the total resistance, is about 3% of the relative peak size for the Nb sample (b). From [5].

An isolated semiconducting SWNT, contacted by electrodes, is lying on a doped silicon substrate coated with SiO_2. Through the substrate, a gate can be applied to the SWNT. This is the knob the authors use to induce superconductivity in their devices. Zhang *et al.* fabricated three devices with nanotube diameters between 1 and 3.3 nm and lengths between 1 and 2 μm. The contacts are either superconducting (bilayer of 5 nm of Pd and 200 nm of Nb) with $T_c = 9.1$ K or normal (50 nm of Pd). While for the vast majority of the gate voltages (V_g) the differential resis-

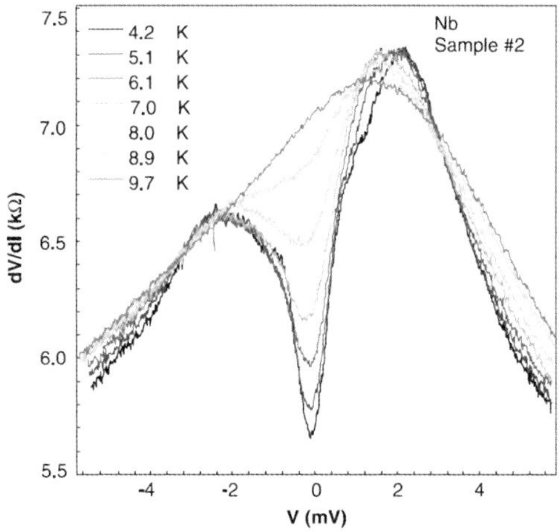

Figure 9.3 Differential resistance dV/dI vs. bias voltage V for a SWNT sample #2 of diameter $D < 1.8$ nm with Nb electrodes over a range of temperatures around T_c of Nb (9.2 K). The magnitude of the Andreev dip decreases with increasing T and disappears above T_c. From [5]. For a color version of this figure, please see the color plates at the beginning of the book.

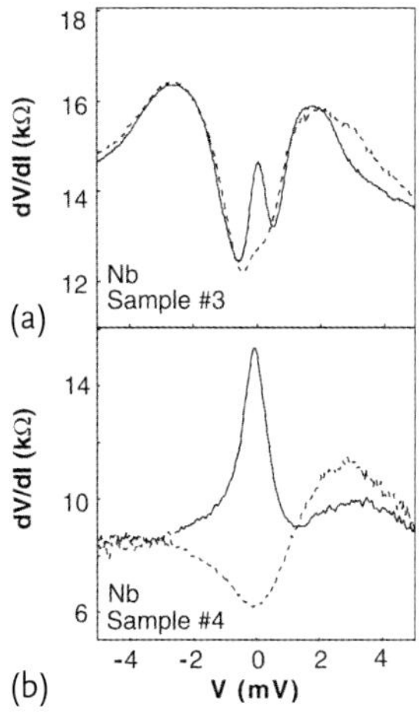

Figure 9.4 (a) Differential resistance dV/dI vs. bias voltage V for a SWNT sample #3 of diameter $D < 1.8$ nm with Nb electrodes measured at 4.2 K (dashed line) and 2 K (solid line), showing a narrow peak emerging from the center of the Andreev dip at lower temperatures ($V_g = -40$ V). (b) At a temperature of 40 mK, dV/dI may show either a peak or a dip, even at large negative gate voltage. This is different from what is found at 4.2 K, where dV/dI will always have a dip at sufficiently negative gate voltage. The low temperature traces differ by only a few volts (dashed curve, $V_g = -38$ V; solid curve, $V_g = -40$ V), but show opposite curvatures around zero bias. The overall behavior at temperatures well below 4 K is not understood. From [5].

tance (dV/dI) decreases from room temperature down to $\approx$ 30 K and then slowly increases down to the lowest temperatures measured, for some small range of V_g, dV/dI has a sharp downturn for $T \lesssim 12$ K. In correspondence with these sharp downturns, dV/dI has a dip as a function of the source-drain voltage. Notice that in absence of the downturn, in the same range of source-drain voltages, dV/dI has a peak. All three samples exhibit these characteristics (Figure 9.5).

One difference that exists between the sample with superconducting contacts and the other is that in the first case, the ZBA starts to appear around 30 K, while in the others, it appears below 15 K. The authors exclude several possibilities regard-

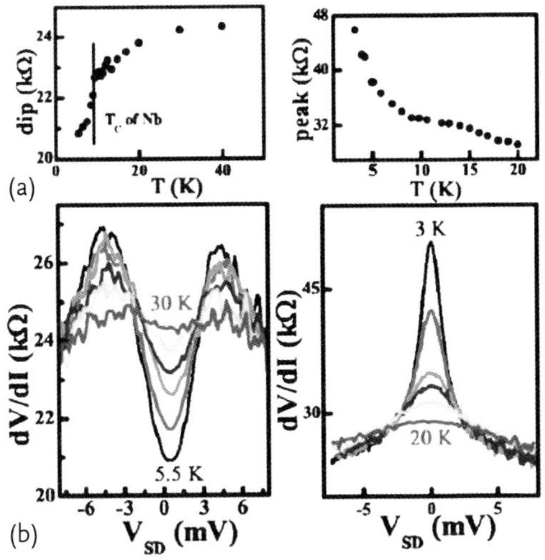

Figure 9.5 Differential resistance as a function of bias and temperature for CNFET (carbon nanotube field-effect-transistor) no. 1. $V_g = 47$ V for the dip on the left. $V_g = 48$ V for the peak on the right. The temperatures for the curves shown in (c) are 5.5, 8.5, 9.5, 15, 20, and 30 K. The temperatures for the curves shown in (d) 3, 4, 7, 9, 15, and 20 K. From [10]. For a color version of this figure, please see the color plates at the beginning of the book.

ing the appearance of the ZBA. Obviously, the ZBA cannot be attributed exclusively to the presence of superconducting contact. The authors are able to exclude resonant tunneling through imperfection in the SWNT (shape of differential resistance vs. T), Kondo effects (would not fit the shape of the ZBA) and Fano resonances (Coulomb blockade is not visible). The explanation suggested by the authors is that the SWNT themselves are superconducting within the small ranges of voltages where the ZBA is observed. This happens because the net effect of the gate voltage is to shift the Fermi energy of the nanotube. For sufficiently large values of V_g, E_F will be shifted to values of energy corresponding to the Van Hove singularities in the nanotube density of states (see e.g., [11]). Not only are they able to reasonably reproduce the dip of the differential resistance as a function of the temperature, for devices with and without superconducting contacts, but this hypothesis is supported by the spacing of ZBAs observed at different gating voltages; this spacing is consistent with the distance in energy between different Van Hove singularities, associated with the 1-dimensional subbands, as calculated for the physical parameters of the SWNT under study.

9.2.3
Superconductivity in Ropes Mediated by the Environment

One of the nanotube systems from which a lot of results have been obtained is the 4-Å (0.4 nm) nanotubes. Discovered in 1999, they are the smallest nanotubes and are grown inside the linear channel of $AlPO_4$-5 zeolite crystal (IUPAC code:AFI).

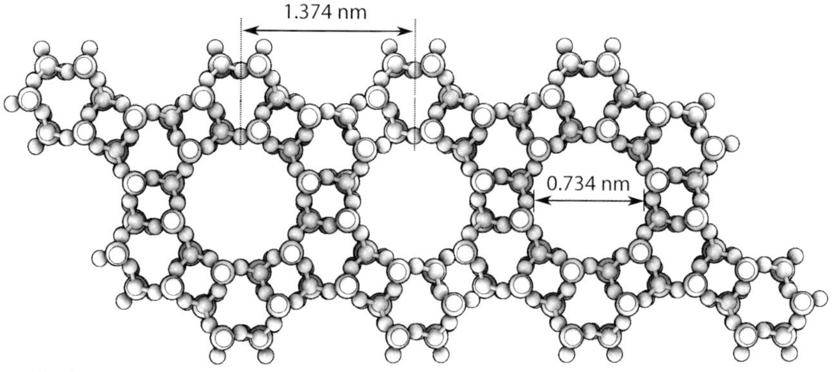

Figure 9.6 The framework structure of the AFI crystal viewed along the *c*-axis. The structure has been rendered with Qutemol [12].

The channel width is about 7 Å (0.7 nm) and the distance between channels is about 14 Å (1.4 nm) (Figure 9.6). The first number determines the radius of the nanotube, while the second determines the distance among nanotubes which turns out to be relevant for explaining observed superconductivity. Given the small diameter of the channel, only extremely small nanotubes can be grown inside. Still, there are three types of nanotubes which have such a small diameter: they have chirality (4,2), (5,0), and (3,3), and *ab initio* calculations show that the first one is semiconducting while the other two are metallic. It must be noted that during the growth process, all three chiralities are grown, but as it turns out, only one is relevant to the superconductivity. The first observation of superconductivity in this system dates back to 2001 [13] when the magnetic susceptibility of a pure AFI crystal and an AFI crystal with embedded nanotubes was measured. The main observation regarded the isotropy of the susceptibility signal: while the AFI pure crystals were isotropic, the SWNT@AFI was not. The results are shown in Figure 9.7 after subtracting the signal from the pure AFI, together with simulation based on phenomenological calculation using the Ginzburg–Landau formalism. An interesting aspect of this calculation is that it allowed one to identify which of the three chiralities is involved in the superconductivity; in fact, only the effective electron mass of the (5,0) is consistent with the value used in the calculation ($0.36\,m_e$ with m_e the bare electron mass).

This experimental work has also spurred a lot of interest in assessing the possibility of superconductivity in these nanotubes. However, all calculations have suggested that the critical temperature for a normal to superconductor transition in nanotubes would be either quite lower than the 10 K observed in [13] or Peierls instability would render superconductivity impossible. However, a theoretical investigation from Gonzales and Perfetto [14] (Section 2.3) uncovered the possibility of superconductivity in the *array* of nanotubes depending on the dielectric constant of

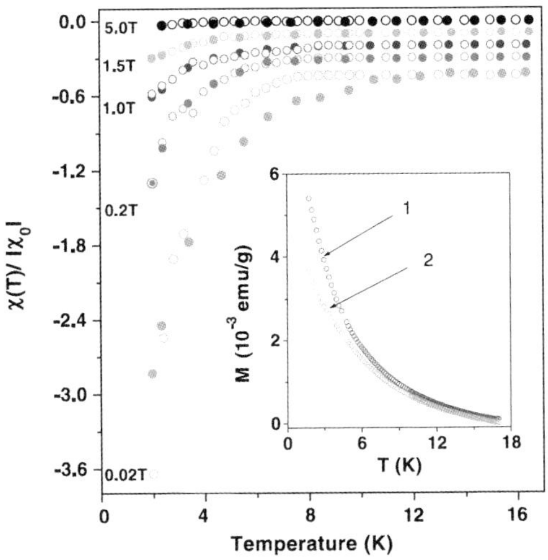

Figure 9.7 Normalized magnetic susceptibility of the SWNTs plotted as a function of temperature for five values of the magnetic field. The curves are displaced vertically for clarity. Values shown are for theory (open symbols) and experiment (filled symbols). χ_0 denotes the value of the susceptibility at $T = 1.6\,\text{K}$ and magnetic field $= 0.2\,\text{T}$. The experimental value of χ_0, when normalized to the volume of the SWNTs, is -0.015 (in units where $B = 0$ denotes $\chi_0 = -1$). The scatter in the theory points reflects statistical fluctuations inherent in the Monte Carlo calculations. (Inset) Temperature dependence of magnetization density for zeolite AFI crystallites (curve 1) and for AFI crystallites with SWNTs in their channels (curve 2). Both curves are measured at 2000 Oe. From [13]. For a color version of this figure, please see the color plates at the beginning of the book.

the environment. While this was not clearly recognized in the original experiment, it turns out that the interaction between nanotubes plays a significant role.

A key improvement to test the experimental results came from the marked improvement of the fabrication technique [15]; again, this is a case where improvements in the fabrication have brought about marked improvements in the experimental results.

The original fabrication consisted of two steps: growth of AFI crystals and then growth of nanotubes. In the original process, TPA (trypropylamince) molecules were embedded in the AFI matrix as precursor; after growing the AFI, a heat treatment at 580 °C decomposed the TPA from which the nanotubes can grow. However, the number of carbon atoms available for the formation of a carbon nanotube was not be sufficient to occupy the full channel. In addition to this, carbon atoms could also escape from the AFI channels, thus further decreasing the actual number of carbon atoms available to form the nanotube. Moreover to this, the nitrogen and hydrogen atoms from the TPA decomposition could form NH_3 which corrodes the AFI crystals. Thermal gravimetric analysis of these samples reveals a mass change of only 1.47%, corresponding to a filling factor of 4.5%. Again, we should point out that during the nanotube growth, all three types of nanotubes are grown and only the nanotubes with (5,0) chirality are superconducting. The new approach starts from the AFI crystals, but then removes the TPA by heating the AFI crystals in vacuum for 8 h. Successively, ethylene gas continuously flows as a source of carbon atoms for the formation of carbon nanotubes. With this approach, the group was able to produce a thermogravimetric change of 3.35%, corresponding to a filling factor of 10.3%. With some additional tweaks, the filling factor was raised up to 16.6%.

In the following, we will only discuss the electrical measurement performed on these nanotubes and refer to the review by Wang *et al.* [16] for additional details.

9.2.3.1 **Electrical Experiments**

Immediately after the formation of the AFI crystal, FIB is used to etch the crystal in order to define two electrodes with a gap of $\approx$ 100 nm. As the carbon nanotubes grown with this method are typically 100–200 nm long, this length serves to make sure a (array of) single continuous nanotube is connected.

After the nanotube growth, a bilayer of Ti/Au (Ti is used for adhesion) is evaporated on the AFI, FIB is used to form the electrodes by removing the film in a

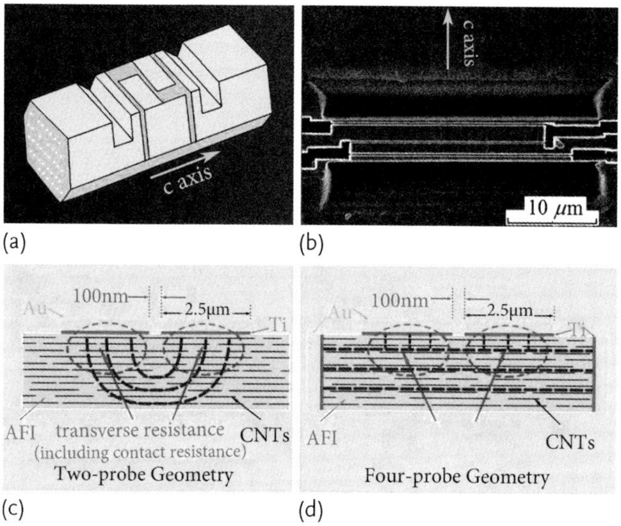

Figure 9.8 (a) Cartoon of the sample contain an array of SWNT in an AFI zeolite crystal. Yellow denotes gold and green denotes the AFI crystal surface exposed by FIB etching. Nanotubes are delineated schematically by open circles. (b) SEM image of one sample. The *c*-axis is along the N-S direction. The thin, light, horizontal line in the middle is the 100 nm separation between the two surface voltage electrodes that are on its two sides. The dark regions are the grooves cut by the FIB and sputtered with Au/Ti to serve as the end-contact current electrodes. (c) and (d) show schematic drawings of the two-probe and four-probe geometries, respectively. Blue-dashed lines represent the current paths. In (d), the two end-contact current pads are 4 mm in depth and 30 mm in width. The difference between the two-probe and the four-probe measurements is the transverse resistance, delineated by the red circles in (c). From [17]. For a color version of this figure, please see the color plates at the beginning of the book.

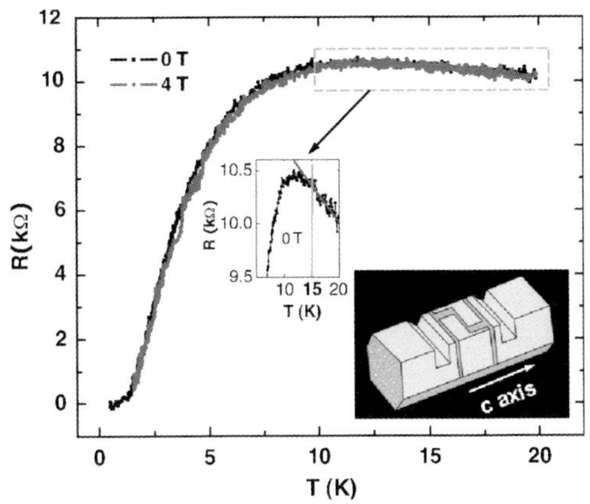

Figure 9.9 Resistance plotted as a function of temperature for the 4-terminal sample, measured at 10 nA. The geometry of the sample is illustrated in the cartoon inset, with yellow indicating gold coating and green the AFI crystal surface. The slice between the 2 grooves is 5 μm thick. The enlarged section shows the transition starts at 15 K. Up to 4 T, there is no appreciable change in the $R(T)$ behavior. From [18].

predetermined pattern and finally annealing at 450 °C is used to reduce the contact resistance. By appropriately modifying the electrode pattern, both two- and four-terminal electrical measurements can be performed (Figure 9.8). These measurements reveal a normal to superconducting transition that starts at 15 K; the measured resistance drops from 11 kΩ to approximately zero in a 10 K span, and additionally, is insensitive to a 4 T magnetic field (Figure 9.9). Both of these are indicative of the transition of an array of 1D superconductors. Additional information is provided by measurements of differential resistance: measurements up to 11 T performed at $T = 0.35$ K clearly demonstrate the presence of a supercurrent as a dip in the dV/dI. In addition, measurements in zero field at higher temperatures reveal a peak in the transition. The peak first appears at 2.7 K, a temperature at which, from the R vs. T data, the system should be in the 1D superconducting state. Recent theoretical work [19] has provided indication that for a thin array of (5,0) nanotubes, Peierls distortion/charge density waves may represent an excited state of the array and destabilize the superconductivity.

An additional sample has revealed measurements consistent with an overall critical temperature of 15 K and has, in addition, provided indication of other mechanisms. As already mentioned, wide transition in the R vs. T curve and independence on the magnetic field is a hallmark of a N/S transition in a 1D superconductor. In a different sample, while the measured T_C remained similar, a sharp transition dependent on the magnetic field was observed around 7.5 K. The observed behavior is found to be consistent with a Kosterlitz–Thouless–Berezinskii-type (KTB) transition: lowering the temperature, the growth of the superconducting condensate favors Josephson coupling in the ab plane (perpendicular to the nanotube growth). This is analogous to a 2D spin system in which vortex–antivortex excitation can be found; these vortices can become bound KTB pairs at sufficiently low temperature. The nanotube system of 1D superconductors becomes a system of 2D superconductors (in the ab plane) weakly linked in the direction of nanotube growth; the magnetic field would decrease the superconducting electron density in the quasi-1D elements which would shift the T_{KTB} transition at lower temperatures and diminish the amplitude of resistance drop associated with the transition. This interpretation is supported by both nonlinear $I-V$ characteristics and temperature dependence of the resistance. Additional details can be found in [13] together with the possible observation of a FFLO state (Fulde, Ferrel Larkin and Ovchnnikov); in this state, a type-II superconductor would sacrifice part of its volume to a normal state in order to increase its upper critical fields.

Superconductivity in this nanotube system is also supported by measurements of the magnetic Meissner effect and thermal specific heat. However, as both of these types of measurements measure the bulk effect, they are extremely difficult in the SWNT@AFI system. The signal of the AFI, in fact, typically dominates the overall signal, and while resolving the effect of the nanotube is quite challenging, the overall picture agrees with the electrical measurements discussed above.

9.3 Majorana Experiments

Another class of 1D devices in which the proximity effect plays an important role in revealing novel physical phenomena includes semiconducting nanowires with strong spin–orbit interaction, as discussed in Section 5.4. In fact, recently, it was suggested ([20–23] and references in Section 5.4) that semiconducting nanowires could be a convenient place to observe Majorana fermions, a special class of real solutions of the Dirac equations which have the extremely intriguing characteristic of the Dirac particles being identical to their antiparticles; in a semiconducting nanowire with strong spin–orbit coupling, in the presence of a normal and a superconducting electrode, it would be possible to observe Majorana fermions (see Section 5.4 for a theoretical overview). An experimental realization of this suggestion has recently been reported by Mourik *et al.* [24] in L.P. Kouwenhoven's group at Delft University, and in other groups [25–27]. Another suggestion for the observation of a Majorana mode comes from Fu and Kane [28], who have predicted its existence at the interface between a conventional superconductor and a superconducting topological surface state; this can be formed by proximitization of a topological insulator by the conventional superconductor. This experiment has recently been reported by Williams *et al.* in D. Goldhaber-Gordon's group at Stanford University [29].

9.3.1 Majorana Experiment in Semiconducting Nanowires

Before reviewing the Majorana experiment, it is useful to review the original experiment on semiconducting nanowires [30]. InAs nanowires are grown via a cat-

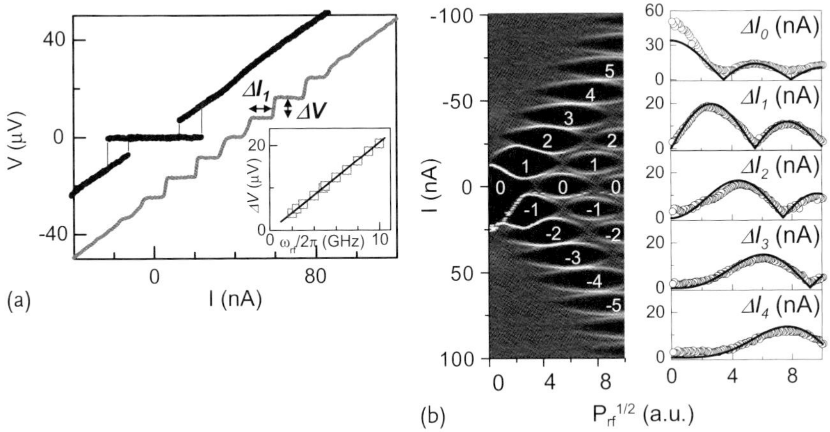

Figure 9.10 (a) $V(I)$ characteristics for device no. 3 at 40 mK, with (red) and without (black) an externally applied 4-GHz radiation (this device has $R_N = 860\,\Omega$ and $I_C = 26$ nA at $T = 40$ mK). The red trace is horizontally offset by 40 nA. The applied microwave radiation results in voltage plateaus (Shapiro steps) at integer multiples of $\Delta V = 8.3\,\mu$V. (Inset) Measured voltage spacing ΔV (symbol) as a function of microwave angular frequency ω_{RF}. The solid line (theory) shows the agreement with the AC Josephson relation $\Delta V = \hbar\omega_{RF}/(2e)$. From [30]. For a color version of this figure, please see the color plates at the beginning of the book.

alytic process: a thin layer of gold is deposited on the surface of a InAs or InP substrate. The substrate is heated to a temperature close to the growth temperature, at which point the gold film agglomerates in nanoparticles. The diameter of the nanowire is determined by the size of the gold nanoparticles and the semiconductor material is provided by pulsed laser ablation of an undoped InAs target. The single-crystalline nanowires thus formed are then transferred onto a heavily doped Si substrate covered with a SiO_2 oxide layer; standard e-beam lithography is used to define the electrode pattern which is then deposited by e-beam evaporation of a Ti/Al (10 nm/120 nm) bilayer. Before the metal deposition, a light etching in buffered hydrofluoric acid is used to improve the contact. While no thermal annealing is performed, the contact transparency is extremely high, with nanowire resistances ranging from 0.4 to 4.5 kΩ. As the nanowire acts as a weak link between the two superconducting electrodes, it is necessary, in fact, that the interface resistance is extremely small, and thus Cooper pairs from the electrodes can diffuse into the semiconductor. Additionally, the length of the nanowire determined by the distance between the electrodes must be comparable if not smaller than the phase coherence length for electron propagation in the nanowire. One of the main results of the paper is summarized in Figure 9.10.

Here, the black line shows the $V(I)$ curve at 40 mK for the nanowire. The hysteresis at low currents can clearly be seen. To demonstrate that the behavior of this nanowire is similar to the behavior of a Josephson junction, the authors apply microwave radiation to demonstrate the presence of Shapiro steps [31]. The phase locking between the applied microwave field and the Josephson frequency

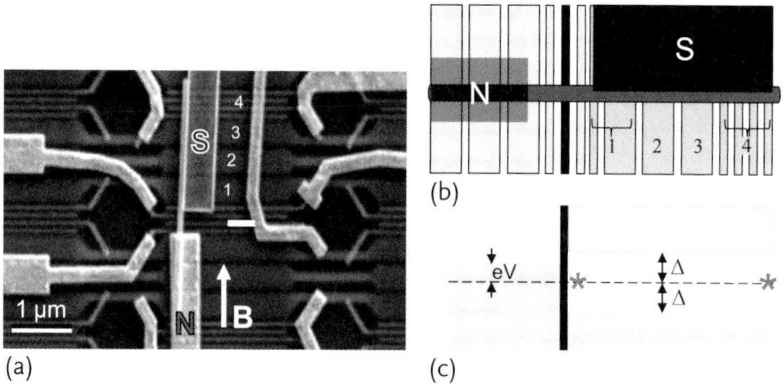

Figure 9.11 (a) Implemented version of theoretical proposals, using InSb SmNW and Nb superconductor. Scanning electron microscope image of the device with normal (N) and superconducting (S) contacts. The S contact only covers the right part of the InSb nanowire. The underlying gates, numbered 1 to 4, are covered with a dielectric. (Note that gate 1 connects two gates, and gate 4 connects four narrow gates; see (b).) (b) schematic of our device. (c) an illustration of energy states. The black rectangle indicates the tunnel barrier separating the normal part of the nanowire on the left from the wire section with induced superconducting gap, Δ. (In (a), the barrier gate is also shown in white.) An external voltage, V, applied between N and S drops across the tunnel barrier. (Red) stars, again, indicate the idealized locations of the Majorana pair. Only the left Majorana is probed in this experiment. Adapted from [24]. For a color version of this figure, please see the color plates at the beginning of the book.

produces voltage plateaus at $V_n = n\hbar\omega_{\mathrm{RF}}/(2e)$ with n as an integer, e as the electron charge, and $\hbar$ as the Planck constant. The inset in Figure 9.10 shows the agreement between the measured and predicted ΔV as a function of the microwave frequency.

This demonstration of the proximity effect in a semiconducting nanowire is a necessary condition for the study of Majorana fermions. Based on the theoretical proposal [20–23], Mourik *et al.* [24] devised an experiment to observe their existence: an InSb nanowire[4)] grown with a process similar to the one described above is contacted by a normal (bilayer of Ti/Au) and a superconducting (NbTiN) electrode. A set of gates is also deposited on a substrate in order to provide additional knobs to the experiment. Device and schematics are reproduced in Figure 9.11.

The main results of the paper are shown in Figure 9.12a,b. By tuning one of the fine gates between the normal and superconducting electrodes, the authors create a tunneling barrier in the nanowire, which allows the measurement of the nanowire differential conductance. The Majorana fermions, while having zero charge and energy, are observed in tunneling spectroscopy measurements as states at zero energy while changing magnetic fields and gate voltages over wide ranges. To this extent, it is important to notice that the superconducting electrode only partially covers the nanowire; this is intentional so that the effect of the underlying gates is not completely screened. The traces in Figure 9.12a show the evolution of the differential conductance as a function of the magnetic field applied along the nanowire axis. While at small fields, only the peaks corresponding to the quasiparticle den-

4) InSb has larger spin–orbit coupling than InAs.

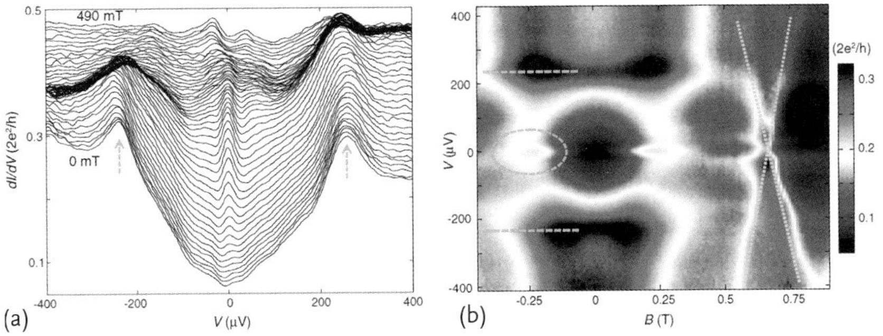

Figure 9.12 Magnetic field-dependent spectroscopy. (a) dI/dV vs. V at 70 mK taken at different B fields (from 0 to 490 mT in 10 mT steps; traces are offset for clarity, except for the lowest trace at $B = 0$). Data are from device 1. Arrows indicate the induced gap peaks. (b) Colorscale plot of dI/dV vs. V and B. The ZBP is highlighted by a dashed oval; green dashed lines indicate the gap edges. At ≈ 0.6 T, a non-Majorana state is crossing zero bias with a slope equal to $\approx$ 3 meV/T (indicated by sloped yellow dotted lines). Traces in (a) are extracted from (b). From [24]. For a color version of this figure, please see the color plates at the beginning of the book.

sity of states are visible (± 250 μeV), at fields between 100 and 400 mT a zero bias peak (ZBP) appears; this is more clearly observed in Figure 9.12b. This peak is identified with the left Majorana states of Figure 9.11c. Other features present in Figure 9.12b are used to verify the experimental conditions; in particular, the orange lines follow a pair of resonances with a slope of the order of $E_z = g\mu_B B/2$, where μ_B is the Bohr magneton and g is the Landé factor for the nanowire. The theoretical expectation is to observe Majorana fermions when $E_z > (\Delta^2 + \mu^2)^{\frac{1}{2}}$, where Δ is the superconducting gap in the nanowire and μ is the Fermi energy. For $\mu = 0$, and $\Delta \approx 250$ μeV, we obtain a magnetic field of about 0.15 T, consistent with the experimental observation.

Additional evidence that the ZBP is due to the presence of Majorana fermions is provided by extensive experimental checks in devices fabricated with leads which were both normal or both superconducting [24], and exploring the dependence on the magnetic field orientation. In the absence of any of the main ingredients, that is, one normal and one superconducting contact, and an external field aligned along the nanowire axis, the magnetic field-dependent spectroscopy displays different characteristics and, while a ZBP can also appear, its presence is not consistent with the existence of Majorana fermions [24]. While this work does not address the topological properties of Majorana fermions, it provides an unequivocal signature of their existence and is a first step toward more sophisticated experiments that can demonstrate the usefulness of Majorana fermions in topological quantum computing [29, 32, 33, 35].

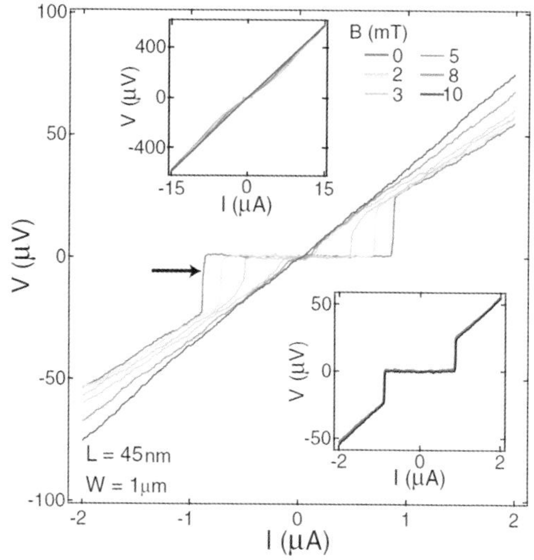

Figure 9.13 (main) V vs. I for Al-Bi_2Se_3-Al S-TI-S devices of dimensions $(L, W) = (45\,\text{nm}, 1\,\mu\text{m})$ for $B = 0, 2, 3, 5, 8, 10$ mT and at a temperature of 12 mK. At $B = 0$, I_C is 850 nA, which is reduced upon increasing B. For this device, the product $I_C R_N = 30.6\,\mu\text{V}$, which is much lower than theoretically expected for conventional JJs. (Upper-left inset) I–V curves overlap for all values of B at $V \geq 2\Delta/e \sim 300\,\mu\text{V}$. (Lower-right inset) Sweeps up and down in I show little hysteresis, indicating that the junction is in the overdamped regime. Adapted from [36]. For a color version of this figure, please see the color plates at the beginning of the book.

9.3.2 Majorana Experiment in Hybrid Superconductor-Topological Insulator Devices

In a recent experiment, Williams *et al.* [36] developed the original proposal by Fu and Kane [28] and demonstrated features that are compatible with the existence of the Majorana mode in an S-TI-S structure, where TI stands for topological insulator (Bi_2Se_3), rather than in a S-SmNW-S structure. After preparing flakes of Bi_2Se_3 by mechanical exfoliation [34] standard e-beam lithography is performed in order to deposit superconducting aluminum electrodes whose adhesion is guaranteed by a thin layer of Ti. The measurements performed on these two devices are voltage–current trace and the differential resistance as a function of the magnetic field.

One main observation is that the product $I_c R_N$, with I_c and R_N the weak link critical current and normal resistance respectively, is inconsistent with the Ambegaokar–Baratoff formula [6, 7] $I_c R_N = \pi\Delta/(2e)$ with Δ denoting the gap of the superconducting electrodes (Figure 9.13). The second observation is that upon application of a magnetic field in the plane perpendicular to the Bi_2Se_3 flake, the differential resistance as a function of the magnetic field does not show a Fraunhofer pattern which would correspond to the current–phase relationship (CPR) for a less exotic weak link.

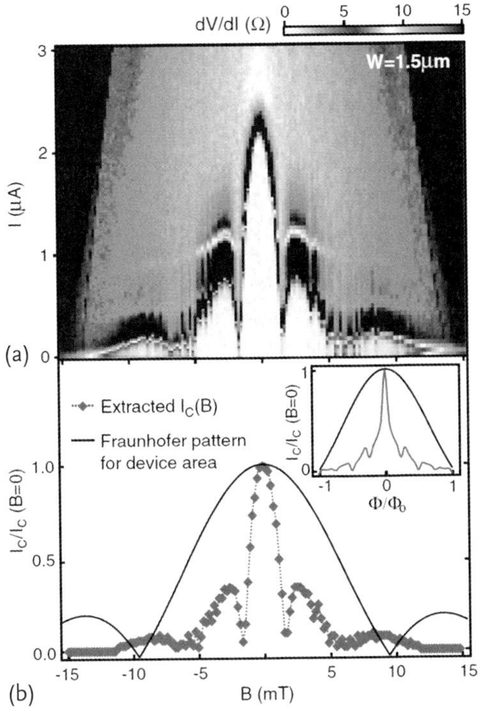

Figure 9.14 (a) Differential resistance dV/dI as a function of B and I showing an anomalous magnetic diffraction pattern for a $W = 1.5\,\mu m$ Al-Bi_2Se_3-Al S-TI-S junction. Two features are of note: a smaller-than-expected value of B_C at 1.70 mT and a nonuniform spacing between minima at values $B = 1.70$, 6.50, 11.80 mT. (b) (Main) $I_C(B)$ (dashed line with circles) extracted from dV/dI in (a) is compared to the expected Fraunhofer pattern for the junction (solid line) where a reduction of the scale of the pattern and the nonuniform spacing are evident. (Inset) a comparison of the simulated Fraunhofer pattern for a sinusoidal and an empirically determined, peaked CPR (current–phase relation). The narrowing of the diffraction pattern and the aperiodic minima observed in (a) are captured by this CPR. From [36].

Including the presence of a 1D Majorana wire (i.e., two Majorana modes at the two S-TI interfaces), and assuming that the supercurrent is determined by the physics of the junction near the zero energy – dominated by the Majorana modes – which in turn produces a peaked CPR, the author can qualitatively reproduce the experimental results as shown in the inset of Figure 9.14. It should be stressed that in a control sample in which the Bi_2Se_3 is replaced with a graphite flake, both the $I_c R_N$ product and the Fraunhofer diffraction pattern are close to the expectation for a weak link.

9.4
Superconducting Nanowires as Single-Photon Detectors

Superconducting nanowires are showing a lot of promise as single-photon detectors. The field is quite vast and specialized, and we will only discuss the aspects which are more relevant for nanowires of extremely reduced size. Our discussion will follow Natarajan *et al.* ([37], and references therein) to which we refer for a more in-depth review of the topic. The principle of operation of a superconducting (nanowire) single-photon detector (SSPD or SNSPD) is relatively simple and was demonstrated by Goltsman *et al.* in 2001 [38]. A superconducting nanowire held at temperature well below its critical temperature is current-biased at a current just below the nanowire's critical current (I_c), or more accurately its switching current (I_s). A photon incident on the nanowire will release enough energy to drive a small section of the nanowire normal, thus creating a hotspot [39] (Figure 9.15a-ii). While the size of the hotspot is smaller than the nanowire width, the current flowing through the nanowire will tend to flow around the normal region (Figure 9.15a-iii). The local current density around the hotspot will thus increase, locally exceeding I_c, which, in turn, will increase the size of the hotspot. This process will continue until the hotspot is as large as the nanowire (Figure 9.15a-iv). At this point, Joule heating will continue increasing the size of the normal region (Figure 9.15a-v) until the current is stopped (by the external circuit) and the nanowire can cool down (Figure 9.15a-vi) so that the SNSPD is ready for detecting another incident photon.

The simplified equivalent circuit used to understand the detection is shown in Figure 9.15b and a simulation of its time response, for typical parameters, in Figure 9.15c. The switch is initially closed so that all of the current flows through it; the absorption of a photon is equivalent to opening the switch. The response of the circuit is limited by the time constant $\tau_1 = L_k/(Z_0 + R_n(t))$ of the order of 1 ns for the parameters in Figure 9.15c, where $Z_0 = 50\ \Omega$, the kinetic inductance $L_k = \mu_0 \lambda^2 \mathsf{L}/\mathsf{A}$ with λ as the penetration depth, L as the nanowire length and A denoting the cross-sectional area of the nanowire. When the current is diverted through R_n, the nanowire starts to cool down: at this point, the switch is closed and the time constant will be $\tau_2 = L_k/Z_0$. The total dead time in which the detector is not sensitive to the incoming photon is $\tau = \tau_1 + \tau_2 \approx \tau_2 = L_k/Z_0$. From the last expression and the circuit diagram, it is obvious that to reduce the dead time, it is necessary to reduce the kinetic inductance (by changing its dimension or material) or placing additional resistors in series with the device.

A SNSPD is typically characterized by its efficiency. This, in turn, is related to: (1) coupling with the incident radiation, (2) detector material and geometry, and (3) the probability that an incident photon is converted into a signal. Other important metrics for a detector are: (1) noise that can produce false counts (usually measured as the dark count rate (DCR) (2) jitter, that is, the time before the arrival of a photon and the generation of the output signal, and (3) recovery time.

Typical SNSPD nanowires are of the order of 100 nm in width; from this point of view, they are quite far from the reduced size nanowires which we have touched upon in this work. However, it has been recognized that narrow (20–30 nm wide)

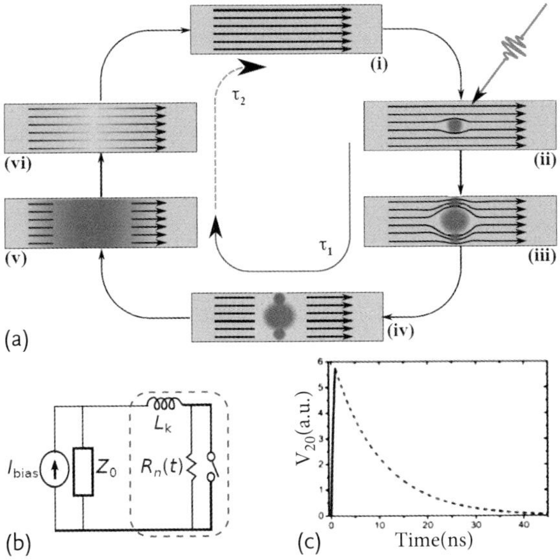

Figure 9.15 The basic operation principle of the superconducting nanowire single-photon detector (SNSPD) (after Goltsman Reference 7 in [37], Semenov *et al.* Reference 19 in [37] and Yang Reference 20 in [37]): (a) A schematic illustrating the detection cycle. (i) The superconducting nanowire maintained well below the critical temperature is direct current (DC)-biased just below the critical current. (ii) When a photon is absorbed by the nanowire, thus creating a small resistive hotspot. (iii) The supercurrent is forced to flow along the periphery of the hotspot. Since the NbN nanowires are narrow, the local current density around the hotspot increases, exceeding the superconducting critical current density. (iv) This, in turn, leads to the formation of a resistive barrier across the width of the nanowire Reference 7 in [37]. (v) Joule heating (via the DC bias) aids the growth of the resistive region along the axis of the nanowire Reference 20 in [37] until the current flow is blocked and the bias current is shunted by the external circuit. (vi) This allows the resistive region to subside and the wire becomes fully superconducting again. The bias current through the nanowire returns to the original value (i). (b) A simple electrical equivalent circuit of a SNSPD. L_k is the kinetic inductance of the superconducting nanowire and R_n is the hotspot resistance of the SNSPD. The SNSPD is current-biased at I_{bias}. Opening and closing the switch simulates the absorption of a photon. An output pulse is measured across the load resistor Z0 (Reference 21, 22 in [37]). (c) A simulation of the output voltage pulse of the SNSPD (approximating the pulse shape typically observed on an oscilloscope after amplification). Values of $L_k = 500$ nH and $R_n = 500\ \Omega$ have been used for this simulation (for simplicity, the R_n is assumed fixed, although a more detailed treatment Reference 20 in [37] shows $R_n(t)$). The solid blue line is the leading edge of the SNSPD output pulse, whilst the dotted red line is the trailing edge of the output pulse. The time constants relate to the phases of the detection cycle in (a). From [37].

nanowires may present substantial advantages in terms of registering probability and operation in the midinfrared range.

In the following, we will discuss, in detail, recent experimental results on SNSPD with ultranarrow nanowires. Typical SNSPDs use NbN wires of about 100 nm width and 3–5 nm thickness which are typically arranged in meander in order to cover the large illuminated area. However, this comes with the drawback of an

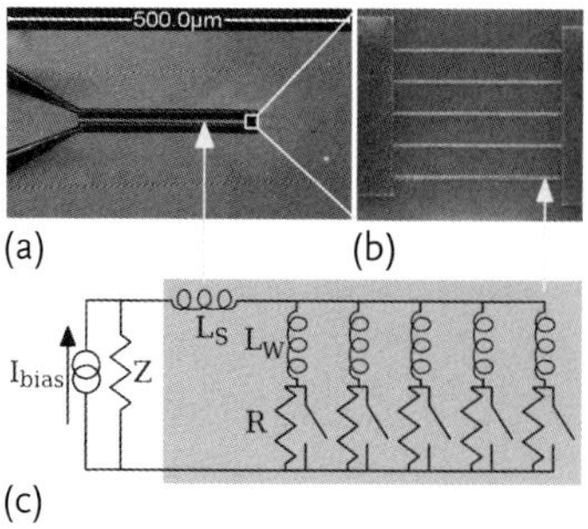

Figure 9.16 The proposed parallel detector. (a) Electron microscope image of the detector showing the coplanar transmission line used for DC biasing and signal readout, the series inductor L_S (indicated by the arrow) and the parallel structure (in the box). (b) Enlargement of the structure of the five parallel wires: each wire is 5 μm long and 100 nm wide. (c) Electrical model of the parallel detector used to simulate its response. The gray box represents the detector. From [40].

extremely large inductance, mostly due to the kinetic inductance of the wire. As discussed above, this increases the detector recovery time and also reduces the maximum count rate. A clever workaround [40] for reducing the kinetic inductance of the detector consisted of designing a circuit with five identical nanowires in parallel (electrical schematics in Figure 9.16). This circuit, dubbed the superconducting nanowire avalanche photodetector (SNAP), is biased near the switching current I_s of the nanowires: the arrival of a photon will cause one section of the nanowire to turn normal, thus redirecting the current in the other nanowires. If the biasing current is large enough, the increase in current will switch the remaining nanowires as well. Since the total inductance is $L_{det} = (1 + 1/N)L_W$, the output signal will be larger and the recovery time will be smaller than a meandered nanowire covering the same surface.

Since detectors based on narrow nanowire are expected to be more efficient in detecting midinfrared photons and are able to operate at a lower bias than detectors based on larger nanowires, Marsili *et al.* [41] implemented an SNSPD with a 30 nm wide NbN nanowire. Based on this result, they also developed a SNAP based on 30-nm wide nanowires. In particular, the ability of these detectors to work at lower bias operation, with respect to a 90 nm wide nanowire detector, is a real advantage.

The fabrication starts with the deposition of a 4–5 nm of NbN. Gold contacts are fabricated using optical lithography and deposited using an e-beam evaporator. Electron beam lithography is used to pattern negative resist hydrogen silsesquioxane (HSQ); the transfer of this pattern in the film using reactive ion etching forms the nanowire, thus completing the detector [41]. Since the circuit density is quite high, proximity effect correction during e-beam lithography [42] plays a key role in the successful fabrication of the device.

The first result by Marsili *et al.* (Figure 9.17) is the demonstration of an SNSPD based on a 30 nm nanowire. This is a single meandering nanowire with a lateral width of 30 nm and a thickness of 4–5 nm. The first result is that a 90 nm wide SNSPD, free of constriction, has a lower detection efficiency than the 30 nm wide SNSPD. Additionally, constrictions in 30 nm-SNSPD devices produce a relatively small effect when compared with a 90 nm-SNSPD. Intuitively, this should not be surprising as the detector performance depends on its bias current and if the switching current of the constriction is much smaller than the switching current of the detector, the detector performance would suffer. Any constriction will lower the operating bias of the detector because if the constriction switches to the normal state, the detector operation will be impossible. Assuming a constriction to be a reduction of the nanowire superconducting cross-sectional area with respect to the nominal value ($\sigma_c = \sigma_n(1 - I_{SW}/I_C)$ implies that if $I_{SW} = I_C$, the area of the constriction is zero), we can see how a constriction in the 90 nm-SNSPD such that $\sigma_c = 27\,\text{nm}^2$ implies that the constriction has a switching current close to 90% of the critical current. If $\sigma_c = 90\,\text{nm}^2$, the constriction will switch at a current equal to $0.75\,I_c^{90\,\text{nm}}$. If we take a 30 nm-SNSPD with the same fractional constriction, that is, $\sigma_c = 27\,\text{nm}^2$ and $I_{SW} = 0.75\,I_c^{30\,\text{nm}}$, we can see how the impact of such a constriction is almost negligible (Figure 9.17).

This should not be surprising if we recall the working of an SNSPD displayed in Figure 9.15. The arrival of a photon will drive a superconducting region of size comparable to the superconducting coherence length, in the normal state. If the nanowire is quite larger than the coherence length, in order to have a high probability of registering the photon, the bias current will have to be very close to the nanowire switching current. It is natural to see how the performance of such a detector is dramatically affected by a constriction.

Note that all the above experiments were performed with a light at 1550 nm wavelength. As it is well-known that longer wavelengths shift the detection efficiency curve of Figure 9.17 at higher currents, these results bode well for the operation of these detectors at longer wavelengths [43–45].

However, since the response of an SNSPD is proportional to its bias current, low bias operation inevitably produces a low signal to noise ratio. To overcome this problem, Marsili *et al.* fabricated a superconducting nanowire avalanche photodetector (SNAP) to exploit its advantages with respect to SNSPD. The experimental results for SNAP with 2, 3 and 4 nanowires in parallel are shown in Figure 9.18.

The detection efficiency curves are shown in Figure 9.18 and show plateaus for 3-SNAP and 4-SNAP. While the *n*-SNAP behaves as expected if biased at a current at which the detection efficiency is independent of the illumination level (I_{AV}), the presence of the plateau regions points to a different behavior for lower current ($I_b < I_{AV}$). Contrary to the simplified explanation provided above, at this bias, the arrival of a photon (or dark count) has the effect of "arming" the detector by altering the current distribution in the remaining nanowires. The arrival of a subsequent photon "triggers" the detector, thus making these detectors suitable for detection of correlated photons. This explanation is consistent with an electrothermal model [46], which takes into account the heat flow between the detector and substrate. While the reduction in nanowire size seems to be a fruitful path to follow, in order to increase the performance of SNSPDs, the overall detector geometry needs to be properly designed. After working on the polymethylmethacrylate cold-

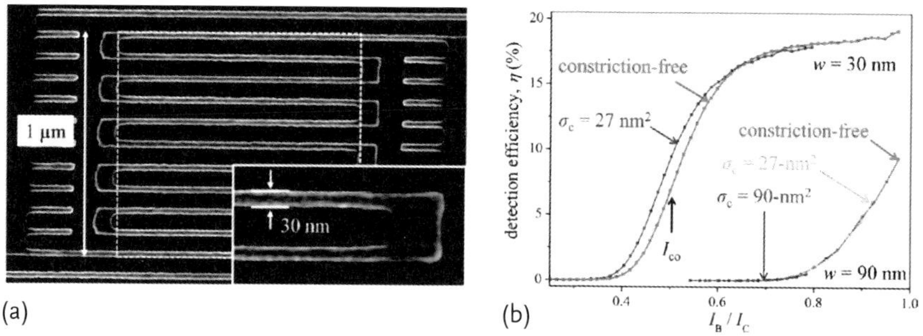

Figure 9.17 SNSPDs based on 30 nm wide NbN nanowires. (a) Scanning electron microscope (SEM) images of an SNSPD hydrogen silsesquioxane (HSQ) mask on NbN. The nanowires are 30 nm wide and the pitch is 100 nm (inset), covering an active area of 1.03 μm × 1.14 μm (dashed frame). See Supporting Information for the device fabrication process. (b) Device detection efficiency at $\lambda = 1550$ nm as a function of normalized bias current for constricted and constriction-free SNSPDs based on 30 nm wide and 90 nm wide nanowires. We assumed the constriction to be a reduction of the nanowire superconducting cross-sectional area with respect to the nominal value. We quantified the device constriction state by estimating the area of the nonsuperconducting part of the nanowire cross section as $\sigma_C = \sigma_n(1 - I_{SW}/I_C)$, where σ_n is the nominal nanowire cross section estimated from the nanowire width (measured by SEM) and thickness (estimated from the material deposition time and rate), I_{SW} is the device switching current, defined as the bias current at which the device switches from the superconducting to the normal state, and I_C is the device critical current, experimentally defined as the highest measured I_{SW} of the devices fabricated on the same film for the ultranarrow nanowire SNSPDs ($I_C = 7.2$ μA) and extracted from kinetic inductance vs. I_B measurements for the 90 nm nanowire-width SNSPDs ($I_C = 18.8$–20.1 μA). The device detection efficiency was calculated as $\eta = H(\mathrm{CR} - \mathrm{DCR})/N_{ph}$, where CR is the count rate measured when the SNSPD was illuminated, DCR is the count rate measured when the SNSPD was not illuminated, H is a normalization factor (see Supporting Information), and N_{ph} is the number of photons per second incident on the device active area. The I_{co} (defined to be at the inflection point of the η vs. I_B curves) of the 30 nm wide nanowire-width SNSPDs is marked with a black arrow (see Supporting Information). From [41].

development technique (Section 6.10.3), and encouraged by the suggestion of the higher fill factor (i.e., the ratio between the total area encompassed by the detector and the active area), Berggren's group set out to demonstrate an SNSPD with 90% fill factor [47]: a detector of area 3 μm × 3 μm with a 88 nm wide line separated by 12 nm gaps. However, the testing of such a detector revealed unexpected results: the switching current of this detector was smaller than the switching current of the detector with lower fill factor. The explanation to this somewhat surprising effect lies in the geometry of SNSPDs. The nanowire meander is composed by a nanowire region of 88 nm width connected with a wider region ($\approx$ 190 nm) in correspondence with a turn. The turns, while they have a larger cross-sectional area and thus a critical current larger than the nanowire, can negatively impact the detector performance because the current can "crowd" at the edge of the turn [48–50]; the density of current lines increases near the shortest current path (inside of a turn). Optimizing the design has improved the performance of the detector in terms of bi-

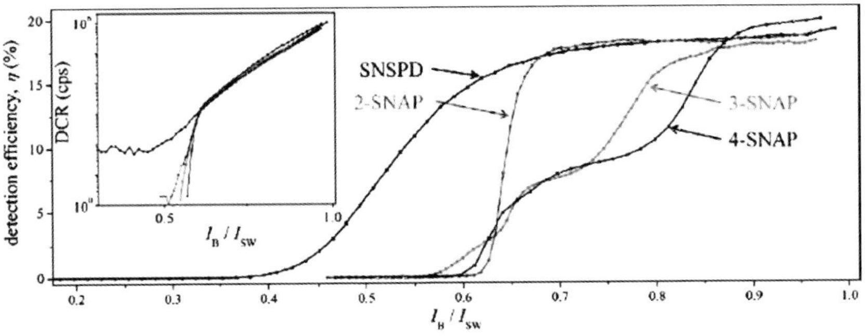

Figure 9.18 Detection efficiency and jitter of NbN SNAPs (superconducting nanowire avalanche detectors) vs. normalized bias current. (a) η at 1550 nm wavelength and DCR (inset) vs. normalized bias current (ratio of bias current and switching current of each device, I_B/I_{SW}) for an SNSPD, a 2-SNAP, a 3-SNAP, and a 4-SNAP based on 30 nm wide nanowires (in black, red, green, and blue, respectively). The devices were designed by integrating in parallel the same 1.47 μm × 230 nm section $N = 1, 2, 3, 4$ times. The series inductance of the N-SNAPs was designed to satisfy the condition $L_S(N-1)/L_0 = 10$, where $L_0 = 13$ nH is the kinetic inductance of one section (estimated from the fall time of the SNSPD response pulse). The detector switching currents were $I_{SW} = 7.2$ μA (SNSPD); 13.4 μA (2-SNAP); 18.1 μA (3-SNAP); 28.4 μA (4-SNAP). The photon fluxes (expressed in photons per second, photons/s) on the detector active area were $\mu = 5.5$ Mphotons/s (SNSPD); 12 Mphotons/s (2-SNAP); 19 Mphotons/s (3-SNAP); 25 Mphotons/s (4-SNAP). Adapted from [41].

asing current and reduced the dark count [50]. The dark count is an additional cause of low performance in detectors. As the pulses created by dark counts are similar to pulses generated by the absorption of photons [51] it is plausible to imagine that dark counts are generated by a similar mechanism: a hotspot in the detector becomes large enough to drive a section of the nanowire, and successively the entire nanowire, normal (Figure 9.15). Mechanisms leading to the creation of normal region in the superconductor are quantum and thermal phase slips, quantum and thermal generation of vortices and vortex–anti-vortex creation [51–53]. While experimental evidence is so far incomplete, the most plausible explanation are related to thermal excitation of vortices [52, 53] or vortex–anti-vortex creation [53]. However it must be noted that, while the continuous reduction in nanowire width might make phase slips relevant, the temperature at which the detector operate is too low for purely thermally phase slips and too high for quantum processes.

References

1 Iijima, S. (1991) Helical microtubules of graphitic carbon. *Nature*, **354**, 56–58.

2 Franklin, A.D. and Chen, Z. (2010) Length scaling of carbon nanotube transistors. *Nature Nanotechnology*, **5**, 858–862.

3 Avouris, P. (2007) Electronics with carbon nanotubes. *Physics World*, **20**, 40.

4 Kasumov, A.Y., Deblock, R., Kociak, M., Reulet, B., Bouchiat, H., Khodos, I.I., Gorbatov, Y.B., Volkov, V.T., Journet, C., and Burghard, M. (1999) Supercurrents

Through Single-Walled Carbon Nanotubes. *Science*, **284**, 1508.

5 Morpurgo, A.F., Kong, J., Marcus, C.M., and Dai, H. (1999) Gate-controlled superconducting proximity effect in carbon nanotubes. *Science*, **286** (5438), 263–265.

6 Ambegaokar, V. and Baratoff, A. (1963) Tunneling between superconductors. *Physical Review Letters*, **10**, 486–489.

7 Ambegaokar, V. and Baratoff, A. (1963) Tunneling between superconductors. *Physical Review Letters*, **11**, 104–104.

8 Reulet, B., Kasumov, A.Y., Kociak, M., Deblock, R., Khodos, I.I., Gorbatov, Y.B., Volkov, V.T., Journet, C., and Bouchiat, H. (2000) Acoustoelectric effects in carbon nanotubes. *Physical Review Letters*, **85**, 2829–2832.

9 Eichler, A., Deblock, R., Weiss, M., Karrasch, C., Meden, V., Schönenberger, C., and Bouchiat, H. (2009) Tuning the Josephson current in carbon nanotubes with the Kondo effect. *Physical Review B*, **79** (16), 161407.

10 Zhang, J., Tselev, A., Yang, Y., Hatton, K., Barbara, P., and Shafraniuk, S. (2006) Zero-bias anomaly and possible superconductivity in single-walled carbon nanotubes. *Physical Review B*, **74** (15), 155414.

11 Ashcroft, N.W. and Mermin, N.D. (1976) *Solid State Physics*, 1st edn, Brooks Cole.

12 Tarini, M., Cignoni, P., and Montani, C. (2006) Ambient occlusion and edge cueing for enhancing real time molecular visualization IEEE transactions on visualization and computer graphics, *IEEE Educational Activities Department*, **12**, 1237–1244.

13 Tang, Z.K., Zhang, L., Wang, N., Zhang, X.X., Wen, G.H., Li, G.D., Wang, J.N., Chan, C.T., and Sheng, P. (2001) Superconductivity in 4 angstrom single-walled carbon nanotubes. *Science*, **292**, 2462–2465.

14 González, J. and Perfetto, E. (2005) Coulomb screening and electronic instabilities of small-diameter (5,0) nanotubes. *Physical Review B*, **72** (20), 205406.

15 Lortz, R., Zhang, Q., Shi, W., Ye, J.T., Qiu, C., Wang, Z., He, H., Sheng, P., Qian, T., Tang, Z., Wang, N., Zhang, X., Wang, J., and Chan, C.T. (2009) From the cover: Superconducting characteristics of 4-A carbon nanotube-zeolite composite. *Proceedings of the National Academy of Science*, **106**, 7299–7303.

16 Wang, Z., Shi, W., Lortz, R., and Sheng, P. (2012) Superconductivity in 4 angstrom carbon nanotubes – a short review. *Nanoscale*, **4**, 21–41.

17 Wang, Z., Shi, W., Xie, H., Zhang, T., Wang, N., Tang, Z., Zhang, X., Lortz, R., Sheng, P., Sheikin, I., and Demuer, A. (2010) Superconducting resistive transition in coupled arrays of 4 Å carbon nanotubes. *Physical Review B*, **81** (17), 174530.

18 Lortz, R., Zhang, Q., Shi, W., Ye, J.T., Qiu, C., Wang, Z., He, H., Sheng, P., Qian, T., Tang, Z., Wang, N., Zhang, X., Wang, J., and Chan, C.T. (2009) Superconducting characteristics of 4-A carbon nanotube-zeolite composite. *Proceedings of the National Academy of Science*, **106**, 7299–7303.

19 Zhang, T., Sun, M.Y., Wang, Z., Shi, W., and Sheng, P. (2011) Crossover from Peierls distortion to one-dimensional superconductivity in arrays of (5,0) carbon nanotubes. *Physical Review B*, **84** (24), 245449.

20 Sau, J.D., Lutchyn, R.M., Tewari, S., and Das Sarma, S. (2010) Generic new platform for topological quantum computation using semiconductor heterostructures. *Physical Review Letters*, **104** (4), 040502.

21 Alicea, J. (2010) Majorana fermions in a tunable semiconductor device. *Physical Review B*, **81** (12), 125318.

22 Lutchyn, R.M., Sau, J.D., and Das Sarma, S. (2010) Majorana fermions and a topological phase transition in semiconductor–superconductor heterostructures. *Physical Review Letters*, **105** (7), 077001.

23 Oreg, Y., Refael, G., and von Oppen, F. (2010) Helical liquids and Majorana bound states in quantum wires. *Physical Review Letters*, **105** (17), 177002.

24 Mourik, V., Zuo, K., Frolov, S.M., Plissard, S.R., Bakkers, E.P.A.M., and Kouwenhoven, L.P. (2012) Signatures of Majorana fermions in hybrid su-

perconductor–semiconductor nanowire devices. *Science*, **336**, 1003.

25 Deng, M.T., Yu, C.L., Huang, G.Y., Larsson, M., Caroff, P., and Xu, H.Q. (2012) Observation of Majorana fermions in a Nb-InSb nanowire-Nb hybrid quantum device. *arXiv:1204.4130*.

26 Rokhinson, L.P., Liu, X., and Furdyna, J.K. (2012) The fractional a.c. Josephson effect in a semiconductor-superconductor nanowire as a signature of Majorana particles. *Nature Physics*, **8**, 795–799.

27 Das, A., Ronen, Y., Most, Y., Oreg, Y., Heiblum, M., and Shtrikman, H. (2012) Evidence of Majorana fermions in an Al–InAs nanowire topological superconductor. *arXiv:1205.7073*.

28 Fu, L. and Kane, C.L. (2008) Superconducting proximity effect and Majorana fermions at the surface of a topological insulator. *Physical Review Letters*, **100** (9), 096407.

29 Bernevig, B.A. and Zhang, S.C. (2006) Quantum spin Hall effect. *Physical Review Letters*, **96** (10), 106802.

30 Doh, Y.J., van Dam, J.A., Roest, A.L., Bakkers, E.P.A.M., Kouwenhoven, L.P., and De Franceschi, S. (2005) Tunable supercurrent through semiconductor nanowires. *Science*, **309**, 272–275.

31 Shapiro, S. (1963) Josephson currents in superconducting tunneling: The effect of microwaves and other observations. *Physical Review Letters*, **11**, 80–82.

32 Kane, C.L. and Mele, E.J. (2005) Z_2 topological order and the quantum spin Hall effect. *Physical Review Letters*, **95** (14), 146802.

33 Kane, C.L. and Mele, E.J. (2005) Quantum spin Hall effect in graphene. *Physical Review Letters*, **95** (22), 226801.

34 Geim, A.K. and Novoselov, K.S. (2007) The rise of graphene. *Nature Materials*, **6**, 183–191.

35 Kitaev, A.Y. (2003) Fault-tolerant quantum computation by anyons. *Annals of Physics*, **303**, 2–30.

36 Williams, J.R., Bestwick, A.J., Gallagher, P., Hong, S.S., Cui, Y., Bleich, A.S., Analytis, J.G., Fisher, I.R., and Goldhaber-Gordon, D. (2012) Unconventional Josephson effect in hybrid superconductor-topological insulator devices. *Physical Review Letters*, **109** (5), 056803.

37 Natarajan, C.M., Tanner, M.G., and Hadfield, R.H. (2012) Superconducting nanowire single-photon detectors: physics and applications. *Superconductor Science Technology*, **25** (6), 063001.

38 Goltsman, G.N., Okunev, O., Chulkova, G., Lipatov, A., Semenov, A., Smirnov, K., Voronov, B., Dzardanov, A., Williams, C., and Sobolewski, R. (2001) Picosecond superconducting single-photon optical detector. *Applied Physics Letters*, **79**, 705.

39 Skocpol, W.J., Beasley, M.R., and Tinkham, M. (1974) Phase-slip centers and nonequilibrium processes in superconducting tin microbridges. *Journal of Low Temperature Physics*, **16**, 145–167.

40 Ejrnaes, M., Cristiano, R., Quaranta, O., Pagano, S., Gaggero, A., Mattioli, F., Leoni, R., Voronov, B., and Gol'tsman, G. (2007) A cascade switching superconducting single photon detector. *Applied Physics Letters, AIP*, **91**, 262509

41 Marsili, F., Najafi, F., Dauler, E., Bellei, F., Hu, X., Csete, M., Molnar, R.J., and Berggren, K.K. (2011) Single-photon detectors based on ultranarrow superconducting nanowires. *Nano Letters*, **11**, 2048–2053.

42 Yang, J.K.W., Dauler, E., Ferri, A., Pearlman, A., Verevkin, A., Goltsman, G., Voronov, B., Sobolewski, R., Keicher, W.E., and Berggren, K.K. (2005) Fabrication development for nanowire ghz-counting-rate single-photon detectors. *IEEE Transactions on Applied Superconductivity*, **15**, 626–630.

43 Korneeva, Y., Florya, I., Semenov, A., Korneev, A., and Goltsman, G. (2011) New generation of nanowire NbN superconducting single-photon detector for mid-infrared. *IEEE Transactions on Applied Superconductivity*, **21**, 323–326.

44 Semenov, A., Engel, A., Hübers, H.W., Il'in, K., and Siegel, M. (2005) Spectral cut-off in the efficiency of the resistive state formation caused by absorption of a single-photon in current-carrying superconducting nano-strips. *European Physical Journal B*, **47**, 495–501.

45 Semenov, A.D., Goltsman, G.N., and Korneev, A.A. (2001) Quantum detection by current carrying superconducting film. *Physica C Superconductivity*, **351**, 349–356.

46 Marsili, F., Najafi, F., Herder, C., and Berggren, K.K. (2011) Electrothermal simulation of superconducting nanowire avalanche photodetectors. *Applied Physics Letters*, **98** (9), 093507.

47 Yang, J.K.W., Kerman, A.J., Dauler, E.A., Cord, B., Anant, V., Molnar, R.J., and Berggren, K.K. (2009) Suppressed critical current in superconducting nanowire single-photon detectors with high fill-factors. *IEEE Transactions on Applied Superconductivity*, **19**, 318–322.

48 Hortensius, H.L., Driessen, E.F.C., Klapwijk, T.M., Berggren, K.K., and Clem, J.R. (2012) Critical-current reduction in thin superconducting wires due to current crowding. *Applied Physics Letters*, **100** (18), 182602.

49 Henrich, D., Reichensperger, P., Hofherr, M., Ilin, K., Siegel, M., Semenov, A., Zotova, A., and Vodolazov, D.Y. (2012) Geometry-induced reduction of the critical current in superconducting nanowires. *Physical Review B*, **86**, 14450.

50 Akhlaghi, M.K., Atikian, H., Eftekharian, A., Loncar, M., and Majedi, A.H. (2012) Reduced dark counts in optimized geometries for superconducting nanowire single photon detectors. *Optics Express*, **20**, 23610.

51 Kitaygorsky, J., Dorenbos, S., Reiger, E., Schouten, R., Zwiller, V., and Sobolewski, R. (2009) HEMT-based readout technique for dark- and photon-count studies in NbN superconducting single-photon detectors. *IEEE Transactions on Applied Superconductivity*, **19**, 346.
June 2009.

52 Bulaevskii, L.N., Graf, M.J., Batista, C.D., and Kogan, V.G. (2011) Vortex-induced dissipation in narrow current-biased thin-lm superconducting strips. *Physical Review B*, **83** (14), 144526.

53 Bartolf, H., Engel, A., Schilling, A., Il'in, K., Siegel, M., Hübers, H.-W., and Semenov, A. (2010) Current assisted thermally activated flux liberation in ultrathin nanopatterned NbN superconducting meander structures. *Physical Review B*, **81** (2), 024502.

10
Concluding Remarks

We hope we have presented the reader with a flavor of the richness of superconductivity effects in 1D. A lot more could have been written, but due to space constraints, we have limited our discussion to what we feel are the most consequential and fruit-bearing theoretical and experimental approaches.

As we have already emphasized, the development of novel fabrication techniques is tightly intertwined with the recent rapid progress in this field. In particular, we have seen how the initial development of new techniques has produced a renewed interest in the field, which in turn has stimulated other groups to invest time and resources in probing and understanding 1D superconductivity from both a theoretical and an experimental point of view.

On a fundamental level, and across subdisciplines of physics, the prospects are bright for investigating the connection between many-body systems and high-energy quantum field theories. Important concepts, for example, instantons in effective Euclidean actions to compute tunneling rates, quantum phase transitions, and even the holographic principle, which yields a duality between gravity and Yang–Mills theory, may be relevant to 1D superconductivity. Moreover, the 1D nature begs one to speculate as to whether the issues encountered in superconductivity may one day prove relevant for superstrings, or cosmic strings.

From an experimental point of view, superconductivity in 1D is interesting in itself.

Uniformity and cleanliness, in conjunction with improvements in measurement techniques, should push us closer to answering many of the questions this field has set out to probe. For instance, how does superconductivity get destroyed in 1D nanowires? Does the wire diameter (cross section) play the dominant role through the superconducting stiffness parameter alone, or does disorder enter as well? Current data on the SIT (superconductor-insulator transition) still differ in many respects from theory, for example, in quantitative estimates of the controlling parameter(s), and regarding how the transition is approached, particularly on the superconducting side. Other important questions, such as whether the linear resistance approaches zero toward absolute zero in temperature as well as the functional form, are only hinted at by experiments.

On the applications side, there are several possibilities. Detectors with single-photon sensitivity and tunable wavelength sensitivity can be expected, as the ma-

One-Dimensional Superconductivity in Nanowires, First Edition. F. Altomare and A.M. Chang

terials are improved to reduce noise. Wavelength tuning may be accomplished by using different material systems, current-biasing, or the application of a moderate magnetic field.

Another, exciting aspect is a revolution in computing. This may occur in conventional computing where the push toward exaflops and beyond is hindered by the power dissipation of semiconducting devices as well as in the metallic interconnects. Here, superconducting nanowire interconnects might prove valuable. Another possibly more interesting direction would be the implementation of low noise, high Q, qubits based on the QPS, 1D superconducting nanowire platform.

In fact, in the coming years, one may anticipate (we envision) the fuller development and exploration of the duality concept, between QPS-based and Josephson junction-based devices. The advent of clean, uniform, and reproducible nanowires and material systems which exhibit minimal noise arising from materials issues, for example, in elemental superconductors, has great potential to bring useful devices to the forefront. These QPS devices will complement conventional Josephson junction devices, for example, SQUIDS, Shapiro voltage standards, as well as qubits. In particular, phase-slip qubits which could perform better than their more established cousins have just been demonstrated. Recently, a considerable amount of effort was devoted to the development of a current standard based on phase-slip junctions. This would allow closing the metrological triangle, establishing the unit of measurement of current, voltage, and resistance in terms of universal constants, and thus certifying our understanding of the underlying physical phenomena.

We look with great anticipation and excitement toward the future, and the continued contributions of all involved, that is, in the past, present, and future.

Index